European Fisheries Law

Europa Law Publishing, Groningen 2009

European Fisheries Law

From Promotion to Management

Dr. Till Markus

Europa Law Publishing is a publishing company specializing in European Union law, international trade law, public international law, environmental law and comparative national law.
For further information please contact Europa Law Publishing via email: info@europalawpublishing.com or visit our website at: www.europalawpublishing.com.

Typeset in Scala and Scala Sans, Graphic design by G2K Designers, Groningen/Amsterdam

NUR 828; ISBN 978-90-8952-003-6

Mixed Sources
Productgroep uit goed beheerde bossen, gecontroleerde bronnen en gerecycled materiaal.
FSC
www.fsc.org Cert no. CU-COC-803615
© 1996 Forest Stewardship Council

'Recognizing the significant contribution of sustainable fisheries to food security, income and wealth for present and future generations,...'

United Nations General Assembly, 6 March 2007, (A/RES/61/105)

Preface and acknowledgements

This study is based on an international and interdisciplinary research project that I participated in from 2005 to 2008: 'Integrating Multiple Demands on Coastal Zones with Emphasis on Aquatic Ecosystems and Fisheries, INCOFISH' was financed by the European Community. The book represents the culmination of my work over this period, initially submitted as a doctoral thesis to the Faculty of Law of the University of Bremen (November 2008). The work has been updated to include recent references and developments in the law and was finalised in early February 2009.

I am grateful for this opportunity to thank the many who have contributed to this study.

First and foremost, I would like to thank Prof. Dr. Gerd Winter for his guidance throughout the process. He was a constant source of good ideas, and his constructive comments greatly assisted and influenced my work. I also appreciate his efforts to foster an open-minded and collegial working environment at the Research Centre for European Environmental Law at the University of Bremen. In addition, I am grateful to Paul Nemitz LL.M. (DG MARE), Prof. Dr. Sabine Schlacke, Prof. Dr. Josef Falke, Dr. Evanson Chege Kamau LL.M. and David Lukaßen for their kind support and helpful comments in the dissertation colloquium. I also thank Anna-Maria Hubert for editing and revising this book with so much patience and congeniality. Warm regards also go to all of my colleagues, past and present, at the University of Bremen, the INCOFISH-Project, and the Bremen International Graduate School for Marine Science, particularly Antje Spalink, Jantje Struß, Bevis Fedder, Dennis Kenji Kipker and David Wille.

I would also like to express my sincere and heartfelt thanks to my close friends. Their encouragement and forbearance spurred me to carry on toward my goals. At the risk of almost certainly omitting important people, I would like to thank Andreas Cordes, Florian Obersold, Ellen Gerdes, Franziska Prüger, Sven Höhne, Jan Drews, Miriam Oppenheim, Jörn Eickmann, Claus Lahner, Cita Wischhusen and Xandra Kramer.

I also wish to express my deepest gratitude to those two teachers whose words and deeds have encouraged and inspired me so much over the last years, Daisaku Ikeda and Thich Nhat Hanh.

I am very grateful to my dear Janna Wolff for her companionship and for bringing so much laughter and happiness into my life.

Finally, thank you to my parents and sister who have been there for me throughout the years, believed in me and encouraged me. This book is dedicated to them with deep affection and gratitude.

Bremen, April 2009

Till Markus

Abbreviations

1982 UNCLOS	United Nation Convention on the Law of the Sea
ACFA	Consultative Committees on Fisheries and Aquaculture
CCAMLR	The Commission for the Conservation of Antarctic Marine Living Resources
CECAF	The Fishery Committee for the Eastern Central Atlantic
CFI	Court of First Instance
CFP	Common Fisheries Policy
EAGGF	European Agricultural Guidance and Guarantee Fund
ECJ	European Court of Justice
EEA	European Environmental Agency
EEC	European Economic Community
EEZ	Exclusive Economic Zone
EFF	European Fisheries Fund
EFZ	Exclusive Fishery Zone
ERDF	European Regional Development Fund
ESF	European Social Fund
FAO	Food and Agricultural Organisation of the United Nations
FIFG	Financial Instrument for Fisheries Guidance
F-lim	Point at which fish stocks are threatened by a substantial decline or collapse
FMC	Fisheries Monitoring Centre
FSA	The United Nations Agreement for the implementation of the Provisions of the United Nations Convention on the Law of the Sea of 10 December 1982 relating to the Conservation and Management of Straddling Fish Stocks and Highly Migratory Fish Stocks
GFCM	General Fisheries Commission for the Mediterranean Area
HELCOM	Helsinki Commission (Baltic Marine Environmental Protection Commission)
IATTC	Inter-American Tropical Tuna Commission
ICCAT	The International Commission for the Conservation of Atlantic Tuna
ICES	International Council for the Exploration of the Sea
ICNAF	International Commission for the Northwest Atlantic Fisheries
ICZM	Integrated Coastal Zone Management
IDCP	International Dolphin Conservation Program
IFO	International Fisheries Organisation
IOTC	The Indian Ocean Tuna Commission
IPOA	International Plan of Action
MAGP	Multiannual Guidance Programme
MSC	Marine Stewardship Council
MSE	Maximum Economic Yield
MSY	Maximum Sustainable Yield
NAFO	The North-West Atlantic Fisheries Organisation

NASCO	The North Atlantic Salmon Conservation Organisation
NEAFC	The North-East Atlantic Fisheries Commission
NGO	Non-Governmental Organisation
OSPAR-Convention	The Convention for the Protection of the Marine Environment of the North-East Atlantic
RAC	Regional Advisory Council
SCM-rules	Subsidies and Countervailing Measures
SEA	Single European Act
SEAFO	The South-East Atlantic Fisheries Organisation
STECF	Scientific, Technical and Economic Committee for Fisheries
SWIOFC	South West Indian Ocean Fisheries Commission
TAC	Total Allowable Catch
TAE	Total Allowable Effort
Treaty	Treaty Establishing the European Community
UK	United Kingdom
VMS	Vessel Monitoring System
WCPFC	The Commission for the Conservation and Management of Highly Migratory Fish Stocks in the Western and Central Pacific Ocean
WECAFC	The Western Central Atlantic Fishery Commission
WTO	World Trade Organisation
WWF	World Wide Fund for Nature

Introduction

The European Community's Common Fisheries Policy (CFP) was established to ensure that the exploitation of living aquatic resources in Community waters and by Community fisheries is carried out at sustainable levels. To this end, a complex management regime, which includes conservation, control and enforcement measures, has been established. In contrast to this aim, the Community has frequently promoted production increases and employment in the fisheries industry to stimulate economic development in coastal regions and to stabilise the fishing sector and its markets.

Given that many fish stocks exploited by Community fisheries are overfished and catches as well as employment continue to decline, Community management and promotion policies have generally failed. Conservation measures, such as total allowable catches, effort restrictions, and technical measures often encourage fishing at unsustainable levels; and control and enforcement measures have lacked effectiveness. Promotion, on the other hand, has contributed its share to the management system's ineffectiveness. Subsidies granted under the CFP's structural policy have, *e.g.*, increased fishing and processing capacities of the Community's fisheries industry. High capacity in the sector, arising from a large work force and huge capital investments, demands high catch rates, putting pressure on marine capture resources. Since the Community's management regime is neither perfectly designed, controlled, enforced nor implemented, promotional activities in many cases have fuelled the 'race to fish'. It has only been recently that the CFP has really begun to adjust its support practices to correspond to the situational and legal management requirements stopping aid to the sector for the most problematic purposes. Nevertheless, such subsidisation continues even under the new European Fisheries Fund.

World-wide policy makers tend to treat the promotion and management of fisheries as separate categories, drawing up their policy approaches in relative isolation from each other. This can lead to a lack of coherence in fisheries policy as a whole. In this author's opinion, in order to integrate these policy areas, promotion should aim first and foremost at adjusting consumption and production capacity to the size commensurate with sustainable exploitation limits. The approach is based on two basic assumptions:

First, the exploitation of marine capture resources ultimately depends on the level of available fish stocks. To maintain sustainable exploitation, fish stocks cannot be continuously exploited beyond sustainable levels. Thus, sustainable exploitation must be the initial point of reference for all other policy segments, promotion in particular.

Second, economic theory generally assumes investment in the fishing industry up to the point at which revenues equal production costs. In theory, exploitation rates thus would not exceed sustainable levels. However, economics also teach that the open access nature of fisheries resources not only fuels the fishermen's 'race to fish' ('the tragedy of the commons'), but it also encourages investment in the sector, creating overcapacity. This dynamic, again, is aggravated by the granting of subsidies that alter cost-revenue structures, by reducing production costs and investment risks. Overcapacity, however, poses a major obstacle to the successful implementation of fisheries management objectives.

Accordingly, this book will try to: (a) explain the CFP's complex management and promotional regimes; (b) identify practical and legal problems and failures in both systems; and (c) assess whether CFP promotional measures are consistent with management rules and how consistency can be increased.

The first chapter will explain the environmental and socio-economic background of the CFP, with special emphasis devoted to the state of Community fisheries resources, the multiple demands on the coastal zones, the socio-economical relevance of the fisheries sector and the political process underlying the CFP.

In the second chapter, the institutional and organisational structures of the CFP will be described. This includes an analysis of the competence order established in the Treaty Establishing the European Community (Treaty), and its effect on the CFP. Subsequently, the Community's territorial and substantial competences will be explained. Then, it will be examined the extent to which the Community actually has made use of its competences and how tasks and functions are divided between the Community, its Member States and other relevant stakeholders. The final section will assess how competences of the Community and the Member States overlap with regards to the regulation of fisheries for environmental purposes and how this conflict should be resolved.

Chapter 3 explains the Community's fisheries management regime. A description of the management rules currently in place, such as regarding TACs, effort limitations, technical measures, multi-annual recovery plans, etc. will be provided. Control and enforcement instruments will also be explained. All the measures will be assessed and, where necessary, criticised from a practical or legal point of view. Particular attention will be paid to the legal status of small-scale coastal fisheries.

Chapter 4 provides an in-depth analysis of the development and current state of the promotional regime. The analysis will take into consideration current theoretical models and empirical studies on the effects of subsidies in marine fisheries.

Chapter 5 develops a concept for the integration of promotion and management. To this end, it outlines the state of integration of promotional and management measures throughout the development of the CFP and describes the ramifications of the lack of integration in this area.

Chapter 6 looks at the degree of access that natural and legal persons (including environmental organisations) have to the Community judicature for the review of CFP measures.

Chapter 7 provides an analysis of several additional legal issues that arise under the CFP. The first part of this inquiry will examine the extent to which current legislation and state practice with regard to the CFP accord with international and Community law. The ensuing section will look at whether fishing limitations under Community law impinge on fundamental rights. The final part of the chapter examines the legality of the Commission's quota deduction competences.

The book ends with an outline of the main insights of this study.

Despite all the criticism expressed in this book, it must be noted that the Community's management regime is the furthest developed international (even supranational) fisheries management system in the world. Besides, many management aspects have improved substantially over the years. For example, the stakeholder approach under the new EFF rules and the increasing integration of promotion and management are promising developments that will, hopefully, contribute to sustainable fisheries management. These and other results of the CFP's tentative learning process can certainly be inspiration to fisheries managers world wide, including CFP managers themselves.

CHAPTER I

Environmental, Socioeconomic and Political Background of the European Common Fisheries Policy

Different influences shape the Community fisheries regime. To illustrate the forces directing the CFP, the ensuing section will present a brief overview of the multiple demands and pressures on the Community's coasts, marine waters and fish stocks. Following this description will be an explanation of the political and public discourse on fisheries.

1 Overview of multiple demands and pressures on the EC coastal zones and marine ecosystem

The Community has a coastline of approximately 68,000 km.[1] Its EEZ is the world's largest, covering over 25 million square-km.[2] Activities in European coastal zones are manifold. Next to the marine fisheries sector, tourism, urban settlement, and industrial operations, such as shipping, port activities, wastewater cleansing, oil and gas drilling, dumping, sand and gravel extraction, and wind energy parks affect the marine ecosystems and fisheries. In addition, climate change and the loss of biodiversity alter the life conditions for marine resources. Although the impact of these factors on fisheries is still inconclusive, they are widely perceived to be of relevance.

1.1 Overview of human activities affecting Europe's coasts and marine waters

The previous two biennial reports of the European Environmental Agency (EEA) described the scale and the character of commercial activities and other pressures on Europe's marine waters and coastal environment.[3]

The EEA states that urban settlements on Europe's coasts are comprised of about 280 cities with more than 50,000 inhabitants in each. About 70 million EU citizens live in coastal regions, with approximately 16% of the population living on about 11% of the total land mass.[4] Furthermore, the tourist industry in coastal areas and the acquisitions of second homes for urban residents have increased sharply over the last decades.

Currently, up to 60% of the world's tourists spend their annual holiday in the European Community. Since many tourists visit the coasts, this industry has been one of the prime commercial activities contributing to coastal development. For example, France, Spain and Italy together received about 174 million tourists in 2004, many of them visiting the Mediterranean. The quantity of

---

1 European Commission, *Maritime Facts and Figures*, (Brussels: 2006), p. 3.

2 See EUROPA Glossary, available online at http://europa.eu/scadplus.

3 European Environment Agency, *Europe's environment – The fourth assessment*, (Luxembourg: 2007), pp. 210-250; see also European Environment Agency, *The European environment – State and outlook 2005*, (Luxembourg: 2005), pp. 132-167; 380-391.

4 European Environment Agency, *Europe's environment – The fourth assessment*, p. 240.

tourists is expected to rise by 44% over the next 20 years. Tourism constitutes a major economic factor in coastal regions, in some places providing for more than 40% of the jobs.[5]

About 20% of the Community's industry is located on the coasts, with one third at the North Sea. These enterprises (*e.g.*, conveyance businesses) depend on seaways or ocean resources, such as oil, gas and fish, or water and wind power. During the 1990's the marine transport of passengers and goods within Europe rose steadily, an increase of about one third to about 1270 billion tonne-kilometres.[6]

Urban settlements, tourism, industrial and agricultural activities put pressure on Community fish habitats, contributing to, for example, oil spills, the introduction of alien species through marine transports, chemical pollution and eutrophication.[7] These factors also affect coastal ecosystems as a whole. The enormous increase in infrastructure within the 10 kilometre coastal zone has aggravated existing problems associated with soil sealing and water shortages in coastal regions.[8]

1.2 Overview of the nature and socioeconomic significance of the Community fisheries sector

According to Eurostat, the per capita supply of fish varies greatly throughout the Community's 27 Member States. While the average consumption of an EC-citizen is 20.3 kg per year, each Spaniard, Portuguese and Lithuanian consumes about 43.3, 54.5, and 36.7 kg per year, respectively, while Slovenians, Bulgarians and Romanians eat about 6.5, 3.1 and 2.6 kg per year.[9]

Currently, marine *catching and aquaculture* production amounts to approximately 4% of the world's total. Denmark, Spain, France and the United Kingdom contribute about 53% of the Community's share. Since 1993, total EU-27 production has declined by about 17%. The Community accounts for 6% of the world total *catch* (5.6 million tonnes), a decline over 1993 to 2005 by about 22%. The Community share of the world's *aquaculture* production was only 2% or 1.27 million tonnes.[10]

---

5 European Environmental Agency, *Outlook 2005*, p. 149.

6 *Ibid.*, pp. 150-151.

7 European Environmental Agency *Outlook 2005*, pp. 152-153; European Environmental Agency, *Europe's environment 2007*, pp. 215-217.

8 European Environmental Agency *Outlook 2005*, p. 143; European Environmental Agency, *Europe's environment 2007*, pp. 243, 245.

9 The average yearly per capita supply for Iceland lies at 91.5 kg and 48.7 kg for Norway. In contrast, the average per capita food fish supply on the African continent in 2002 lay at 7.8 kg per year; Eurostat, *Fishery Statistics – Data 1990 – 2006*, (Luxembourg: 2007) pp. 8-61, 58; FAO, *The State of World Fisheries and Aquaculture*, (Rome: 2004), pp. 39-43.

10 Eurostat, *Fishery Statistics – Data 1990 – 2006*, pp. 18-19, 22-23.

The Community has a negative trade balance in fishery products, both in terms of volume and value. France, Germany, Italy, Spain and the United Kingdom are the major importers of Community members. Imports have risen to 6.23 million tonnes in 2006, equalling €17.2 billion.[11]

In 2006, the EU-25 fleet was comprised of 87,004 fishing vessels. However, the mere number of vessels is not an accurate indicator of the overall fishing power. Tonnage and engine power are considered to be more reliable factors. In 2005, these two accounted for 1,955,879 gross registered tonnes (GRT) and 7,068,471 KW for the then 25 Member states. It is necessary to mention that the fleet is structured differently throughout the Community. The fleets of Greece, Portugal, Italy and Finland tend to maintain a large number of small-scale fishing vessels. Countries like Belgium and the Netherlands have a few large ships in terms of engine power and tonnage.[12] The fleets belonging to France, Spain and the UK maintain many small-scale vessels, as well as a few high-powered vessels.[13]

Due to variances in assessment methods across the Member States, employment rates in the fisheries sector are somewhat imprecise.[14] Nevertheless, in 2006 the Commission published a comprehensive study,[15] which found that the total number of people employed in the fisheries sector in 2002/2003 was estimated to be about 421,000 persons, of whom 405,000 were involved in the marine fisheries sector. Approximately 209,000 people had actually worked as fishermen on board fishing vessels (approximately 99,000 on small-scale coastal vessels and 110,000 on off-shore vessels). The number of fishermen has declined since 1996 by 4 to 5%. Aquaculture provided for about 65,400 of these jobs and the processing industry for about 147,000 positions. In 2003, fishing jobs in the Community's 25 Member States were distributed as follows: Spain employed 92,777; France 64,712; Italy 47,957; Greece 37,701; UK 33,534; Portugal 33,229; Germany 16,409; Denmark, 14,060; Poland, 13,617; Ireland 10,584; The Netherlands 9,049; Sweden 3,955; Finland 2,740; Belgium 1,743; Lithuania, 1,433; Latvia, 1,007; Slovenia, 897; Estonia, 549; Cyprus, 327; Malta, 148.[16]

Aquaculture production in the Community rose from 642,000 tonnes in 1980 to 1,374,000 tonnes in 2006, now accounting for 19% of overall volume

[11] *Ibid.*, pp. 44-45.

[12] Commission Communication Com(2001) 135 final Green Paper on the future of the Common Fisheries Policy, vol. II b, p. 5.

[13] *Ibid.* pp. 25-40.

[14] Eurostat, *Fishery Statistics – Data 1990 – 2006*, p. 57.

[15] Pavel Salz, Erik Buisman, Jos Smit and Birgit de Vos, *Employment in the Fisheries Sector*, (Brussels: 2006); For older data, see European Commission, *Regional Socio-economic Studies on Employment and the Level of Dependency on Fishing*, (Brussels: 2000).

[16] Salz *et al.*, *Employment in the Fisheries Sector*, p. 17; European Commission, *Regional Socio-economic Studies on Employment and the Level of Dependency on Fishing, Lot No. 23*, (Brussels: 2000), pp. 30-31.

and 30% of total fishery production.[17] The aquaculture sector basically consists of three groups: freshwater fish farming, marine mollusc farming, and marine fish farming. While marine mollusc farming accounts for about 60% of the Community's aquaculture production and 30% of its overall value, marine fish farming is considered to be the most profitable field, showing significant gains over the past decades. In many cases, aquaculture offers employment for workers who are being squeezed out of the declining capture fisheries sector.

In 1998, the value of the whole production chain (*i.e.*, fishing, aquaculture, processing and marketing) was estimated at €20 billion, approximately 0.28% of the Community's gross domestic product.[18] Despite this relatively small share, many coastal communities rely heavily on fishing as a source of jobs and income. In some areas in Scotland and Spain, the fishery sector provides for more than 10% of the overall employment.[19]

2 State of the marine environment throughout the Community

2.1 State of the fish stocks of Community interest

In 2001 the Commission declared in its 'Greenpaper on the future of the Common Fisheries Policy' that fish resources in Community waters are overexploited, and in some cases seriously threatened:[20]

> The quantities of mature demersal fish in the sea as assessed by the International Council for the Exploration of the Seas (ICES) have, in many cases, declined significantly over the last twenty five years. On average, the quantities were about 90% greater in the early 1970s than in the late 1990s. [...] In recent years, for many stocks quantities of mature fish in the sea have been below or very close to the minimum levels required to provide high probability of sustainability (precautionary level of stock biomass).[21]

---

17 Strategy Paper Com(2002) 511 final for the Sustainable Development of European Aquaculture, pp. 3-4; European Commission, *Facts and Figures on the CFP 2006*, (Luxembourg: 2006), p. 16.

18 Commission Communication Com(2001) 135 final Green Paper on the future of the Common Fisheries Policy, vol. II b, p. 4. In 2005, GDP at current prices stood at €10.817,000 million for EU-25, Eurostat, *Eurostat Yearbook*, (Luxembourg: 2007), pp. 50 and 151.

19 European Environmental European Environmental Agency, *The European environment – State and outlook 2005*, (Luxembourg: 2005), p. 147.

20 Commission Communication Com(2001) 135 final Green Paper on the future of the Common Fisheries Policy, vol. I, pp. 7-8.

21 *Ibid.*, pp. 7-8; Communication Com(2002) 181 final on the Reform of the Common Fisheries Policy ('Roadmap'), pp. 29-31.

For benthic resources, a 'general economic overexploitation' was also observed. Bluefin tuna was also considered to be 'clearly overexploited', suggesting the need for at least a 25% reduction in fishing mortality. The situation was better for the pelagic stocks, which were subject to recovery measures during the 1990s.

At the end of 2002, the CFP was restructured in favour of more sustainable fisheries practices.[22] However, the European Environmental Agency (EEA) Report on Europe's environment draws the following conclusions for 2007:

> Of those stocks that had been assessed, 14% in the Arctic were outside safe biological limits, whilst for the North-East Atlantic and Baltic Seas, this was 26%. Within the North-East Atlantic, the North Sea was the most severely affected with 44% of the assessed commercial stocks outside safe biological limits, followed by the Celtic Sea with 30% outside them [...]. In the Mediterranean Sea, the percentage of assessed commercial stocks outside safe biological limits in 2005 ranged from 10-20% with Aegean and Cretan being in the worst condition. [...] Bluefin tuna stocks both in the eastern Atlantic and Mediterranean Seas have been identified as being near collapse.[23]

Since 1995, overall catches of the 27 Member States have decreased from 8,054,000 to 5,312,000 tonnes live weight.[24] A further problem is that many of the exploited stocks are insufficiently monitored. For example, the EEA states that, in 2006, 81% of Arctic, 67% of Baltic Sea and 54% of North-Eastern Atlantic commercial fish stocks remained unassessed.[25]

2.2 State of the Community marine ecosystem in relation to fish stocks

Overfishing has been widely identified as one of the most important factors causing the depletion of fish stocks. However, as mentioned above, other factors also influence the state of the fisheries resources, both

---

[22] Regulation 2371/2002 on the conservation a sustainable exploitation of fisheries resources under the Common Fisheries Policy, OJ 2002 No. L358/59; Council Regulation (EC) 2370/2002 of 20 December 2002 establishing an emergency Community measure for scrapping fishing vessels; Council Regulation (EC) 2369/2002 of 20 December 2002 amending Regulation (EC) No. 2792/1999 laying down the detailed rules and arrangements regarding Community structural assistance in the fisheries sector.

[23] European Environmental Agency, *Europe's environment 2007*, p. 223; More numbers are provided in Communication Com(2007) 295 final from the Commission to the Council on Fishing Opportunities for 2008 – Policy Statement from the European Commission, p. 5; ICES, *Environmental Status of the European Seas*, (Copenhagen: 2003), pp. 37-42.

[24] Source: Eurostat, available online at : http://epp. eurostat.ec.europa.eu.

[25] European Environmental Agency, *Europe's environment 2007*, p. 223; Communication Com(2007) 295 final from the Commission to the Council on Fishing Opportunities for 2008, p. 5.

directly and indirectly. For instance, climate change, oil spills, industrial and touristic activities, as well as alien species contaminate and put pressure on fish and their habitats. It is also important to remember that overfishing decreases the resilience of stocks against such influences.[26]

The Community's fish stocks must cope with climate change. The Bay of Biscay, the North Sea, the Baltic Sea, the Barents Sea, the Mediterranean and parts of the Arctic Ocean, in particular, have already sustained adverse effects. The melting Arctic ice and inland glaciers and the changing oceanic currents influence the temperature and the salinity of the Community's marine waters. These developments alter the composition of the stocks. The warming of the northern waters generates a relocation of fish and plant species from the south to the north, and sometimes even results in their extinction.[27] This trend is likely to accelerate over time.[28]

Industrial, agricultural, touristic activities together with urban settlements put pressure on the Community fish habitats, introducing harmful substances into the marine ecosystem. Oil extraction, oil discharge, the release of nitrogen into marine waters, and the use and emission of anti-foulants by ships put a great deal of pressure on the marine ecosystem, in particular on the main shipping routes and in drilling areas. For example, recent oil-tanker accidents have damaged the coasts of Spain.[29] Other harmful substances, such as heavy metals, pesticides and hydrocarbons, enter the water from air or land based sources, and accumulate in the bodies of fish, affecting their fertility, growth and health. The concentration of these substances in fish in Community waters has declined over the past 15 years, with the exception of certain substances in the Baltic Sea (*e.g.*, PCBs and DDT).[30] Eutrophication, in particular, adversely affects the Adriatic Sea, Baltic Sea, North Sea, the Black Sea, Caspian Sea, the Kattegat, Norwegian fjords, some areas of the Mediterranean and the North Sea's Wadden Sea.[31]

The invasion of alien species is a major problem in the Mediterranean, the Baltic, Celtic, Caspian, Black, and the North Seas, as well as Iberian Coastal and Norwegian waters. This usually occurs when foreign organisms are harboured in ships' ballast water and then released, spreading all over the world. For exam-

---

26 D. Pauly, V. Christensen, J. Dalsgaard, R. Froese and F. Torres, 'Fishing down the marine food webs' (1998) *Science*, 279, pp. 860-863; European Environmental Agency, *Europe's environment 2007*, pp. 237-239.

27 European Environmental Agency, *Europe's environment 2007*, p. 246.

28 European Environmental Agency, *Outlook 2005*, pp. 134, 136, 139, 146; European Environmental Agency, *Europe's environment 2007*, p. 216.

29 European Environmental Agency, *Outlook 2005*, pp. 134-142; European Environmental Agency, *Europe's environment 2007*, pp. 229-237.

30 European Environmental Agency, *Outlook 2005*, pp. 153-154; European Environmental Agency, *Europe's environment 2007*, pp. 229-237.

31 European Environmental Agency *Outlook 2005*, pp. 152-153; European Environmental Agency, *Europe's environment 2007*, pp. 215-217.

ple, since 2000, over 100 new species have been reported in the Mediterranean Sea. Alien species are more likely to become established where ecosystems are already weakened by pollution or overfishing.[32]

3 Formation of the Common Fisheries Policy

Various political actors and interest groups help to shape the CFP. One must particularly note that the Community is a compound of nations and not a sovereign state. To the extent that legislative duties have not been delegated to the Commission, Member States' fisheries ministers and the Commission jointly negotiate the CFP and adopt pertinent legislation within the Community's central legislative organ, *i.e.*, the Council.[33] They decide on matters concerning fisheries management, structural policy, market organisation and external affairs. Thereby, they are influenced and supported by various scientific bodies, as well as non-governmental interest groups. The following section will describe the current political discourse on fisheries, examining the views of the relevant political players of CFP. Before this, a rough sketch of the political concerns connected with the CFP will be provided.

3.1 The central elements of the Common Fisheries Policy

The CFP is concerned with the sustainable exploitation of living aquatic resources. This requires a broad range of political and legislative actions in a range of policy areas.

To guarantee that Community fishermen do not fish at unsustainable levels, the Community limits fishing opportunities. Currently, it does so by adopting total allowable catches (TACs), effort limitations and technical measures. It also apportions the available resources among its Member States, which then allocate their share to their own fishers. When imposing catch limitations, the Community increasingly takes environmental concerns into consideration, for example, requiring the use of environmental friendly fishing gear or prohibiting fishing in environmentally sensitive areas.

Rules adopted under the CFP also relate to the structure of the Community's fisheries sector. 'Structure' basically refers to the equipment required for the production of fisheries products and the organisation of the production process.[34] The primary aim of such structural policies is to support the sector in adapting its production capacities to correspond to available resources, thereby

---

[32] European Environmental Agency, *Europe's environment 2007*, pp. 237-239.

[33] Arts. 203, paragraph 2 and 202, paragraph 3 of the Treaty, Art. 211, paragraph 4, Art. 37(2), paragraph 2 of the Treaty; see also Art. 202, Art. 37(2), paragraph, Art. 250(1) of the Treaty and Art. 29 Regulation (EC) 2371/02.

[34] R.R. Churchill, *EEC Fisheries Law*, (Dordrecht: 1987), p. 203.

guaranteeing efficient and sustainable production. It also aims at increasing the competitiveness of the sector, and working towards socioeconomic stability and social cohesion within different fishing regions.[35] Just like catch limitations, structural measures increasingly consider environmental aspects.[36] Public financial transfers to the fisheries sectors are considerable under the CFP. From 2000 to 2006, over €4.1 billion had been allocated to the Community's fisheries sector.[37] Financial transfers from 2007 to 2013 lie at approximately €3.8 billion.[38]

Another area of concern is the common organisation of the market for fisheries products. Different objectives are pursued in this regard. Given the unstable supply of fisheries resources, the Community takes measures to stabilise markets.[39] It also aims at matching supply to demand. To this end, the Community intervenes in the market, and maintains common marketing standards, producer organisations, and a system of trade with third countries. The system had been introduced in 1970 and has, to this day, not been modified substantially.[40]

To effectively limit fishing activities, structure the sector and stabilise markets, the Community is also responsible for the external promotion of its fisheries policies (*i.e.*, with non-Member States).[41] It has concluded several international agreements, which allow Community vessels to fish in the waters of third countries.[42] Furthermore, the Community is currently a member of 11 international fisheries organisations (IFOs), which are mainly concerned with

---

35 Council Regulation (EC) 1198/06 on the European Fisheries Fund, OJ 2006 No. L223/1.

36 See Art. 25 (6)(c),(d), (7),(8); Arts. 30 (2)(a-c) and 35 (1)(d); Arts. 37 and 38; Arts. 40 (3)(d),(f); Arts. 43 (2)(c), 44 (1)(b) and (f); Arts. 30 (2)(d), (4)(d) and 38 (2)(c) of Regulation 1198/06.

37 European Commission, *Facts and Figures on the CFP 2006*, p. 25.

38 European Commission, *The European Fisheries Fund 2007 – 2013*, (Luxembourg: 2006), p. 10.

39 Regarding market stability see Reasons (4), (11) and (27); regarding increasing profitability see Reason (6) and Arts. 2 and 3; regarding increasing the variety of supply, see Reason (8) of Regulation 104/2000 on the common organisation of the markets in fishery and aquaculture products, OJ 2000 No. L017/22.

40 See Arts. 7 to 16 of Regulation 2142/70 on the common organisation of the market in fishery (and aquaculture) products, OJ 1970, No. L236/5; Arts. 8 to 17 of Regulation 100/76, OJ, 1976, No. L020/1; Arts. 9 to 18 of Regulation 3796/81, OJ 1981, No. L379/1; Arts. 9 to 21 Regulation 3687/91, OJ 1991, No. L354/1; Arts. 8 to 18 of Regulation 3759/92, OJ 1999, No. L388/1.

41 Cases 3, 4 and 6/76, Kramer and others [1976] ECR 1279, paras. 30-33.

42 Faeroe Islands OJ 1980 No. L226/12, Norway OJ 1980 No. L 226/47, Senegal, OJ 1980 No. L226/16, Guinea Bissau, OJ 1980 No. L226/33, Guinea OJ 1983 No. L111/2, Equatorial Guinea OJ 1984 No. L 188/2, Sao Tomé and Principe OJ 1984 No. L54/2, Madagascar, OJ 1986 No. L73/26, USA OJ 1984 No. L272/1, Seychelles OJ 1987 No. L160/1, Mozambique, OJ 1987 No. L201/1, Gambia, OJ 1987 No. L146/1, Angola, OJ 1987 No. L341/1, Dominican Republic, OJ 1993 No. L299/1, Comoros, OJ 1988 No. L137/24, Mauritius, OJ 1989 No. L159/1, Sierra Leone, OJ 1990 No. L125/27, Cap Verde, OJ 1990 No. 212/1, Ivory Coast, OJ 1990 No. L379/1, Tanzania, OJ 1990 No. L379/24.

the management of fisheries resources on the high seas.[43] Finally, the Community controls the implementation of CFP measures by Member States.[44]

3.2 The structure of the political debate

The political debate on the CFP is largely determined by the legal framework that supports the adoption of legislation under the CFP. Art. 37(2), paragraph 3 of the Treaty establishing the European Community (Treaty) provides that '[...] the Council shall, on a proposal from the Commission and after consulting the European Parliament, acting by a qualified majority, make regulations, issue directives, or take decisions [...]'.

In fact, most political initiatives connected with the CFP find their origin in the Community's executive branch, the Commission. This includes establishing recurrent management measures like, for example, the yearly setting of TACs.[45] The Commission relies on scientific data to a great extent, especially on stock assessments. This data is mainly acquired from the International Council for the Exploration of the Sea (ICES), international fisheries commissions, Member States, and third countries.[46] The scientific data is then analysed and prepared by the Commission's Scientific, Technical and Economic Committee for Fisheries (STEFC) for legislative purposes. In addition, the STECF, stakeholder committees, other Commission departments, as well as the European Parliament are also consulted. When these consultations are over, the Commission draws up a final proposal and forwards it to the Council. At the end of each year, the Council then finally decides on TACs, effort limitations and technical measures for the upcoming year for the commercially most important stocks.[47]

3.3 The interests of the Member States and negotiations in the Council

The Council is the central body within which Member States negotiate the CFP and adopt pertinent legislation.[48] It is comprised of ministerial representatives responsible for fisheries policy within their respective

---

43 See Com(1999) 613 final from the Commission to the Council and the European Parliament – Community Participation in Regional Fisheries Organisations (RFOs).

44 See Arts. 16(1), 23(4), 26(3), 27 of Regulation 2371/02. See also Council Regulation (EEC) 2847/93 establishing a control system applicable to the Common Fisheries Policy, OJ 1993, No. L261/1 as amended.

45 Art. 20 Regulation (EC) 2371/02 on the conservation and sustainable exploitation of the fisheries resources under the Common Fisheries Policy, OJ 2002 No. L358/59.

46 For the agreement between the EC and ICES, see OJ 1987 No. L149/14.

47 See, for example, Council Regulation (EC) 43/09 fixing for 2009 the fishing opportunities and associated conditions for certain fish stocks and groups of fish stocks, applicable in Community waters and, for Community vessels, in waters where catch limitations are required, OJ 2009 No. L22/1.

48 Art. 202, Art. 37(2), paragraph 3 of the Treaty and Art. 29 Regulation (EC) 2371/02.

Member State, who decide on all-important matters concerning fisheries management, structural policy, market organisation and other external concerns, to the extent that these duties have not been delegated to the Commission.[49]

Even though the CFP only constitutes a minor portion of the Community's gross domestic income, it has often proven to be a cumbersome and difficult policy area to manage. There are several reasons for that:

The most important reason is Member States' ruthless promotion of their national economic interests. This basically leads to struggles between the Member States for fishing rights and other benefits within the Council. Two examples will be given:

First, every year the Commission proposes a TAC to the Council, which is, as indicated above, generally based on scientific advice and interdisciplinary consultations. However, in most cases, the Commission's suggested maximum limit is increased by the Council.[50] This is due to Commission's proposals generally conflicting with the Member States' economic interests, as their own national fishing industries require high exploitation rates. National government's short-term interest in getting re-elected cause fisheries ministers to pursue populist policies.[51]

Second, the struggles between the Member States within the Council also became apparent during the European enlargement process in the 1970s. On 30 June 1970, on the eve of the accession of the United Kingdom, Denmark, Norway and Ireland, the 'Community of the six' agreed on the main elements of a CFP, and unilaterally imposed the free access principle on all new Member States.[52] This tactic generated mistrust and intense political conflict, and ultimately resulted in the non-accession of Norway, the limited accession of the Faroe Islands, and Greenland leaving the Community.[53] Fisheries policy was also a contentious issue in the accession negotiations with Spain and Portugal

---

49 Arts. 203, paragraph 2 and 202, paragraph 3 of the Treaty, Art. 211, paragraph 4, Art. 37(2), paragraph 2 of the Treaty.

50 Communication Com(2007) 295 final from the Commission to the Council on Fishing Opportunities for 2008, p. 5; D. Symes, 'The European Community's Fisheries Policy' (1997) 35 *Ocean & Coastal Management*, p. 147; T. Daw and T. Gray, 'Fisheries science and sustainability in international policy: a study of failure in the European Union's Common Fisheries Policy' (2005) *Marine Policy*, 29, pp. 189-197; J-P. Boude, J. Bonceur and D. Bailly, 'Regulating the access to fisheries: learning from the European experience' (2001) 25 *Marine Policy*, p. 315; Communication Com (2002) 135 final Green Paper on the Future of the Common Fisheries Policy Vol. 1, pp. 6-8; Rat von Sachverständigen für Umweltfragen (SRU), *Meeresschutz für Nord- und Ostsee – Sondergutachten*, p. 126; M. Holden, *The Common Fisheries Policy*, (Oxford: 1996), pp. 57-60.

51 E. da Conceicao-Heldt, *The Common Fisheries Policy in the European Union*, (New York: 2004), pp. 39-41.

52 See Art. 2 of Regulation 2141/70, OJ 1970 No. L236/1. See also C. Nonnenmacher, 'Das EG-Fischereiregime als supranationale Ressourcenordnung' (1989) *Europarecht*, 1, p. 35; M. Holden, *The Common Fisheries Policy*, p. 19.

53 H. Schneider, 'Die gemeinsame Fischereipolitik' in Bleckmann (ed.), *Europarecht*, (Köln: 1997), p. 815.

in the 1980s. These difficulties are clearly visible in the Spanish and Portuguese accession agreements, which established highly complicated access and distribution schemes.[54]

The adoption of a comprehensive policy reform in 2002 aimed at resolving the most obvious deficiencies in the CFP[55] and improving current condition of fisheries resources, as well as ending the Council's continued practice of adopting catch limits and management measures at unsustainable levels. In the end, however, it cannot be perceived a breakthrough in sustainable and environmentally friendly fisheries management.[56]

3.4 The Commission's role in the CFP

Under the Community's institutional framework, the Commission has been assigned three basic tasks, which help to shape its own political agenda regarding the CFP. According to the Treaty, the Commission controls, implements and institutes Community policies.[57] Its primary political motivations are to further the European integration process, to act as a mediator for Member States in the Council, and to maintain its own powers.[58] Its position in the political process is also defined according to its bureaucratic properties

54 Limitations existed regarding access and number of Spanish and Portuguese fishing vessels to Community waters , see Arts. 157 to 166, 157 to 160, 164, 165 and Arts. 347 to 353, 351, 352 of the Accession Treaties with Spain and Portugal, see OJ 1985 No. L302/1. A complete integration of Spain and Portugal's fleets followed on 31 December 1993, see Council Regulation (EC) 1275/94 on adjustments to the arrangements in the fisheries chapters of the Act of Accession of Spain and Portugal, OJ 1994 No. L140/1.

55 Council Regulation (EC) 2371/2002; Council Regulation (EC) 2370/2002; Council Regulation (EC) 2369/2002.

56 Communication Com(2007) 295 final from the Commission to the Council on Fishing Opportunities for 2008 – Policy Statement from the European Commission; Commission Com(2007) 828 final annual report from the Commission to the Council and the European Parliament on Member States' efforts to during 2006 to achieve a sustainable balance between fishing capacity and fishing opportunities, pp. 9-10; WWF, *WWF Mid-Term Review of the EU common Fisheries Policy*, (Brussels: 2007); Rat von Sachverständigen für Umweltfragen (SRU), *Meeresschutz für Nord- und Ostsee – Sondergutachten*, p. 126; Commission Staff Working Document SEC(2008) 2760 accompanying the Proposal for a Council Regulation establishing a Community control system for ensuring compliance with the rules of the Common fisheries Policy, p. 11; Court of Auditors, *Special Report 7/2007*; House of Lords – European Union Committee, *The Progress of the Common Fisheries Policy* (London: 2008).

57 Art. 211 of the Treaty. In principle, the Commission is the sole organ of the EC that initiates legislative proposals. However, the Council and Parliament can request that the Commission submit proposals to the Council, pursuant with Art. 211, Art. 208 and Art. 192, paragraph 2 of the Treaty.

58 Marc E. Smyrl, 'When (and how) Do the Commission's Preferences Matter?' (1998) 36 *Journal of Common Market Studies*, pp. 79-99; Conceicao-Heldt, *The Common Fisheries Policy in the European Union*, pp. 46-47.

i.e., continuity, control of expertise and an information asymmetry:[59] Continuity refers to the fact that compared to Council members and Parliamentarians, Commission staff is free from electoral pressure and remains in office for a longer time. This allows the Commission to develop and advance long-term approaches to fisheries management. Expertise monopoly and information asymmetry basically refer to the fact that, compared to other political actors, public administrations in general and the Commission in particular are often in lead of information.[60]

The Commission's strong legislative power derives from its exclusive right to initiate proposals,[61] from the fact that the Council may only overturn its proposal by unanimous vote,[62] and by its right to adopt provisional measures in case Member States do not manage to agree on conservation measures such as the annual TACs.[63] Given that in practice Member States hardly ever reach unanimity in the Council with regards to the adoption of utilisation and conservation measures, most CFP measures can only be established with the Commission giving its consent. Thus, the Commission's political responsibility for the current state of the CFP's overmodest political results should not be underestimated.[64] It actually has been observed in a range of policy areas that the Commission often anticipates Council preferences, mirroring them in its own proposals.[65] Under the CFP, for example, the Commission regularly proposes higher TACs than recommended by ICES, which are then increased by the Council.[66]

Nevertheless, as the executive branch of the Community, the Commission is required to take a more objective view of the CFP than the Council or Parliament. Consequently, the Commission's approach to the CFP tends to be more integrated, comprehensive and conservation-oriented, than the legislation that is finally adopted. The Commission has tried in many cases to integrate other

59 Da Conceicao-Heldt, *The Common Fisheries Policy in the European Union*, pp. 46-47.

60 *Ibid.*, pp. 46-47.

61 Arts. 211 and 250 of the Treaty.

62 See Arts. 211 and 250(1) of the Treaty; G. Garrett, 'International Co-operation and Institutional Choice: The European Community's Internal Market' (1992) 46 *International Organization*, pp. 533-560.

63 The competence to adopt measures in such cases is based on Arts. 10 and 211 of the Treaty, see van Rijn, 'Art. 37 EGV' in H. von der Groeben and J. Schwarze (eds.), *Kommentar zum Vertrag über die Europäische Union und zur Gründung der Europäischen Gemeinschaft*, (Baden-Baden: 2004), p. 1241.

64 *Payne* erroneaously assumes an uniform Member States interest in setting fishing opportunities as high as possible. According to *Payne*, unanimity allows Member States to to overrule the Commission, see Payne, Policy-Making in Nested Institutions: Explaining the Conservation Failure of the EU's Common Fisheries Policy', p. 315.

65 Mark A. Pollack, 'Delegation, agency, agenda setting in the European Community' (1997) 51 *International Organization*, pp. 99-134.

66 See Chapter on Fisheries Management under the CFP in this report.

important aspects into its fisheries policy such as, for example, biodiversity,[67] environmental policies[68] and good governance.[69]

3.5 The political dialogue with the European Parliament

Art. 37(2), paragraph 3 of the Treaty only requires that Parliament be consulted during the legislative process. Moreover, it does not have the right to initiate legislative procedures itself, but may only request that the Commission submit a proposal on matters on which it considers a Community act is legally necessary for the implementation of the Treaty.[70]

Parliament's assent is required, however, whenever the Community enters into agreements with foreign states or international organisations, which have 'important budgetary implication for the Community […]'.[71] Parliament also influences the CFP substantially through its powers regarding the yearly adoption of the Community budget.[72] It also issues regular reports on its own initiatives.[73] Furthermore, throughout the life of the CFP, it has actively promoted special interests, such as the free access principle and, despite the fact it lacks the express power to do so, has even made proposals to the Council.[74] It must also be noted that in the course of the adoption of the Lisbon Treaty, the Parliament's legislative powers will be expanded. In case of ratification, assent will be required in all cases except price fixing, setting of levies, structural aid, and the adoption of quantitative limitations and the allocation of fishing opportunities.[75]

---

67 Commission Communication Com (2001) 162 final – Biodiversity Action Plan for Fisheries, pp. 4-32.

68 Commission Communication Com (2001) 143 final – Elements of a Strategy for the Integration of Environmental Protection Requirements into the Common Fisheries Policy, pp. 3-22.

69 Since the reform in 2002, the CFP is guided by four principles of good governance laid down in Art. 2(2) of Regulation 2371/2002. Regarding transparency and industry participation and environmental stakeholders, see Commission Decision 1999/478/EC of 14 July 1999 renewing the Advisory Committee on Fisheries and Aquaculture, OJ 1999 No. L187/70-73 (amended by Commission Decision 2004/864/EC) and Council Regulation (EC) No. 657/2000 of 27 March 2000 on closer dialogue with the fishing sector and groups affected by the common fisheries policy, OJ 2000 No. L80/7.

70 Art. 192, paragraph 2 of the Treaty.

71 Art. 300(3), paragraph 2 of the Treaty.

72 Art. 272(3)-(8) of the Treaty. The powers of the Parliament are laid down in Section 38, Annexes, IV, VI of the Interinstitutional Agreement of 6 May 1999 between the European Parliament, the Council and the Commission on budgetary discipline and improvement of the budgetary procedure, OJ 1999 No. C172/1-22. Annex IV of the latter Agreement defines expenditures from the Common Agricultural Policy and Expenditure resulting from international fisheries agreements as compulsory and makes them subject to the regular procedure of Art. 272(3)-(8) granting fewer powers to Parliament than Art. 272(9) would, if they were categorized to be non-compulsory expenditures.

73 See online at www.europarl.eu.int/committees/pech_home_en.htm.

74 Carol Nonnemacher, *Die gemeinsame Fischereipolitik der Europäischen Wirtschaftsgemeinschaft*, (Tübingen: 1990), pp. 168-172.

75 See Art. 43(3) of the not yet ratified Treaty on the Functioning of the European Union.

Parliament's Committee on Fisheries, which is composed of parliamentarians from different European parties from different Member States, is the Parliament's responsible forum that deals with the opinions, reports, and requests for proposals. These Parliamentarians are mainly perceived to be 'intermediaries for the fishing industry's territorialized interests in the Community sphere'.[76] The committee's one-sided economical interests as well as a lack of control of its reports through the plenary make the Parliament a weak promoter of sustainability under the CFP. Nevertheless, environmental NGOs also try to exert their influence by lobbying 'green' parliamentarians.[77] Parliament also organises informal 'Intergroups' consisting of members from different political parties and factions with an interest in particular issues. Some Intergroups are concerned with fisheries issues, such as the Intergroups on Fisheries, on Animal Welfare, on Sustainable Development and on Maritime Affairs.[78]

3.6 Public perception and its integration in the CFP

There are no empirical studies specifically dedicated to the topic of public awareness of fisheries issues. Thus, the following section will attempt to describe how affected non-governmental actors perceive fisheries issues and how their perceptions and interests influence the political discourse. The subject of fisheries is an important one amongst Europe's non-governmental actors working in commercial, environmental, development and consumer groups and the media.

Due to the range of institutional structures and manifold interests at national and regional levels, interest representation activities in the commercial fisheries sector take place at the national level. Thus, representation at the Community level largely depends on the Member States acting as agents.[79] However, interest representation also exists at the European level. Associations such as Europêche (Association of the National Organisations of Fishery Enterprises in the European Union) actively represent the interests of fishers, fishing enterprises and shipowners in Brussels.[80] Direct communication with the Commission occurs within the Consultative Committees on Fisheries and Aquaculture (ACFA)[81] and, since 2004, also within the new Regional Advisory

76 C. Lequesne, *The politics of fisheries in the European Union*, (Manchester: 2004) p. 40.

77 *Ibid.*, p. 40.

78 See for example www.ebcd.org/EPISD.html.

79 Lequesne, *The politics of fisheries in the European Union*, pp. 43-53; da Conceicao-Heldt, *The Common Fisheries Policy in the European Union*, (New York: 2004), pp. 42-43; see also M. Shackleton, 'Fishing for a Policy? The Common Fisheries Policy of the Community', in Wallace, Wallace and Webb (ed.), *Policy-Making in the European Community*, (Chichester: 1983), p. 357.

80 See online at www.europeche.org.

81 Legislation establishing and modifying the ACFA Committee: Commission Decision 71/128/EEC; Commission Decision 1999/478/EC; Commission Decision 2004/864/EC.

Councils (RAC).[82] In addition, the fishing industry also tries to influence the CFP via the Fisheries Committee of the European Parliament.

In general, the European fishing industry tends to regard the CFP as hierarchical ('top down'), and out of touch with realities of the business.[83] They often complain that their own knowledge about the state of fisheries resources is not given enough weight by the Commission's scientific advisors.[84] In many cases, their views may be characterised as being one-sided, and determined by their own (often 'short-term') economic interests. In pursuit of these interests, industry members often lobby governments at the regional and national levels. They also provide useful informational and other services to their members.[85]

The processing sector has helped to establish private regimes for the labelling of sustainably-caught fish products. Unilever, the world's largest buyer of fish, has joined with the WWF to set up an organisation that provides an eco-labelling regime for fish products called the Marine Stewardship Council (MSC).[86] The MSC approves fish products that have been caught and processed according to sustainable and environmentally sound methods. These standards have been developed by the Marine Stewardship Council together with scientists and stakeholders from all over the world. Every day, the Council assesses fisheries and their products from around the world.

Environmental, development and consumer NGOs pay much attention to overfishing and other threats to marine living resources.[87] NGOs are often very critical of the CFP. They conduct research, review Community legislation, issue reports and opinions to the Community institutions and organise public campaigns and educational programs, in order to draw attention to problems in the industry. Formal communication with the Commission mainly takes place through the ACFA and the Regional Advisory Councils. In the reform of the ACFA in 1999, environmental, development and consumers groups were given 3

---

[82] See Art. 31 of Regulation 2371/02 and Council Decision (EC) 585/04 establishing Regional Advisory Councils under the Common Fisheries Policy, OJ 2004 No. L256/17; as amended by Council Decision (EC) 409/07, 2007 No. L155/68.

[83] Holden, *The Common Fisheries Policy*, pp. 1-2.

[84] D. Garrod, The Common Fisheries Policy – Now', in Holden (ed.), *The Common Fisheries Policy*, (Oxford: 1996), pp. 270-271.

[85] Examples for Germany, Scotland and the Netherlands are Bundesverband der Deutschen Fischindustrie und des Fischgroßhandels, see online at www.fischverband.de; The Scottish Fishermen's Association, see online at www.sff.co.uk; and the Productschap Vis, see online at www.pvis.nl.

[86] See online at http://eng.msc.org/. See also Tavis Potts and Marcus Haward, 'International Trade, Eco-Labelling, and Sustainable Fisheries – Recent Issues, Concepts and Practices' (2006) *Environment, Development, and Sustainability*, pp. 91-106.

[87] Friends of the Earth Europe online at www.foeeurope.org; Greenpeace Europe Unit online, available at http://eu.greenpeace.org/issues/oceans.html; WWF, available online at www.panda.org; Oceana, available online at www.oceana.org. A network of smaller, more locally oriented NGOs is Seas at Risk, see online at www.seas-at-risk.org.

of the 21 seats in the assembly. In the Regional Advisory Councils, they are allotted not more than one-third of the seats.[88] Communication with the Parliament is informal, often through individual parliamentarians, especially members of the Green Parties as well as through Intergroups.[89]

The media throughout Europe regularly pays a lot of attention to the fisheries problem. For example, major German, English, Dutch and Spanish newspapers (excluding the rainbow press) continuously issue articles on the poor state of marine resources, the ensuing social problems from the decline in fisheries, and on the CFP in general. English and German public service broadcasters, particularly BBC and 3Sat, aired a series of programs concerned with the state of Community fisheries from 2005 to 2008. It remains to be seen whether the media's interest will continue and whether increased media interest will actually reach consumers and policy makers.

88 Art. 31 (2) Council Regulation 2002/2371/EC; Art. 5 (3) Council Decision 2004/585/EC; Arts. 2, 3 (1)(2), 7 Commission Decision 1999/478/EC and Art. 1 Commission Decision 1999/478/EC.

89 Lequesne, *The politics of fisheries in the European Union*, p. 40.

CHAPTER 2

Institutional and Organisational Structures

The following chapter will analyse the institutional and legal structure of fisheries governance in the European Community. To this end, attention will be paid to the choice of the legal basis for the policy under the Treaty, as well as the scope of powers granted by the Treaty in relation to different CFP areas. It will describe the substantial and territorial powers claimed by the Community in secondary legislation and outline the division of competences between Member States and the Community. The focus will then shift to the relationship between the CFP and the Community's environmental policies.

1 Community competences under the Treaty

The Treaty provides the legal foundation on which the Community builds its fisheries policy. It regulates the distribution of competences between the Community and the Member States, and allocates competences to the Community organs. In accordance with its rules, the Community has adopted legislation in the area of fisheries management (including control and enforcement), structural policies, market organisation and external relations, *i.e.*, international fisheries access agreements or fisheries management agreements.

1.1 Legal bases

According to the fundamental 'principle of conferred powers', the Community may act only where it has been authorised to do so under the Treaty.[1] Thus, each legislative act of the Community must be based on a Treaty provision that confers upon it legislative powers.[2] Different policy areas are supported by different legal bases under the Treaty. In addition, Community organs participate in the legislative process only insofar as powers have been conferred on them under the Treaty.[3] Thus, the choice of the legal basis for a particular policy has important institutional and political consequences.[4] In order to safeguard the principle of conferred powers, the choice of the legal basis must be founded on objective elements and open for judicial review by the ECJ.[5] Reviewing the legislators' choices, the ECJ considers the aims, objectives and content of the measure in connection with the powers conferred by a specific Treaty provi-

1 Arts. 5(1) and 249(1) of the Treaty.

2 Armin von Bogdandy and Jürgen Bast, 'EGV Art. 5', in Grabitz and Hilf (ed.), *Das Recht der Europäischen Union*, (München: 2005), para. 7.

3 Art. 7(1) of the Treaty.

4 Holly Cullen and Andrew Charlesworth, 'Diplomacy by other means: the use of the legal basis litigation as a political strategy by the European Parliament and Member States' (1999) 36 *Common Market Law Review*, pp. 1243-1270.

5 Damian Chalmers, Christos Hadjiemmanuil, Giorgio Monti and Adam Tomkins, European Union law: text and material, (Oxford: 2006), p. 143.

sion.[6] The Court determines the limits of the powers granted under the Treaty by interpreting the language of the provision in question and analysing it 'in the light of its purpose and place in the scheme of the Treaty'.[7]

There are several provisions in the Treaty that explicitly mention fisheries either in combination with or under those sections on the Common Agricultural Policy (CAP).[8] Consequently, the Community might simply choose to base its fisheries policy on the provisions on agriculture in Arts. 32 to 38 of the Treaty – particularly Art. 37(2), third paragraph. In fact, however, the CFP is composed of different policy groupings. It includes measures helping the fishing sector to adapt to available resources; it intervenes in the market, and increasingly targets environmental objectives. It could be argued, for example, that in the context of fisheries policies, conservation measures primarily protecting non-target species or marine habitats ought to be based on Art. 175(1) of the Treaty, *i.e.*, the legal basis for environmental measures.[9] While Art. 175(1) refers to the co-decision procedure, providing the Parliament with the competence to finally reject legislative proposals, Art. 37(2), third paragraph limits Parliament's powers to the contribution of non-binding opinions. In addition, it is likely that the Council of Environmental Ministers would adopt more conservation oriented fisheries legislation than the Council of Fisheries Ministers has done throughout the last twenty-five years.

In its early days, there was concern about finding the right legal bases to support a Community fisheries policy. Among chosen provisions were Arts. 26, 36, 37, 150 and 308 (previously Arts. 7, 28, 42, 43, 127 and 235).[10] In principle, under Community law it is possible to base legislation on several Treaty provisions. However, this is not permitted where such provisions refer to different legislative procedures in which Community organs are provided with varying

---

6 Case C 155/91 Commission v. Council [1993] ECR I-939, paras. 5-21; Case 165/87, Commission v. Council [1988] ECR 5545, para. 5; Case 176/03, Commission v. Council [2004] ECR I-11671, para. 51.

7 Cases 188-190/80, France, Italy and United Kingdom v. Commission [1982], ECR 2545, para. 6.

8 Arts. 3(1)(e) and 32(1) of the Treaty.

9 See arguments in C-405/92, Mondiet v. Armement Islais [1993] ECR I-6133, para. 17; Wolff, Fisheries and the Environment, pp. 26 27, 172; Proelß, *Meeresschutz*, pp. 311-21; More general, see G. Winter, *Towards a Legal Clinic for Fisheries Management* (forthcoming: 2009), pp. 39-40.

10 Particularly in relation to the structural measures, regulations have been based on varying Treaty provisions. However, none of these provisions required a different legislative procedure in the sense that the powers of the Community and its organs varied from Art. 37(2), third paragraph. See, for example, Regulation 2141/70 on structural measures. The latter was based on Arts. 36, 37 and 308 (ex Arts. 7, 42, 43, 235). See also Regulation 2088/85 on the Integrated Programme for the Mediterranean being based on Arts. 37, 150 and 308 (ex Arts. 43, 127 and 235). Later regulations have merely referred to Arts. 36 or 37 (ex Arts. 42 and 43) or solely to Art. 37 (ex Art. 43). Using different provisions to support this legislation arose from legal questions in relation to the Treaty and its meaning. For a comprehensive overview of legislative bases prior to 1988, see H. Schneider, *Die gemeinsame Fischereipolitik der Europäischen Gemeinschaften*, (Köln: 1988), pp. 21-23.

powers.[11] After the adoption of the Single European Act, no fisheries regulation has ever referred to the Treaty section on the environment, found in Art. 175(1) (previously Art. 130 s); and since 1983 all of the basic regulations arising from the CFP have primarily been based on Art. 37 (ex 43).[12] In order to find objective grounds justifying Art. 37 as the CFP's main legal basis for CFP measures, one has to take a closer look at the structure of Arts. 32 to 38 as well as of Art. 6 of the Treaty, which sets out the so-called 'integration principle'.

The exact language used by Arts. 3(e) and 32 to 38 of the Treaty indicates that the scope of the fisheries policy is somewhat limited. Art. 32(1) and (3) combined with Annex I of the Treaty state that fish, crustaceans and molluscs are considered to be 'agricultural products', and thus are subject to Arts. 33 to 38, which regulate the Community's agricultural policy. The legal basis for secondary fisheries legislation could thus be Art. 37(2), third paragraph. However, Art. 32 only reads, 'the common market shall extend to agriculture and trade with agricultural products'. Based on this wording, the CFP likely only applies to a common market that extends to trade in fisheries products, but not necessarily a structural or a comprehensive fisheries management policy.[13] However, such arguments were primarily advanced by Member States who sought to maintain their powers to govern fisheries issues. Today, neither Member States nor legal scholars argue in favour of such a narrow interpretation.[14]

Although it may be desirable to further include the Council of Environmental Ministers into the decision making process, there are several legal and political arguments which favour relying on Art. 37(2), third paragraph, under the current Treaty as the sole legal basis for a comprehensive CFP. At present, the integration principle, laid down in Art. 6 of the Treaty, must provide a sufficient basis for the Community to include conservation aspects into the CFP.

The first legal argument in favour of this view derives from the ECJ's understanding of Art. 102 of the 1972 Act of Accession of Denmark, Ireland and the United Kingdom.[15] Art. 102 of the Act of Accession states that 'from the sixth year after the accession at the latest (*i.e.*, the end of 1978),[16] the Council, acting on a proposal from the Commission, shall determine conditions for fishing with a view to ensuring protection of the fishing grounds and conservation of the biological resources at sea'.[17] The ECJ has ruled that even though Art. 102 was

---

[11] Cases C-164-165/97, Parliament v. Council [1999] ECR I-01139, para. 14.

[12] Some regulations have been based on Arts. 26 and 36, as well, see Regulation 104/2000.

[13] Proelß, *Meeresschutz*, pp. 312-314.

[14] Nonnenmacher, *Die gemeinsame Fischereipolitik*, p. 93.

[15] Case 61/77, Commission v. Ireland [1978] 417, paras. 28-32.

[16] Brackets included by the author.

[17] From a dogmatic perspective, scholars and different decisions of the ECJ have taken diverging views on whether Art. 102 of the Act of Accession merely complements Art. 32 et seq. (ex Art. 38 et seq.) of the Treaty or whether it forms an independent legal basis regarding a comprehensive CFP. Since Art. 102 was only in force until the end of 1978, this question shall not be answered here. However, for an in-

only valid until the end of 1978,[18] in relation to the protection and conservation of fish stocks, Art. 102 must be seen as an 'addition' to the Treaty provisions on agriculture.[19] The decision clearly indicates that the ECJ views Art. 37(2), third paragraph of the Treaty as the legal basis for a comprehensive fisheries policy, including fisheries management, the protection of fishing grounds and the conservation of fisheries resources. The Court has explicitly stated that 'fisheries are treated as agriculture under the terms of Art. 32 (previously Art. 38)' of the Treaty.[20]

Second, prior to the entering into force of the Maastricht Treaty in 1993, Community activities listed in Art. 3(1) of the Single European Act merely provided that the Community should include 'a common policy in the sphere of agriculture'. The Maastricht Treaty later added 'and fisheries' to this provision. Thus, the Member States did not choose to insert a separate chapter or title on fisheries. Since Art. 3 serves as a main interpretive provision for subsequent policies and other Treaty provisions, a broader interpretation of the existing provisions on agriculture, *i.e.*, Art. 32 to 38, ought to be applied.[21] Furthermore, the adoption of amendments under the Maastricht Treaty is generally interpreted as a matching of Treaty provisions with the real state of Community integration.[22] Thus, inserting 'fisheries' into Art. 3(1)(e) which governs agriculture, as well as not adopting more specific provision on fisheries or assigning it to Art. 175(1) of the Treaty, can be seen as an expression of the Member States' political will to continue its CFP under the provisions on agriculture.[23]

Third, even though the objectives of the CAP under Art. 33 have primarily been designed to guide the agricultural policy sector, particularly 'to increase productivity', the remaining objectives are applicable to a comprehensive fisheries policy. For example, Art. 33(1)(a) refers to 'promoting technical progress' and 'ensuring the rational development of agricultural production and the optimum utilisation of the factors of production, in particular labour'. Art. 33(1)(b) mentions 'a fair standard of living for the agricultural community, in particular by increasing the individual earnings'. It can thus be argued that the latter three objectives call for the adoption of common structural interventions into the fisheries production process.[24] In addition, Art. 33(1)(d) provides for the 'stabilisation of markets' and empowers a common market organisation to intervene

---

depth discussion on this issue, see Churchill, *EEC Fisheries Law*, pp. 29 to 30 and Nonnenmacher, *'Die gemeinsame Fischereipolitik'*, pp. 104-105.

18 Churchill, *EEC Fisheries Law*, pp. 30 and 46 in reference to an interpretation of Art. 102 by the ECJ in joined Cases185-204/78, Officier van Justitie v. Van Dam en Zonen [1979] ECR 2345.

19 Case 61/77, Commission v. Ireland [1978] 417, para. 30.

20 *Ibid.*, para 28.

21 Proelß, *Meeresschutz*, p. 313.

22 Armin von Bogdandy, 'EGV Art. 3', in Grabitz and Hilf (ed.), *Das Recht der Europäischen Union*, (München: 2005), p. 2.

23 Fischer, *'Die gemeinsame Fischereipolitik'*, p. 1.

24 Schneider, *Die gemeinsame Fischereipolitik der EG*, pp. 52 and 53.

in the market when necessary.[25] Last but not least, Art. 33(1)(d), which 'assur[es] the availability of supplies', in the context of fisheries, supports a comprehensive fisheries management or conservation system.[26]

Fourth, regarding conservation measures which go beyond protecting commercially exploited stocks, Art. 6 of the Treaty provides that 'environmental protection requirements must be integrated into the definition and implementation of the Community policies and activities referred to in Art. 3, particularly with a view to promoting sustainable development'. The principle is based on the assumption that environmental policies cannot be viewed as separate from other policies.[27] As stated above, Art. 3(1)(e) mentions fisheries policy, and therefore environmental protection requirements must be integrated into the fisheries policy framework. Due to the adoption of the integration principle under the Single European Act in 1987, policy objectives in areas other than the environment have in fact broadened under Community law.[28] This, of course, also applies to the CFP,[29] which can thus touch upon environmental issues.[30] In a case regarding the limitation on the use of driftnets, the ECJ has ruled that as long as fisheries are the main object of regulation, environmental issues may be regulated under the provisions on agriculture.[31] Thus, according to this decision, the Community may adopt a broad range of environmental measures based on Art. 37(2), third paragraph.

There are also *political* arguments that support the approach of basing the CFP solely on Art. 37(2), third paragraph. The first argument is that any division of the Community's legislative competence regarding the CFP would disintegrate fisheries management. Involving environmental ministers and the Parliament under Art. 175(1) in the legislation process only on specific questions, *i.e.*, only where environmental conservation aspects would be touched, may disturb the establishment of an expeditious and coherent legislative approach to fisheries management. The Community has first-hand experience in this regard, since there has already been a disintegration of structural and management policies under the existing regime. Another argument supporting the use of Art. 37(2), third paragraph, as the sole basis of CFP is that the Community has more legislative competences in relation to fisheries policies than in rela-

---

25 *Ibid*, p. 54; Kapteyn and van Themaat, Introduction to the Law of the European Communities, pp. 1128-1129.

26 Wolff, *Fisheries and the Environment*, p. 23, Schneider, *Die gemeinsame Fischereipolitik der EG*, p. 21.

27 Krämer, *EC Environmental Law*, p. 21.

28 Jans, *European Environmental Law*, (Groningen: 2008), p. 19; This does not lead to environmental policy losing its character as a stand-alone policy under Art. 174 et seq. of the Treaty, see C. Calliess, 'Art. 6 EGV', in Calliess and Ruffert (eds.), *Kommentar zu EU-Vertrag und EG-Vertrag*, (München: 2007), p. 403.

29 Booß, 'Umweltschutz und Fischerei', p. 1526-1527; Nonnenmacher, Die gemeinsame Fischereipolitik, pp. 93-94; Proelß, Meeresschutz, pp. 312-314; van Rijn, ,Fischereipolitik', p. 1250.

30 Jans, *European Environmental Law*, p. 19.

31 Case C-405/92, Mondiet v. Armement Islais [1993] I-6133, para. 24.

tion to environmental issues. The competence to govern the conservation of sea resources lies exclusively with the Community. Taking into account the international dimension of fisheries management, the CFP's supranational approach is advantageous in comparison to the less integrated approach under the Community's environmental policy.

1.2 Territorial scope

According to Art. 299(1) and (2), the Treaty applies in all Member States as well as the Azores, Canary Islands, Madeira as well as French oversees departments. Overseas territories, in particular, Greenland, listed in Annex II of the Treaty, are excluded.[32] Art. 299(1) refers to the Member States as legal entities, but does not explicitly mention their territories. Nevertheless, the use of the term *Member States* implies that the Treaty applies to all territories under their sovereign control, including their inland waters, ports and territorial seas,[33] the latter being determined by Art. 2 of 1982 UNCLOS.

Since sovereignty does not extend to the marine areas beyond the territorial sea, neither the EEZ nor to the high seas would be included here. Thus, Art. 299(1) of the Treaty does not justify the application of the Treaty beyond the territorial sea.

However, the ECJ has made clear that whenever rule-making competences are conferred on the Community by the Treaty, these competences extend to maritime areas insofar as Member States have similar powers under public international law.[34] Given the fact that Member States are empowered by public international law to exploit fisheries resources on the high seas as well as to exploit fisheries resources exclusively within their own EEZ, the Community's competence to regulate their use and their conservation applies to these areas accordingly. Thus, fisheries governance in the EEZ and the high seas is not based on Art. 299, but on Art. 37(2), third paragraph, of the Treaty.

---

[32] Van Rijn, 'Fischereipolitik', pp. 1251-1252.

[33] Case C-286/90, Anklagemydigheden v Poulsen and Diva Navigation Corp. [1992] ECR I-6019, para. 24; see also M. Schröder, 'Art. 299 EG', in von der Groeben and Schwarze (ed.), *Kommentar zum Vertrag der Europäischen Union und zur Gründung der Europäischen Gemeinschaft*, (Baden-Baden: 2004), p. 1554; Van Rijn, 'Fischereipolitik', pp. 1251-1253; Fischer, 'Fischereipolitik', pp. 5-6; Proelß, *Meeresschutz*, p. 278. Czybulka questions the Community's competence to exclusively regulate fisheries within their territorial sea. He argues that Member States have not yet conferred their 'aquitorial' powers of the territorial sea to the Community, see Czybulka 'Forschungsbedarf im marinen Fischereirecht', in Bauer, Czybulka, Kahl and Vosskuhle (ed.), *Wirtschaft im offenen Verfassungsstaat*, (München: 2006), pp. 808, 811 and 824.

[34] Case 3, 4 and 6/76, Kramer and others [1976] ECR 1279, paras. 30/33. The decision says that it 'nonetheless follows from Article 102 of the Act of Accession, from Article 1 of the said regulation (*i.e.*, Regulation 2141/70 – brackets inserted by the author) and moreover from the very nature of things that the rule-making authority of the Community *ratione materiae* also extends, insofar as the Member States have similar authority under public international law – to fishing on the high seas'.

Based on this decision, Churchill and others have correctly concluded that the CFP rules apply to: (a) all vessels engaged in fishing activities in the territorial sea or the EEZ of one of the territories in which the CFP applies, (b) all vessels registered in the Member States fishing on the high seas (as well as to those fishing on the continental shelf for sedentary species),[35] and (c) all vessels registered in the Member States fishing in third country waters.[36]

1.3 Vertical competence order

Many authors assume that the Community has an *exclusive* competence in the area of fisheries.[37] This assumption seems to be based on several ECJ decisions in which the court has stated that the Community has a *'full and definite power'* to adopt *'measures relating to the conservation of the resources of the sea'*.[38] This assumption deserves further consideration in view of the significant repercussions that exclusivity would have on the powers of Member States. The following subsection will assess the reach of Community legislative powers under the treaty provisions on agriculture and fisheries and whether Member States retain any legislative powers in this area. To this end, it is necessary to take a close look at the exact wording of the above-mentioned ECJ decisions. It will also be important to examine the scope of secondary legislation regarding the CFP's structural and market policies. Competences in the areas of control and enforcement as well as with regards to trade policies will also be assessed.

1.3.1 Exclusive and shared competences

The Treaty does not elaborate on the different categories of competences that exist in Community law. In fact, only one category is explicitly mentioned in this regard: the group of 'exclusive competences' that are set out in Arts. 5 and 106(1) of the Treaty.[39] Moreover, the ECJ has failed to develop a clear doctrine on what 'exclusive competence' really means. Most scholars agree that the existence of an exclusive competence in Community law prohibits Member

35 See Art. 77 (1) and (4) of 1982 UNCLOS.

36 Churchill, *EEC Fisheries Law*, p. 68. Regarding situation (c), Churchill notices that CFP rules are concurrent with the third country's fisheries rules. The third country, of course, has the sole right to enforce rules within its waters; see also Van Rijn, 'Fischereipolitik', p. 1253, fn. 50; and also Wolfgang Graf Vitzthum, 'Begriff, Geschichte und Rechtsquellen des Seerechts', in Graf Vitzthum (ed.), *Handbuch des Seerechts*, (München: 2006), p. 57.

37 This has also been critically recognised by Detlef Czybulka, 'Forschungsbedarf im marinen Fischereirecht', p. 822.

38 Case 804/79, Commission v. United Kingdom, [1981] ECR 1045, para. 1; Case 141/78 France v. United Kingdom [1979], para. 1 (Summary); Case C-405/92, Mondiet [1993] ECR I-6133, para 12.

39 Art. 3(1)(d) of the Treaty on the Functioning of the European Community provides that 'The Union shall have exclusive competence in the following areas: the conservation of marine biological resources under the common fisheries policy'.

States from legislating in that particular area.[40] However, the ECJ made clear that Member States can still legislate in these areas as 'trustees of the common interest' where the Community legislator has failed to act.[41] In such cases, Member States are not acting within their own competence but rather viewed as fulfilling their obligation in Art. 10 of the Treaty.[42]

Some authors argue that there is another form of exclusive competence called 'secondary exclusive competence'.[43] According to these authors, certain secondary legislation has become so comprehensive that there is in fact very little or no room for Member States to legislate. This is often referred to as '*terrain occupé*' or '*pre-emption*'. However, the Member States can also still legislate in these areas when acting as 'trustees of the common interest'.[44]

All non-exclusive competences are regarded as *shared*. Legal scholars working in this area have vigorously debated the exact meaning of 'shared'.[45] Most scholars agree, however, that for all subject matters falling within this category, the Community's exercise of legislative powers does not exclude Member States from adopting legislation. *All* authors accept that any legislation put in place by the Member States is subject to the principle of the primacy of Community law.[46]

Thus, given that Member States are not generally excluded from lawmaking where the Community enjoys an exclusive or shared competence, it is necessary to take a close look at the scope of the competences connected to the different areas of the CFP.

---

40 Hans D. Jarass, 'Die Kompetenzverteilung zwischen der Europäischen Gemeinschaft und den Mitgliedstaaten' (1996) 121 *Archiv des öffentlichen Rechts*, 2, pp. 185-189; A. Bogdandy and J. von Bast, 'The European Union's Vertical Order of Competences: The Current Law and Proposals for its Reform' (2002) 39 *Common Market Law Review*, p. 241.

41 Case 804/79, Commission v. United Kingdom, [1981] ECR 1045, para. 30.

42 Jarass, Die Kompetenzverteilung zwischen der Europäischen Gemeinschaft und den Mitgliedstaaten', p. 186

43 Regarding the CFP's common market organisation, some authors put forward three reasons which suggest the existence of an exclusive competence in this area: First, Art. 34(2) of the Treaty obliges the Community to co-ordinate the various national market organisations. Second, only the Community may effectively achieve this goal. Third, secondary legislation pre-empts Member States from adopting legislation within the entire political field; see Booß, 'Art. 32 EGV', p. 15; Fischer, 'Die gemeinsame Fischereipolitik', p. 4. Kopp, 'Art. 37 EGV', p. 552.

44 Hans D. Jarass, ,Die Kompetenzverteilung zwischen der Europäischen Gemeinschaft und den Mitgliedstaaten' (1996) 121 *Archiv des öffentlichen Rechts*, 2, pp. 185-189

45 There are authors who argue that, in principle, all legislative powers are *shared* between the Community and its Member States, see Andreas Furrer, Die Sperrwirkung des sekundären Gemeinschaftsrechts, (Baden-Baden: 1994), pp. 90 et seq.; E.D. Cross, 'Pre-emption of Member States law in the European Economic Community: a framework analyses, p. 447-472.

46 De Búrca and de Witte, 'The Delimitation of Powers Between the EU and its Member States', pp. 210.

1.3.2 Fisheries management (internally and externally)

The establishment of exclusive Community powers for fisheries management started with the adoption of Art. 102 of the Treaty of Accession of 1972. The Treaty of Accession, constituting primary Community law, provided that from 31 December 1978 the Council had to 'determine conditions for fishing with a view to ensuring the protection of fishing grounds and the conservation of biological resources of the sea'.[47] The ECJ's decision in the *Kramer* case in 1976 concretised Art. 102, stating that 'the Community has at its disposal, on the internal level, the power to take any measures for the conservation of the biological resources of the sea'.[48] The ECJ added that 'the only way to ensure the conservation of the biological resources of the sea both effectively and equitably is through a system of rules binding on all the States concerned, including non-member countries'. Furthermore, it found that the 'competence extends – insofar as the Member States have a similar authority under public international law'[49], *i.e.*, in relation to fishing activities on the high seas. Accordingly, the ECJ concluded that the Community also had the competence to enter into international commitments for the conservation of resources of the sea.[50] Based on this competence, the Community negotiates and concludes international agreements according to the procedure laid down in Art. 300 of the Treaty.[51]

Regarding the competence of the Member States, the Court subsequently declared in *Commission v. United Kingdom* that from the day of the expiry of Art. 102 of the Treaty of Accession (*i.e.*, 1 January 1979), the Community had the full and definite power to adopt, 'as part of the CFP, measures relating to the conservation of the resources of the sea. Thus, the Member States are no longer entitled to institute any of their own conservation measures in the waters under their jurisdiction. The adoption of such measures, with the restrictions which they imply as regards fishing activities, is a matter, as from that date, of Community law'.[52] In this context, the term 'conservation' includes the allocation of resources by the Community.[53]

---

47 OJ 1972, No. L73/1.

48 Case 3, 4 and 6/76, Kramer and others [1976] ECR 1279, paras. 30-33.

49 *Ibid.*, paras. 30-33.

50 *Ibid.*, paras. 30-33.

51 Paul Nemitz, 'Fischereipolitik – Nach Art. 34 EGV', in Lenz and Borchardt (ed.), *EU- und EG-Vertrag*, (Köln: 2006), p. 583.

52 Case 804/79, Commission v. United Kingdom, [1981] ECR 1045, para. 18 in which the ECJ found that the Community had not yet adopted any comprehensive fisheries management regulation; until the Community had done so, Member States were entitled to maintain their already existing management measures; see Churchill, '*EEC Fisheries Law*', pp. 88-89. Argumentations underlining the competence to solely regulate fishing activities can be found in Case C-146/89, Commission v. United Kingdom [1991] ECR I-3533, paras. 23, 28, 29, 53 and 59 of Baseline Decision and Case C-246/89, Commission v. United Kingdom, [1991] ECR I-4585, paras. 13-17.

53 Case 3, 4 and 6/76, Kramer and others [1976] 1279, paras. 30-33.

The Court thus explicitly excluded Member States from the adoption of conservation measures in maritime waters. Insofar, the Community's competence should here be referred to as being exclusive. However, as mentioned above, the boundaries of an exclusive competence, *i.e.*, the conservation of the biological resources of the sea, may be difficult to draw.[54] In this regard, the ECJ has ruled, for example, that it is up to the Member States to decide, in accordance with international law, which vessels may fly their flags or how to define their baselines, as long as such decisions do not interfere with existing CFP conservation legislation.[55] Another case concerned a legal matter which clearly lay *within the scope* of fisheries conservation.[56] A lacuna in the existing law created a problem with regards to the implementation of respective conservation rules. Exclusivity could have implied that Member States were not entitled to fill the legislative gap with national measures. However, the ECJ has confirmed that Member States have the power to fill lacunae in Community conservation legislation.[57] The relevant case concerned a Dutch fisherman who had been prosecuted before a Netherlands criminal court. He was charged with having infringed Art. 14(3) of the former basic Regulation 171/83, by fishing for sole and plaice with beam trawls within the twelve nautical mile zone along the Netherlands coast in a vessel exceeding seventy GRT.[58] The Netherlands court found that it was not clear from the provisions of Regulation 171/81 how GRT were to be calculated. It claimed that the existing Netherlands rules determining GRT were ambiguous in the sense that they gave rise to discriminatory treatment between different fishing vessels. The ECJ confirmed that Community law did not, in fact, provide a method for calculating GRT. It concluded, that since there are no such Community rules, 'in the present state of Community law',[59] these must be determined by national law. According to *Churchill*, any action taken by Member States in this area must be non-discriminatory, relate solely to nationals, may only apply in waters under national jurisdiction or sovereignty and be compatible with the objectives of Art. 2(1) of Regulation 2371/02.[60] Of course, the Community retains the power to fill existing lacunae, making measures taken by Member States void where it considers this to be appropriate.[61]

54 Von Bogdandy and Bast, 'The European Union's Vertical Order of Competence', p. 241.

55 Case C-146/89, Commission v. United Kingdom [1991] ECR I-3533, paras. 23, 28, 29, 53 and 59 and Case C-246/89, Commission v. United Kingdom, [1991] ECR I-4585, paras. 13-17.

56 Churchill, *'EEC Fisheries Law'*, pp. 94-110; Nonnenmacher, *'Die gemeinsame Fischereipolitik'*, p. 140.

57 Case C-146/89, Commission v. United Kingdom [1991] ECR I-3533, paras. 23, 28, 29, 53 and 59 and Case C-246/89, Commission v. United Kingdom, [1991] ECR I-4585, paras. 13-17.

58 Case 86 and 87/84, Bout [1985] ECR 941, para. 2.

59 *Ibid*, para. 7.

60 Churchill, *'EEC Fisheries Law'*, pp. 94-110.

61 *Ibid.*, p. 94.

1.3.3 Structural policies and market organisation

In the above-mentioned rulings the ECJ has made clear that the Community's exclusive competence regards the conservation of the biological resources of the sea. Member States are thus not *a priori* precluded from adopting measures in the sector of structural policy and market organisation and the principle of subsidiarity as set out in Art. 5(2) of the Treaty is applicable.[62] This is particularly true for the fisheries structural policy which is part of the general structural policy.[63] However, where the Community has adopted measures, Member States' competence ends and existing national rules are superseded.[64]

The substantial scope of the existing CFP structural legislation is indeed comprehensive. At the start of the common structural policy under the CFP, the Community only required Member States to co-ordinate their respective national structural policies.[65] Over the last 38 years, the Community increasingly centralised its structural policies and substantially raised its financial commitment. Prior to the Reform in 2002, structural measures concerned the construction and modernisation of the fleet, fleet adjustments, aquaculture, fisheries agreements, joint ventures, the transfers of vessels to third countries, fishing ports, the processing and marketing sector, social measures, and the even grant of financial aid for inland fishing. After the 2002 Reform, aid for the modernisation and renewal of the fleet was severely cut back. Currently, the power of Member States to actively influence structural policies mainly concentrates on the distribution and channelling of the Community's structural aid. Nevertheless, Community law still allows for national subsidies in areas such as, for example, education and scientific research. However, Art. 7(1) of Regulation 1198/06 on the European Fisheries Fund declares Community's rules on national aid, *i.e.*, Arts. 87, 88 and 89 of the Treaty, to be applicable to Member States payments (excluding those payments which co-finance Community aid schemes). Under no circumstances are these public transfers allowed to transgress these rules, nor are they permitted to conflict with the objectives of the CFP. The powers of Member States to allocate aid to their nationals is determined more closely in the Commission Regulations and Guidelines on state aid

---

62 Martin Gellermann, Geschützte Meeresflächen in der deutschen ausschließlichen Wirtschaftszone (AWZ) – Rechtsfragen der Einrichtung und des Schutzes mariner Natura 2000 Gebiete', in Hendler, Marburger, Reinhardt and Schröder (ed.), *Jahrbuch des Umwelt- und Technikrechts 2005* (Berlin: 2005), p. 167; Thiele, 'Art. 37 EGV', p. 704.

63 Since 1993, 'the structural aspects of the CFP have been integrated as a full structural policy under the rules governing the structural funds' [...]; whereas implementation should continue under the structural funds through the Financial Instrument for Fisheries Guidance', see Reason (9) of Regulation 1260/99, OJ 1999 No. L161/1; see also Nemitz, 'Fischereipolitik – Nach Art. 34 EGV', p. 582.

64 F. Kopp, ,Art. 37 EGV', in Streinz (ed.), *EUV/EGV*, (München: 2003), p. 551; Booß, ,Art. 37 EGV', p. 15.

65 See Art. 6 et seq. of Regulation 2141/70, OJ 1970 No. L236/1.

to fisheries and aquaculture.[66] Summing up, even though structural policies on fisheries may not be exclusive, they are comprehensive and far reaching.[67]

The Community's organisation of the market in fisheries products is built on four basic pillars. The first pillar is the creation of common marketing standards and provisions for the establishment of a minimum level of consumer information. The second pillar establishes producer organisations as means for implementation. These organisations are established for the purpose of ensuring that the conditions for the sale of the products of their members are improved. To this end, they are competent to intervene in the markets. The third pillar establishes a common and stabilised price system guaranteeing a minimum price for fish products. The last pillar builds a system of trade with third countries, aiming at providing a custom tariff system that guarantees adequate supply as well as protection of the markets.[68] Member States' competences are restricted in the sense that they may not modify this system, particularly with regards to the competences of producer organisations. Despite the fact that the Community competence in the area of market organisation is not exclusive, it is far reaching.

1.3.4 Implementation, control and enforcement

In principle, Member States implement and enforce Community law.[69] This is due to the limited conferral of powers in this area.[70] Art. 10 of the Treaty provides that Member States have the obligation to implement and enforce Community law effectively.[71] Its first sentence reads: 'Member States shall take all appropriate measures, whether general or particular, to ensure fulfilment of the obligations arising out of this Treaty or resulting from actions taken by the institutions of the Community'. Despite the fact that the Community has obtained far reaching legislative powers with regards to the CFP, executive powers primarily lie with Member States. In reality, Member States' implementation and enforcement rules and practices differ.[72] Control and enforcement has not always been taken seriously, and the strength of the

---

66 Commission's Guidelines for the Examination of State Aid to Fisheries and Aquaculture, OJ 2004, No. C 229/5; Commission Regulation (EC) 1595/04 on the application of Articles 87 and 88 of the EC Treaty to State aid to small and medium-sized enterprises active in the production, processing and marketing of fisheries products, OJ 2004, No. L291/3.

67 Fischer, 'Die gemeinsame Fischereipolitik', p. 4. See also, Booß, 'Art. 37 EGV', p. 15.

68 See, Council Regulation 104/2000.

69 Wolfgang Pühs, *Der Vollzug von Gemeinschaftsrecht* (Berlin: 1997), p. 74;

70 *Ibid.*; The implementation of Community law subjects to the principles of the conferral of powers, subsidiarity and proportionality, see Siegfried Magiera, 'Durchsetzung des Europarechts', in Schulze and Zuleeg (ed.), *Europarecht*, (Baden-Baden: 2006), p. 440.

71 Armin von Bogdandy, ,Art. 10 EGV`, in Grabitz and Hilf (ed.), *Das Recht der Europäischen Union*, (München: 2002), p. 11 et. seq.

72 The Community has, nevertheless, adopted a comprehensive control regime, see Regulation 2847/93 establishing a control system applicable to the common fisheries policy, OJ 1993 No. L261/1.

responses to violations of fisheries laws varies.[73] This has irritated fishermen, who often feel they are being treated unfairly in comparison to their colleagues from neighbouring Member States.[74] In response, the Community has made efforts to promote the alignment of national measures to be taken in response to violations, despite its limited competence in this area.[75]

In addition, even though the competence to implement, control and enforce legislation lies primarily with Member States, the CFP provides a system of 'dual vigilance'.[76] While the Member States ensure control, inspection and enforcement and provide for the judicial or administrational response to infringements of fisheries legislation, the Commission, on behalf of the Community, controls national compliance efforts and tries to ensure that they are carried out properly.[77] In order to increase the effectiveness of the Commission's monitoring efforts, in 2002, the Commission was equipped with particular inspection rights under Art. 27 of Regulation 2371/02. Now, it may carry out inspections and inquiries of its own accord. However, its power does not extend beyond that of the inspectors of Member States and it has no police powers.[78] Where the Commission acknowledges infringements of the duty to implement and enforce Community law, particularly with regards to the enforcement of TACs, it may initiate an infringement procedure with the ECJ under Art. 226, take preventive measures, suspend financial aid, or, where quotas have been exceeded, reduce the quotas allotted to a Member State in upcoming years. Apart from that, the Community has equipped its own inspectors with independent inspection powers under two international treaties on the management of high seas fisheries resources. Under the Joint International Scheme of Inspection, as adopted by the NAFO and the Council in 1988, Community inspectors and the inspectors of Member States carry out the same functions. Art. XXIV of the CCAMLR allows Community inspectors to control vessels operating in the Convention area.[79]

---

73 Commission Staff Working Document SEC(2008) 2760 accompanying the Proposal for a Council Regulation establishing a Community control system for ensuring compliance with the rules of the Common fisheries Policy, p. 11; Court of Auditors, *Special Report 7/2007*; Long and Curran, '*Enforcing the Common Fisheries Policy*', pp. 66-72; Astrid Berg, '*Implementing and Enforcing European Fisheries Law*', p. 44.

74 European Commission, 'A level playing field for better enforcement of CFP rules', *Fishing in Europe*, pp. 3-6; Lequesne, *The politics of fisheries in the European Union*, p. 73.

75 See Art. 25 (2) to (5) of Regulation 2371/02.

76 Long and Curran, '*Enforcing the Common Fisheries Policy*', pp. 63-63.

77 See Arts. 23 (4) and 26 (3) of Regulation 2371/02.

78 See Art. 27(1) of Regulation 23712/02.

79 Long and Curran, '*Enforcing the Common Fisheries Policy*', pp. 61-62, 74. See Council Regulation (EEC) 436/92 amending Regulation (EEC) 1956/88 adopting provisions for the application of the scheme of joint international inspection adopted by the Northwest Atlantic Fisheries Organisation, OJ 1992 No. L54/1.

1.3.5 Trade policy

Community trade policy with regards to fisheries products basically concerns (a) trade between Member States and (b) trade between Member States and third countries.[80] The division of competences regarding trade in fisheries products is primarily determined by the Treaty provisions establishing and governing the internal market as well as those provisions concerning the customs union.

Regarding trade between Member States, Article 32(2) provides that 'save as otherwise provided in Articles 33 to 38, the rules laid down for the establishment of the common market shall apply to agricultural products'. Thus, provisions on the free movement of goods and those prohibiting quantitative restrictions determine the intra-Community trade in fisheries products.[81] These rules abolished custom duties, charges having an equivalent effect as well as quantitative restrictions on fisheries products from other Member States, and prohibited Member States from re-introducing similar measures.[82] Moreover, according to Art. 90 of the Treaty, Member States shall not, directly or indirectly, impose any internal taxation of any kind on fisheries products of other Member States in excess of that imposed directly or indirectly on similar domestic products, nor is it permitted to institute any internal taxation so as to afford indirect protection to other products. Finally, Member States are not allowed to circumvent the Community competition rules in Arts. 81 to 86 as well as respective implementing measures (adopted under Art. 83). The same applies to secondary legislation under Arts. 14 and 95, adopted to further the creation of the internal market.

Regarding trade between the Community and third states,[83] the Community has the competence to fix a common custom tariff and adopt its own export policy. The common custom tariff can be fixed either (a) autonomously by the Council (Art. 26 of the Treaty), (b) in the course of tariff and trade agreements with third states (Art. 133 of the Treaty), or (c) in association agreements with third states (Art. 310 of the Treaty). The Community has established integrated tariffs for the import of fisheries products.[84] Member States must refrain from fixing individual tariffs and amending existing Community tariffs.[85] Exist-

---

80 Churchill, '*EEC Fisheries Law*', pp. 255 et seq.

81 See Arts. 23 to 24 and 28 to 31 of the Treaty. See also, R. Priebe, 'Gemeinsamer Markt für die Landwirtschaft', in Grabitz and Hilf (ed.), *Das Recht der Europäischen Union – Kommentar*, (München: 2003), p. 15; ECJ Case 3, 4 and 6/76, Kramer and others [1976] ECR 1279, paras. 53-54.

82 See Arts. 25 and 29 of the Treaty as well as Arts. 23-24, 28, 30-31 of the Treaty; Arts. 31-38, 42 of the 1972 Act of Accession, 64, 35 of the 1979 Act of Accession.

83 According to Title V of Regulation 104/2000, trade in fisheries products with third countries is part of the market organisation.

84 The Community's custom tariffs are based on Regulation (EEC) 2658/87 on the tariff and statistical nomenclature and on the Common Custom Tariff, OJ 1987, No. L256/1.

85 Reimer Voß, 'Art. 26 EGV', in Grabitz and Hilf (ed.), *Das Recht der Europäischen Union*, (München: 1999), p. 3.

ing tariffs can be divided into two categories: regular tariffs applying to all non-Community countries, and preferential tariffs applying to selected third countries such as the ACP-Countries. While regular tariffs for fish and fisheries products range from 0% to 26%, preferential tariffs are often 0%.[86] To protect the domestic processing industry, tariffs for processed and preserved products are generally higher than those for raw fisheries products (so-called tariff escalation).[87] To influence the Community market positively and protect it against non-European competitors, tariffs are regularly adjusted in line with market developments.[88] In the past, the Community implemented measures which supported the export of fisheries products to third countries by granting export funds. This practice ceased in the early 1980s.[89]

1.4 Horizontal competence order

Many legal measures concerning fisheries can be relevant to other policy areas such as, for example, the establishment of the internal market or environmental protection and vice versa. In particular, such measures may qualify to be based on the respective competences set out in Art. 95(1) and 175(1) of the Treaty. Legislating on the basis of these provisions is distinct from the procedure under Art. 37 of the Treaty, in that parliamentary co-decision is required and the principle of subsidiarity must be met. Given that these competences are aim-based, they are substantial and very broad in scope. However, legislative acts are not allowed to be based on different Treaty provisions if this means that different legislative procedures would be invoked.

To demarcate horizontal competences, the ECJ will consider the aims, objectives and content of measures in question.[90] The Court sees these criteria as objective factors. Where the aims, objectives and content of a measure relate to different policy areas, the Court must determine the predominant purpose of the measure in accordance with its *centre of gravity approach.*[91] It basically asks 'whether the measures in question relate principally to a particular field of

---

86 European Commission online at http://ec.europa.eu/trade/issues/sectoral/agri_fish/fish/.

87 See Art. 30 of Regulation 104/2000; International Centre for Trade and Sustainable Development (ICTSD), *Fisheries, International Trade and Sustainable Development*, pp. 27, 28 and 31; See also A. von Moltke *et al.*, 'Fisheries Subsidies', pp. 101-114.

88 See Art. 28 of the market regulation (Regulation 104/2000). Exceptions and suspensions are adopted within innumerable measures in response to the demands of the Community markets; see, for example, Proposal for a Council Regulation opening and providing for the management of autonomous Community tariff quotas for certain fishery products for the period 2007 to 2009, Com(2007) 289 final.

89 Churchill, '*EEC Fisheries Law*', pp. 269-271.

90 Case C 155/91 Commission v. Council [1993] ECR I-939, paras. 5-21; Case 165/87, Commission v. Council [1988] ECR 5545, para. 5; Case 176/03, Commission v. Council [2004] ECR I-11671, para. 51.

91 *Ibid.*

action, having only incidental effects on other policies, or whether both aspects are equally essential'.[92]

The Court gave its view of the specific content of the fisheries policy in the *Mondiet* decision where it had to decide whether a certain measure had to be based on environmental or fisheries Treaty provisions. This case required the Court to review the choice of the legal basis for a driftnet regulation. Based on the wording of the preamble and the preceding reasonings of the disputed regulation, it held that the regulation was 'adopted *primarily* in order to ensure the conservation and rational exploitation of fishery resources and to limit the fishing effort'.[93] According to the ECJ, the regulation forms an integral part of the common agricultural policy, which objectives include 'ensuring the rational development of production' and 'assuring the availability of supplies' (see Art. 33 of the Treaty).[94] Environmental considerations were regarded merely as 'contributory factors' that had been validly adopted on the base of the Treaty provisions on agriculture as well as the integration clause.[95]

The decision makes it clear that the Court sees the content of fisheries policy as closely linked to fisheries management and the objectives laid down in the Treaty provisions on agriculture, *i.e.*, regulating fishing activities to ensure exploitation and conservation, 'ensuring rational development of production' and 'assuring availability of supplies'. The decision also indicates that as long as Community measures relate to fishing activities, the structure of the production sector, and market organisation of fisheries products, they must be based on the Treaty provision on agriculture and fisheries. The current practice of basing CFP legislation on the provisions on agriculture is in line with this ruling.[96]

The Court's approach to horizontal demarcation has been criticised for being vague and creating uncertainties.[97] In order to eliminate these uncertainties, it has been suggested that where several competence bases are to be considered, the more specific, task-oriented Treaty provisions (such as Art. 37(2)) would

---

92 Case C-164/97 and C-165/97, Parliament v. Council [1999] ECRI-623, para. 14.

93 Emphasis added; see Case C-405/92, Mondiet v. Armement Islais [1993] I-6133, para. 24; Wolf, 'Fisheries and the Environment', p. 171. The Court of First Instance has only recently referred to fisheries and environment being two different kinds of policies, see Case T-37/04, The Autonomous Region of the Azores v. Council, paras. 133, 144 et seq., 164 et seq., available online at http://curia.europa.eu; Regarding the 'centre of gravity theory' in general see Cases C-164-165/97, Parliament v. Council [1999] ECR I-01139, para. 14. Regarding the distinction of environmental policies and the Common Commercial- or Agricultural Policy see Case 62/88 Greece v. Council [1990] ECR I-1527, paras. 5-23; Case C-336/00, Huber [2002] I-7699, paras. 29-37; Case C-164-165/97, Parliament v. Council [1999] ECR I-1139, paras. 8-20;.

94 Case C-405/92, Mondiet v. Armement Islais [1993] I-6133, para. 24.

95 *Ibid.*, paras, 26 to 28.

96 Only in the area of market organisation, legislation is also based on Arts. 26 of the Treaty, see Regulation 104/2000.

97 Wolff, *Fisheries and the Environment*, p. 172; Proelß, *Meeresschutz*, pp. 316-317; Chalmers *et al.*, *European Union Law*, pp. 143-144.

prevail as *leges speciales* over the more general provisions, which are based on aims such as Art. 95 or Art. 175 of the Treaty.[98] Overall, however, this approach does not appear to bring substantial change to the current legislative practice with regard to the CFP.[99]

2 Substantive and territorial scope of CFP secondary legislation

The Community has adopted various regulations on fisheries management, structural policies, and common market organisation and has also concluded international agreements in this area.[100] Art. 1 of the current CFP Regulation 2371/02 sets out the substantial scope of the CFP, and presents a good overview of the areas in which the Community has made use of its above-described competences under the Treaty. According to Art. 1, the CFP should achieve the following:

1. It shall cover the conservation, management and exploitation of living aquatic resources, aquaculture, and the processing and marketing of fishery and aquaculture products where such activities are practised on the territory of the Member States or in Community waters or by Community fishing vessels or, without prejudice of the primary responsibility of the Member States.
2. It shall provide for coherent measures concerning:
   a) conservation management,
   b) limitation of the environmental impact of fishing,
   c) conditions of access to waters and resources,
   d) structural policy and the management of fleet capacity,
   e) control and enforcement,
   f) aquaculture,
   g) common organisation of the markets, and
   h) international relations.

According to Art. 32(1), the CFP shall extend to the trade in fisheries products as well as to products of first-stage processing directly related to these products. As corals and sponges are also being fished, they ought to be seen as fisheries products as well.[101] The same argument applies to cetacea, such as whales or

98 Proelß, *Meeresschutz*, pp. 316-317.

99 *Ibid.*, p. 317.

100 International agreements are negotiated in accordance with Art. 300 of the Treaty.

101 However, since this is not mentioned in Annex I, regulatory powers regarding the trade of such products remains a matter of dispute; see Fischer, 'Die gemeinsame Fischereipolitik', p. 5; van Rijn, 'Fischereipolitik', p. 1251; Booß, 'Umweltschutz und Fischerei', p. 1527; Churchill, '*EEC Fisheries Law*', pp. 53-55.

dolphins. Regulations on the protection of whales, for example, have been inconsistently adopted under the Community's environmental policies.[102]

In principle, secondary legislation has the same territorial scope as the Treaty.[103] The Community has adopted several regulations, providing management measures for different marine areas. The basic Regulation applies to all Community waters. Recognising the specific environmental and political context in the Mediterranean and the Baltic Sea, the Community has adopted tailor-made conservation and management policies.[104] In the Mediterranean, Regulation 1967/2006, concerning management measure for the sustainable exploitation of fishery resources in the Mediterranean Sea,[105] institutes restrictions on gear, and sets out fishing zones for the seven Member States adjacent to the sea.[106] The Baltic Sea management system is laid down in Council Regulation 2187/2005 for the conservation of fishery resources through technical measures in the Baltic Sea, the Belts and the Sound.[107]

3 Distribution of tasks – centralisation v. decentralisation

Even though the Community has an exclusive competence with regards to the conservation of marine resources and comprehensive powers regarding the governance of the fisheries sector, Member States hold certain administrative and legislative powers. Dogmatically, these powers partly derive from (a) the re-delegation of Community activities in this area to Member States, (b) the fact that matters have intentionally or unintentionally remained unregulated by the Community, and (c) from the Member States' competence to adopt measures on environmental protection.

---

[102] See, for example, Regulation 348/81 on common rules for imports of whales or other cetaceans products, OJ 1981 No. L 39/1 .

[103] Schröder, 'Art. 299 EG', p. 1565.

[104] Special *technical measures* also exist with regards to the Antarctic waters, see Council Regulation (EC) 600/04 laying down certain technical measures applicable to fishing activities in the area covered by the Convention on the conservation of Antarctic marine living resources, OJ 2004 No. L97/1.

[105] Council Regulation (EC) 1967/2006 concerning management measure for the sustainable exploitation of fishery resources in the Mediterranean Sea, amending Regulation (EEC) No. 2847/93 and repealing Regulation (EC) No. 1626/94, OJ 2007 No. L36/6. See also Commission Communication Com (2002) 535 final to the Council and the European Parliament laying down a Community Action Plan for the conservation and the sustainable exploitation of fisheries resources in the Mediterranean Sea under the Common Fisheries Policy, Com(2002) 535.

[106] Currently, Cyprus, France, Greece, Italy, Malta, Slovenia and Spain border the Mediterranean.

[107] Council Regulation (EC) 2187/2005 for the conservation of fishery resources through technical measures in the Baltic Sea, the Belts and the Sound, amending Regulation (EC) 1434/98 and repealing Regulation (EC) 88/98, OJ 2005 L349/1.

3.1 Re-delegated powers

Regulation 2371/02 entitles Member States to take specific measures on the conservation of fisheries resources. Since the Community has exclusive competence in this area, this assignment should be viewed as a re-delegation of powers.[108] The same applies where the Community has gained competence on the grounds of its legislative activity in other areas of fisheries governance (*'terrain occupé'*).

3.1.1 Member States emergency measures

According to Art. 8 of Regulation 2371/2002, Member States are allowed to take so-called 'emergency measures' in waters falling under their sovereignty or jurisdiction. They may do so where there is evidence of a serious and unforeseen threat to the conservation of living aquatic resources or the marine ecosystem as a result of fishing activities. This applies only in cases where any undue delay would result in a damage that would be difficult to repair. Measures shall not extend beyond three months. Member States are also required to notify the Commission, other Member States and the Regional Advisory Councils of their intention to take measures. These parties are then allowed to submit comments to the Commission on the proposed measures within five working days; and Member States may then challenge the Commission's decision at the Council within ten working days. The Council then has one month within which it may overrule the Commission's decision.

3.1.2 Member States conservation and management measures within the 12 nm mile zone

According to Art. 9(1), Member States have the right to take measures for the conservation and management of fisheries resources and to minimise the effect of fishing on the conservation of marine ecosystems within twelve nautical miles of their baseline. These measures may, in principle, apply to all fishing vessels.[109] However, they are allowed to do so only where the Community has not adopted specific measures addressing conservation and management in this area. Measures taken by Member States have to be non-discriminatory, compatible with the objective in Art. 2 of the Regulation 2371/2002, and must be as stringent as existing Community legislation. Where the vessels of other Member States are affected, these Member States, the Commission and the Regional Advisory Councils have to be consulted, pursuant to Art. 9(2). In cases where measures apply to fishing vessels of other Member

[108] Jarass, 'Die Kompetenzverteilung zwischen der Europäischen Gemeinschaft und den Mitgliedstaaten', p. 186; Churchill, '*EEC Fisheries Law*', p. 92; Long and Curran, *Enforcing the Common Fisheries Policy*, p. 59.

[109] Reason (11) of Regulation 2371/2002.

States, the same procedure applies as for the implementation of emergency measures by Member States.[110]

3.1.3 Member States measures applying to vessels flying their flag

Art. 10 allows Member States to unilaterally adopt conservation measures for the conservation of stocks in waters under their sovereignty or jurisdiction (*i.e.*, within their territorial sea and their EEZ). However, Member States measures must apply solely to fishing vessels flying the flag of the particular Member State. Measures must also be compatible with the objective in Art. 2(1) and must be as stringent as existing Community legislation. Art. 10 thus provides the possibility for Member States to subject their national fishers to more stringent conservation measures.

3.1.4 Member States option to adopt technical management measures

Art. 45(2) of Council Regulation (EC) 850/1998 for the conservation of fishery resources through technical measures for the protection of juveniles of marine organisms provides that Member States may take non-discriminatory conservation measures in respect of waters under their jurisdiction.[111] They may do so in cases where any delay would result in damage that is difficult to repair. The adoption of measures must be communicated to the Commission, which may require their cancellation or amendment. The Commission's decision may be referred to the Council within 10 working days of receiving the notification of the Commission. Member States may, according to Art. 46, also take technical measures if they strictly concern local stocks, which are of interest solely to the fishermen of the Member State involved. These measures may either supplement Community legislation or go beyond its minimum requirements, provided that they apply solely to the fishermen of the Member State and are compatible with Community law.

3.1.5 Right to reserve fishing in the 12 nm zone to Member States nationals

Art. 17(1) of Regulation 2371/02 provides that the vessels of all Member States have *equal access* to resources within Community waters. This right, however, can be restricted by Member States within waters up to 12 nautical miles from their baselines under their sovereign control or jurisdiction.[112]

110 Art. 9(2) in combination with Art. 8(3) to (6) of Regulation 2371/02.

111 OJ 1998, No. L125/1.

112 The principle of free access is also limited by Community measures listed in Arts. 4 to 10 of Regulation 2371/02. Particularly important in this respect are the annual allocations of stock-specific TAC quotas.

Art. 17(2) allows Member States to restrict fishing to vessels that 'traditionally fish in those waters from ports on the adjacent coasts'.[113] Art. 17(2) subsequently provides an exemption to the exemption, granting fishermen from other Member States fishing rights in accordance with the maintenance of neighbourhood relations or traditional fishing rights as laid down in Annex I of Regulation 2371/02.

It remains somewhat unclear, however, what the terms 'traditional' and 'ports of the adjacent coasts' in Art. 17(2) exactly mean. How long must fishing rights be exercised in order to constitute traditional fishing? Which areas does 'ports of the adjacent coasts' include?[114] Looking at the purpose and at the development of the wording of the exemption rule may provide answers to these questions.

The purpose of the exemption rule has changed throughout the history of the CFP. It was first established during the accession negotiations in 1972. In 1970, the former six Member States had agreed on the equal access principle in order to receive access to the large marine territories of the future acceding Member States. Against this background, the acceding states, *i.e.*, the UK, Ireland and Denmark, insisted upon an exemption to protect their coastal fisheries. Art. 100 of the Act of Accession of 1972 was essentially a political compromise, allowing the acceding states to restrict fishing in a six-mile zone off their coasts to `vessels which fish traditionally in those waters and which operate from ports in that geographical area'. Ten years later, Art. 6 of the first general management regulation, Regulation 170/83, also included this exception. Reason 8 of that regulation stated that the exception was maintained 'to enable the inshore fishing industry to cope with the new fishing conditions resulting from the institution of 200-mile fishing zones'.[115] The Regulation that is currently in force mentions different purposes for the exception rule in Reason 14. It states that the rules restricting access to the 12 nm zones of Member States 'have operated satisfactorily benefiting conservation by restricting fishing effort in the most sensitive part of Community waters and preserving traditional fishing activities on which the social and economic development of certain coastal communities is highly dependent.' The Commission also states that the basic objective of the 12 nm rule is to 'protect fisheries resources by reserving access to small-scale coastal fisheries activities which in general put less pressure on the stocks in these zones which often harbour nurseries and to protect the traditional fishing activities of coastal communities, thus helping to maintain their

---

Quotas are assigned to specific areas that subdivide Community waters, *i.e.*, the so-called ICES-areas. Quotas can only be fished in these specific areas. ICES areas can be viewed online at www.ices.dk.

113 Traditional fishing rights in the EC are listed in Annex I of Regulation 2371/2002.

114 Similar questions arose also with regards to Art. 6 of Regulation 170/83; see Churchill, *EEC Fisheries Law*, p. 135.

115 The ECJ has acknowledged that it recognises the Community's competence to adopt this exception rule on the basis of Art. 37 of the Treaty; see Case 103/91 Spain v. Council [2005] ECR I-2267, para. 30.

social fabric'.[116] Thus, both the wording and rationale for the exemption have changed over time. The rationale has shifted from merely protecting coastal fisheries to also protecting fish stocks from large-scale vessels. Furthermore, the current provision uses the terms 'ports of the adjacent coasts' instead of 'ports of that geographical area', and thus does not specifically refer to the Member States' 12 nm zone.

Considering the slight adjustments to the wording of the exemption rule currently in force in Art. 17(2), as well as the aim of protecting small-scale fisheries and the coastal environment from destructive large-scale fishing, Art. 17(2) should be interpreted as enabling national legislators to limit the *fishing power* of fishing vessels, and to act in consideration of the importance of fishing activity to a coastal region's economy as well as the impact on the coastal environment.[117]

However, most Member States do not give priority rights to *local* fishers traditionally fishing from ports of the adjacent coasts, but basically allow *all* vessels flying *their* flag to fish in the twelve nm zone.[118] This approach, however, may conflict with Art. 12 of the Treaty, which forbids discrimination on the grounds of nationality.[119] The principle of free access represents an attempt to incorporate Art. 12 into the CFP; it does not imply free access, but instead, means non-discriminatory access to marine resources in Community waters.[120] Exceptions to the principle of equal access are possible, but they must be objectively justified.[121] Thus, where Member States merely base their access restriction to their 12 nm zone on the grounds of nationality, their implementation practice may be considered to be incompatible with Art. 12 of the Treaty.[122]

3.1.6 Allocation of quotas assigned to the Member States

While the Council sets TACs and allocates quotas annually, Member States themselves determine the method of distributing the assigned

[116] Communication Com (2002) 135 final Green Paper on the Future of the Common Fisheries Policy, Vol. 1, pp. 24.

[117] Churchill, *EEC Fisheries Law*, pp. 135-138; Nonnenmacher, *Die gemeinsame Fischereipolitik*, p. 220.

[118] Van Rijn, 'Fischereipolitik', p. 1254; Paul Nemitz, 'Fischereipolitik – Nach Art. 34 EGV', p. 576; Churchill, *EEC Fisheries Law*, pp. 135; Nonnenmacher, *Die gemeinsame Fischereipolitik*, p. 220; Fischer, 'Fischereipolitik', p. 7.

[119] Van Rijn, 'Fischereipolitik', p. 1254; Fischer, 'Fischereipolitik', p. 7.

[120] Case 91/03 Spain v. Council [2005] ECR I-2267, paras. 32-56.

[121] In 2003, Spain brought a case before the ECJ, claiming that access restrictions for Spanish vessels to French waters provided in Nr. 6 of Annex I of Regulation 2371/02 violated Art. 12 of the Treaty. The Court accepted that Art. 12 could be restricted for attaining the protection of the sensitive coastal marine areas; see Case 91/03 Spain v. Council [2005] ECR I-2267, paras. 43, 50-55; Case 201-202/85; Klensch v. Secrétaire d' État [1986] ECR, 3477, para. 9; Art. 17(2) shall remain in force until 2012 before it will be reviewed, Art. 17(2), second paragraph.

[122] Van Rijn, 'Fischereipolitik', p. 1254; Fisher, 'Fischereipolitik', p. 7; Churchill, *EEC Fisheries Law*, pp. 135-138.

fishing opportunities to vessels being registered within their territory and which fly their flags.[123] Insofar, they are free to pursue their own political and regulatory aims.[124] The Commission must be notified of the proposed allocation method. Further information regarding Member States' distribution mechanisms will be provided below.

3.1.7 Member States tasks regarding the market organisation

The activities of Member States in the area of market organisation are basically limited to the financial and administrative support of producer organisations and the ability to ensure they fulfil their tasks. Producer organisations form the backbone of the common market organisation.[125] They have the competence to implement catch plans, and promote the concentration of supply and stabilise prices.[126] Within this area of competence, the Member States are only allowed to require producers, which are not members of the producer organisations, to apply those rules producer organisations adopt.[127] Member States may also grant financial support to the organisations to offset the costs arising from their formation and of their production and marketing planning.[128]

3.1.8 Member States tasks regarding structural policies

While the Community precisely determines the conditions under which aid is granted to fishermen, Member States are primarily responsible for the implementation of Community structural measures and the determination of how aid is allocated.[129] This procedure has been referred to as the principle of 'shared responsibility'.

Under the current system, Member States must submit to the Commission a *national strategic* plan on their fisheries industry. These plans serve as the basis for dialogue between the Commission and Member States in regards to structural support schemes from 2007 to 2013.[130] Member States are to consult with relevant stakeholders when drawing up their plans. The plans should set out national priorities, objectives, estimates of financial resources required,

---

123 See Arts. 20(3) and 3(d) of Regulation 2371/02.

124 Fischer, 'Fischereirecht', p. 14.

125 Reason (9) of Regulation 104/2000.

126 See Art. 5 (1) Regulation 104/2000.

127 See Art. 7 (1) of Regulation 104/2004; Case 207/84, de Boer v. Produktschap voor Vis en Visprodukten [1985] ECR 3203, paras. 32-33.

128 See Art. 15 (1) of Regulation 2792/1999 and Art. 10 Regulation 104/2000.

129 Reason 6 of Regulation 2082/93. Prior to 1993, in many cases, the Commission itself decided on individual aid applications, see, for example, Case 514/93, Cobrecaf and other v. Commission, [1995] ECR, II- 621.

130 Art. 15(1) of Regulation 1198/06.

and timelines for putting measures into effect.[131] Member States have also been required to draw up an *operational programme* for implementing the policies that will be co-financed by the EFF. The programme has to be in line with the national strategic plan objectives and include, *inter alia*, tables which detail the financial contributions of the EFF and the Member States and where they will be allocated. The Commission then evaluates, *e.g.*, whether the programmes contribute to EFF objectives[132] and the guiding principles for the operational programmes laid down in Art. 19 of Regulation 1198/06, and whether they take into account the national strategic plans. If, in its opinion, the programme is incongruent with CFP and EFF rules, it will ask the Member State to amend its programme accordingly. If the proposal is approved, the Commission will adopt an approving decision.[133] Thus, the EFF leaves Member States a substantial margin of discretion as to how funds are actually allocated. For example, Germany's current (approved) operational programme states that about €12 million of EFF and national contributions will be set aside for the adaptation of its fleet.[134] This money is designated for modernising approximately 300 vessels.[135] However, when granting aid, Germany must follow the strict conditions (e.g., participation rates and the capacity ceilings) laid down in the EFF regulation.

As explained above, the ability of Member States to grant *additional* state aid is also limited. According to Art. 88(3) of the Treaty, Member States must notify the Commission of their intention to hand out state aid. To reduce administrative workload, the Commission has established a Regulation that exempts aid granted to small- and medium-sized enterprises from the notification obligations under certain conditions. It also lays down guidelines outlining the criteria it applies when evaluating Member States state aid proposals.[136]

3.2 Unregulated fisheries issues

The Community has wilfully refrained from regulating in specific areas in order to leave particular competences to the Member States.

The activities of non-commercial fisheries have an increasing economic importance, and a rising impact on the marine environment – particularly with regard to fish stocks. They may, to a certain extent, affect regular commercial

---

131 Art. 15(2) of Regulation 1198/06.

132 Art. 4 of Regulation 1198/06.

133 Art. 17(6) of Regulation 1198/06.

134 See German Operational Programme, available online at the homepage of the DG Fisheries, pp. 72, 80 and 120.

135 *Ibid.*, p. 81.

136 Commission Regulation (EC) 1595/2002 on the application of Articles 87 and 88 of the EC Treaty to State aid to small and medium-sized enterprises active in the production, processing and marketing of fisheries products, OJ 2004 No. L291/3. The Commission has also published 'Guidelines for the Examination of State Aid to Fisheries and Aquaculture', OJ 2004 No. C229/5.

fishing activities; this implies that, in principle, they ought to be covered by the CFP.[137] However, the main regulation, *i.e.*, Regulation 2371/02, does not explicitly refer to non-commercial activities.[138] Only Art. 17 of Regulation 1967/2006 concerning management measures for the sustainable exploitation of fishery resources in the Mediterranean Sea requires Member States to ensure that non-commercial fishing activities do not jeopardise CFP conservation and management efforts.

Another area of Community responsibility, which remains largely unregulated at the Community level, is the management of sweet water fisheries in inland waters,[139] where only a few market rules and structural measures currently exist.[140]

4 Regulating fisheries for environmental purposes

National, Community and international laws establish environmental protection requirements that either directly or indirectly affect fisheries legislation.[141] From a practical and systematic perspective, it is important that fisheries legislation is compatible with the objectives and requirements of such environmental laws, and that environmental legislation does not impair the existing fisheries legislation.[142]

However, under current Community law, incompatibilities are likely to arise. This is due to conflicting competences between fisheries and environmental legislation. While the Community holds an exclusive competence for the conservation of fish resources, Member States are primarily competent and required under international, Community and national law to adopt rules on environmental protection. Conflicts may particularly arise where these competences overlap in relation to the establishment of marine conservation areas, *i.e.*, Natura 2000 sites.[143] For example, according to Art. 6(2) of the Habitats Directive, adopted under the environmental competence order in Art. 175 of the Treaty:

---

137 Fischer, 'Fischereipolitik', p. 5 and Art. 2(2) Regulation 3760 and Art. 1(2) of Regulation 1626/1994.

138 Only Art. 2(2) of the former basic Regulation 3760/92 required Member States to ensure that non-commercial fishing activities do not jeopardise CFP conservation and management efforts.

139 Fischer, 'Fischereipolitik', p. 8; Art. 2 of Regulation 3760/1992 and Arts. 1 and 30 Regulation 2371/2002. Regulation 2371/2002 does not explicitly reiterate this limitation. There is no indication that the existing practice will be changed, van Rijn, 'Fischereipolitik', p. 1252-1253 at cc) and footnote 46.

140 See Art. 1 Regulation 104/00 and Art. 13 of Regulation 2792/99.

141 Commission Communication Com (2001) 143 to the Council and the European Parliament – Elements of a Strategy for the Integration of Environmental Protection Requirements into the Common Fisheries Policy, pp. 6-8.

142 Cyrille de Klemm, 'Fisheries Conservation and Management and the Conservation of Biological Diversity', in Hey (ed.), *Developments International Fisheries Law*, (the Hague: 1999), p. 447.

143 Hans D. Jarass, *Naturschutz in der Ausschließlichen Wirtschaftszone – Völkerrechtliche, EG- rechtliche und verfassungsrechtliche Probleme der Ausweisung von Meeresschutzgebieten*, (Baden-Baden: 2001), pp. 31-34;

> Member States shall take appropriate steps to avoid, in the special areas of conservation, the deterioration of natural habitats of species as well as disturbance of the species for which the areas have been designated, in so far as such disturbances could be significant in relation to the objectives of this Directive.[144]

Since the Habitats Directive applies within all Community waters (including the EEZ),[145] and fishing is one of the major concerns regarding the protection of Natura 2000 sites, Member States may be required to regulate, confine or even prohibit fishing activities within certain EEZ areas in order to meet their environmental obligations under Community law.[146] However, the Commission has argued that the Community is solely competent to regulate fishing activities – even where the purpose is environmental (*i.e.*, for the implementation of international and Community environmental obligations).[147]

The conflict has two aspects. First, it must be decided whether the Community or the Member States are competent to regulate fisheries for environmental purposes. Second, a mechanism has to be established according to which the decision ought to be adopted.

It will be suggested that, despite the exclusive Community competence in the area of *fisheries conservation*, Member States are still competent to regulate fisheries activities for the purpose of *environmental protection*. However, in order to resolve the competence conflict and establish an integrated approach to fisheries and environmental conservation, Member States and the Community should *co-ordinate* and integrate their fisheries and environmental conservation legislation.

Kyrill Schwarz, 'Fishing management in the Exclusive Economic Zone (EEZ) of the Federal Republic of Germany', in Ritterhoff, Gubbay and Zucco (ed.), *Marine Protected Areas and Fisheries – BfN-Skript 122*, (Bonn: 2004), p. 46.

[144] See also Art. 4 of the Birds Directive; protective areas of the Habitats and the Birds Directive will together form Natura 2000 sites network.

[145] Detlef Czybulka, 'Die Geltung der FFH-Richtlinie in der Ausschließlichen Wirtschaftszone' (2001) 12 Natur und Recht, 1, pp. 19 et seq.; Kyrill Schwarz, ‚Fishing management in the Exclusive Economic Zone (EEZ) of the Federal Republic of Germany', p. 47.

[146] Gellermann, 'Geschützte Meeresflächen in der deutschen Ausschließlichen Wirtschaftszone (AWZ)', pp. 164-169; Czybulka, ‚Forschungsbedarf im marinen Fischereirecht', pp. 831-832.

[147] Commission Communication Com (2001) 143, p. 7-8; see also Commission argumentation in John Clorley, 'Relationship of the Common Fisheries Policy and the Habitats and Birds Directives' in Ritterhof, Gubbay and Zucco (ed.), Marine Protected Areas and Fisheries – BfN-Skript 122, (Bonn: 2004), p. 60.

4.1 Vertical Competences with regards to environmental policies

The Community is required to take actions in the sphere of the environment.[148] Environmental policy is not an exclusive Community competence. According to Art. 174(1) of the Treaty, the Community's environmental policy shall only *contribute* to the pursuit of specific environmental objectives. The Community and Member States thus share the competence to legislate on environmental issues. If the Community decides to implement its own environmental policy in a certain area, it has to respect the subsidiarity principle laid down in Art. 5 of the Treaty. According to this principle, if the Community has not adopted legislation on the preservation or protection of the environment, Member States are free to do so within the limits of Community law. Consequently, there is a dynamic relationship between Community environmental law and that of the Member States.[149]

4.2 Horizontal competences

In order to clearly identify the different objects of regulation, as well as the common core of fisheries and environmental policy, it is necessary to untangle the meanings of the terms 'environmental conservation', 'fisheries' and 'fisheries management'.

The environment covers all flora and fauna, as well as the material base in which they evolve, interact and inter-depend.[150] Thus, the conservation of the marine environment has a broad objective, namely, the protection of marine flora and fauna, their ecosystems and the way they interact. Accordingly, the objectives that underlie environmental policies are pursued from a particular starting point: the protection of the marine environment as a whole.

Fisheries involve the catching, taking, or harvesting of living aquatic resources from the marine environment.[151] Due to over-exploitative tendencies of fisheries, limiting fishing activities must be seen as a precondition, *i.e.*, an intrinsic part of fisheries itself. Limitations on fishing activities that are based on a solid rationale can be referred to as fisheries management. Fisheries policies centre on fishing activities and the consequences such activities have on the future availability of resources.[152] In many cases, fisheries policies include supplemental actions in the areas of structural and trade measures, control and enforcement as well as market organisation.

---

148 Art. 3 of the Treaty.

149 Krämer, *EC Environmental Law*, p. 103.

150 A. Lorz, ‚Fischerei und Naturschutz – eine rechtliche Betrachtung' (1982) *Natur und Recht*, 1, p. 5; Uta Eser and Thomas Potthast, *Naturschutzethik*, (Baden-Baden: 1999), pp. 14-15.

151 Lorz, 'Fischerei und Naturschutz – eine rechtliche Betrachtung', p. 1.

152 Cyrille de Klemm, 'Fisheries Conservation and Management and the Conservation of Biological Diversity', p. 447.

To a limited extent, fisheries management and marine environmental conservation serve the same objective: preventing the disruption of the ecological equilibrium by destructive (fishing) practices.[153] By limiting fishing activities, the environmentally harmful effects of such activities – ranging from the over-fishing of commercially exploited stocks to the by-catching of non-target species, and threats to marine mammals and sea birds – and the destruction of marine and coastal ecosystems are automatically reduced.[154] Thus, there is some overlap between fisheries management and environmental protection. This is particularly the case where fisheries measures are simultaneously directed at protecting commercially unexploited (*i.e.*, non-target species) birds, mammals, reptiles, habitats, coastal areas, etc., or where they take into consideration other environmental factors.[155] On the other hand, certain Community measures such as the Birds Directive, the Habitats Directive, the Regulation on the Protection of Species of Wild Fauna and Flora as well as some international environmental agreements such as the Biological Diversity, Bern, HELCOM, OSPAR and Barcelona Conventions have been adopted to achieve particular environmental aims.[156] These instruments sometimes overlap with fisheries management regimes, some of them protecting certain species, some of them requiring fisheries conservation measures. The effects of this overlap can be either direct or indirect.[157]

The distinction between environmental and fisheries policies also exists within Community law. The ECJ has emphasised this difference in the *Mondiet* decision, where it defined the discrete legislative area of fisheries from the aim-based and almost unlimited scope of environmental policies. In *Mondiet*, the Court had to determine whether a regulation on drift netting had to be based either on the Treaty's environmental or fisheries provisions. The regulation was challenged on the basis that it had been adopted for ecological reasons, namely, the protection of non-target species. Referring to its *centre of gravity theory*, the

---

[153] *Ibid.*; Commission Communication Com (2001) 143, p. 4.

[154] Booß, 'Umweltschutz und Fischerei', pp. 1524-1525.

[155] Wolff, *'Fisheries and the Environment'*, p. 177.

[156] Council Directive (EEC) 79/409 of 2 April 1979 on the conservation of wild birds, OJ 1979, L 103/1 as amended; Council Directive (EEC) 92/43 of 21 May 1992 on the conservation of natural habitats and of wild fauna and flora, OJ 1992, L206/7; OJ 1997, L61/1 and Commission Regulation (EC) 939/97 laying down detailed rules concerning the implementation of Council Regulation (EC) 338/97 on the protection of species of wild fauna and flora by regulating trade therein, OJ 1997, L140/9; Convention on Biological Diversity, OJ 1993, L309/1 as amended by OJ 2002, L201/48; Convention for the Protection of the Marine Environment of the Baltic Sea Area, OJ 1994, L73/19; Convention for the Protection of the Marine Environment of the North-Atlantic, OJ 1998, L104/1 as amended by OJ 2000, L118/44; Convention for the protection of the Mediterranean Sea against pollution, OJ 1977, L240/1. Agreement for cooperation in dealing with pollution of the North Sea by oil and other harmful substances, OJ 1984, L188/7.

[157] Wolff, *Fisheries and the Environment*, pp. 182-195; Commission Communication Com (2001) 143 final, pp. 6-8.

Court found that the regulation had to be based on the Treaty provision on agriculture, because it was 'adopted *primarily* in order to ensure the conservation and rational exploitation of fishery resources and to limit the fishing effort'.[158] The regulation placed less emphasis on environmental concerns.[159] Its rules are thus an integral part of the common agricultural policy, the objectives of which under Art. 33 of the Treaty include 'ensuring the rational development of production and assuring the availability of supplies'.[160] The environmental considerations were regarded merely as 'contributory factors'.[161]

In addition, the ECJ's decisions on the exclusive competence only signal that the Community is competent to adopt 'measures relating to the conservation of the *resources of the sea*'.[162] These rulings do not explicitly preclude the Member States from adopting measures relating to '*the conservation of the marine environment*'.[163]

However, it is difficult to determine the exact point at which the ECJ will find that the centre of gravity has shifted from fisheries to environmental policy.[164] Besides, Art. 6 of the Treaty allows CFP legislation to take into account the effects of fishing activities on marine ecosystems.[165] *Nina Wolff* correctly states that where measures aim at fisheries and environmental protection, the integration principle shifts the centre of gravity towards the fisheries policy. Consequently, measures that are directly linked to fishing activities should be based on the provision on agriculture, 'even if the underlying environmental objectives are of an importance equal to that of the concern for fisheries conservation'.[166]

---

[158] Emphasis added; see Case C-405/92, Mondiet v. Armement Islais [1993] I-6133, para. 24; Wolf, 'Fisheries and the Environment', p. 171. The Court of First Instance has only recently referred to fisheries and environment being two different kinds of policies, see Case T-37/04, The Autonomous Region of the Azores v. Council, paras. 133, 144 et seq., 164 et seq., available online at http://curia.europa.eu; Regarding the 'centre of gravity theory' in general see Cases C-164-165/97, Parliament v. Council [1999] ECR I-01139, para. 14. Regarding the distinction of environmental policies and the Common Commercial- or Agricultural Policy see Case 62/88 Greece v. Council [1990] ECR I-1527, paras. 5-23; Case C-336/00, Huber [2002] I-7699, paras. 29-37; Case C-164-165/97, Parliament v. Council [1999] ECR I-1139, paras. 8-20;.

[159] Case C-405/92, Mondiet v. Armement Islais [1993] I-6133, para. 25.

[160] *Ibid.*, para. 24.

[161] *Ibid.*, para, 28.

[162] Von Bogdandy and Bast, 'The European Union's Vertical Order of Competences', p 241.

[163] Similiar argumentation, see Gellermann, 'Geschützte Meeresflächen in der deutschen Ausschließlichen Wirtschaftszone (AWZ)', pp. 166-169.

[164] Wolff, *Fisheries and the Environment*, p. 170; Proelß, *Meeresschutz*, pp. 315-316.

[165] Art. 2 and 3(i) of Regulation 2371/02.

[166] Nina Wolff, '*Fisheries and the Environment*', p. 172.

4.3 Legal evaluation of the competence conflict

The ECJ has stated that environmental and fisheries policy spheres coexist side by side,– but that they have their own scope and content.[167] Primary law provides for two separate policies that arise from two different vertical competence orders.

It was explained above that where the Community has not taken action so far, the Member States have the right to act, both where the Community enjoys an exclusive or a non-exclusive legislative competence. However, according to the principle of the supremacy of Community law, environmental measures taken by the Member States cannot violate existing Community law. This includes Community's fisheries rules.[168] Given that the Community has established geographical conditions for fishing in Community waters, there will be little room for national legislators to adopt restrictions in specific areas. Potential national measures, for example, that ban certain fishing activities in marine NATURA 2000 areas would have to respect existing Community fisheries legislation.

Given that the Community holds exclusive competences to regulate fishing activities, the principle of subsidiarity does not apply to fisheries law. Accordingly, fishing regulations instituted by individual Member States, which have been adopted for environmental purposes, could simply be replaced by Community fisheries legislation.

4.4 Conflicting principles

The result is unsatisfying. On the one side, Member States have residual competences to regulate fishing activities for environmental purposes. They have not yet transferred exclusive powers to regulate in the area of the environment to the Community. Besides, there may be room to regulate fishing activities where the Community has failed to act. The Commission's position on this basically disregards the competence order, as well as the institutional balance established in the Treaty.[169] In addition, as *Gellermann* correctly states that it would be a contradiction for the Community to adopt a Directive which oblige Member States to guarantee the protection of certain marine areas on the one hand, and not grant them the competence to do so on the other.[170]

On the other side, the unilateral establishment of 'no go areas' or 'no take areas' by Member States could interfere with the Community's hard-earned

---

167 Case C-405/92, Mondiet v. Armement Islais [1993] I-6133, para. 24; for other policy areas see, Case 62/88 Greece v. Council [1990] ECR I-1527, paras. 5-23; Case C-336/00, Huber [2002] I-7699, paras. 29-37; see particularly Case C-164-165/97, Parliament v. Council [1999] ECR I-1139, paras. 8-20.

168 With regards to Community agricultural law, see Krämer, *EC Environmental Law*, pp. 1020-1021.

169 Art. 5, paragraph 1 of the Treaty.

170 Gellermann, 'Geschützte Meeresflächen in der deutschen Ausschließlichen Wirtschaftszone (AWZ)', p. 166.

geographically determined access and distribution regime, as provided under the Regulation 2371/02. In particular, it is possible that the principles of equal access and relative stability, laid down in Art. 17 to 19 and 20(1) of Regulation 2371/02, could be violated.

Where Member States restrict fishing activities by adopting geographical access rules, they run the risk of violating the principle of equal access. The principle of equal access is one of the CFP's cornerstones, introduced in 1970 under Art. 2 of the first structural Regulation 2141/70. It implements Art. 12 (ex Art. 7) of the Treaty, which forbids 'discrimination on the grounds of nationality'.[171] The principle does not imply free access, but instead calls for non-discriminatory access to marine capture resources in Community waters.[172] Any exception to this principle must be objectively justified.[173] The ECJ has decided that, with regards to the principle of equal access, only the Council has the power to determine the conditions of such access and that this situation cannot be modified by the Member States taking unilateral measures.[174] Even though exceptions do exist (such as the twelve nautical mile zone, traditional fishing rights, the Shetland Box[175] or the assignment of fishing quotas to specific ICES areas), all of them were adopted by the Council.

Where Member States are entitled to establish 'no go areas' or 'restricted-take areas' in Community waters, also the principle of relative stability might be transgressed. This principle requires that each year Member States receive the same relative quota of exploitable stocks. It was established in 1983, and forms a central part of the CFP distribution system.[176] The ECJ recognised in *Commission* v. *United Kingdom* that the adoption of geographical access restrictions could have distributory effects.[177] In that case, the United Kingdom had unilaterally shifted its national baselines outward to the sea.[178] This impaired the traditional fishing rights of other Member States, who were fishing within the 12 nm zone. The ECJ noted that the change in fishing activities would clearly be a detriment to these fishers, since fishing areas further to the sea had comparably lower productivity. The Court regarded the conservation, distribution and

---

[171] *Ibid.*, para. 29; Clare Coffey, 'Introduction to the Common Fisheries Policy', p. 295; Van Rijn, 'Fischereipolitik', para. 19; Kapteyn and van Themaat, 'Introduction to the Law of the European Communities', p. 1162; Nonnenmacher, *'Die gemeinsame Fischereipolitik'*, pp. 218-219; see also Case C-146/89, Commission v. United Kingdom [1991] ECR I-3533, paras. 6-37.

[172] Case 91/03 Spain v. Council [2005] ECR I-2267, para. 43 ; see also Nonnenmacher, *'Die gemeinsame Fischereipolitik'*, p. 217.

[173] Case 91/03 Spain v. Council [2005] ECR I-2267, para. 43; 201-202/85 Klensch v. Secrétaire d' État [1986] ECR, 3477, para.9.

[174] Case 804/79, Commission v. United Kingdom, [1981] ECR 1045, para. 29; C-146/89, Commission v. United Kingdom [1991] ECR I-3533, paras. 21-25.

[175] See Arts. Art. 17 (2) and 18 and Annex I of Regulation 2371/02.

[176] See Reasons (4) to (7) of Regulation 170/83, OJ 1983, No. L24/1; Van Rijn, 'Fischereipolitik', p. 1257.

[177] C-146/89, Commission v. United Kingdom [1991] ECR I-3533.

[178] *Ibid.*, paras. 14, 19.

access system laid down in the regulation as establishing a complex balance between traditional fishing rights and rights conferred to fishers from the coastal Member States.[179] It stated that any unilateral amendment of the access regime by Member States could reduce the effectiveness of the basic regulation, as well as endanger the objective of relative stability.[180]

4.5 Possible solutions

All in all, even though Member States retain some competence to alter fisheries regulations for environmental purposes, they should neither unilaterally alter the Community principle of equal access, nor regulate access for fishers from other Member States within waters under their sovereign control or jurisdiction, if this would reduce the effectiveness of the existing conservation, distribution and access system provided in the basic regulation.

Following this, one option might be to oblige Member States to co-ordinate their legislative processes (in accordance with all relevant environmental laws) with the Community and any affected Member States. Under this approach, either the Member States or the Community may finally adopt the required fisheries legislation. This has been successfully practiced when the United Kingdom wanted to establish a Special Area of Protection under the Habitats Directive 185 km northwest of Scotland. The UK asked the Commission to prohibit bottom trawling in order to protect sand and cold-water mounds in that area ('Darwin Mounds'). As a result, the Council agreed to a permanent ban on bottom trawling in the particular area in 2004.[181]

Another option would be to fully confer the power to adopt fisheries regulations for the purpose of environmental protection on the Community. In such a case, the Community would be required to take into consideration and to respect Member States' environmental objectives and obligations when regulating fisheries activities.

5 Participation rights of stakeholders within the CFP legislative process

The Community has established two different kinds of political Committees: the Consultative Committee on Fisheries and Aquaculture

[179] *Ibid.*, paras. 27, 28.

[180] *Ibid.*, paras. 28, 29.

[181] Council Regulation (EC) No. 602/2004 amending Regulation (EC) No. 850/98 as regards the protection of deepwater coral reefs from the effects of trawling in an area north west of Scotland, OJ 2004, No. L97/30; Clorley, 'Relationship of the Common Fisheries Policy and the Habitats and Birds Directives', pp. 58-60.

('ACFA'[182]) and the Regional Advisory Councils[183] (the latter having been established in the course of the 2002 Reforms). These committees grant stakeholders participation rights within the CFP legislative process. However, these rights are limited to the pre-decision phase,[184] as both committees are only *consulted* by the Commission.[185] No provision has been made for compulsory consultation and the advice and opinions provided have no binding legal effect.[186]

5.1 Consultative Committee on Fisheries and Aquaculture

The ACFA Committee is comprised of 21 members; within this group, ten represent commercial interests, eight are experts from the scientific community chosen by the STEFC and three represent consumers, environmental and development interests.[187] The purpose of the Advisory Committee is to communicate with all parties concerned on the development and management of the CFP.[188]

5.2 Regional Advisory Councils

The purpose of the Regional Advisory Councils is similar to that of the ACFA. Both have been established to enable the CFP to benefit from the knowledge and experience of fishermen and other stakeholders, and to take into account the diverse conditions in Community waters.[189] Through the RACs, the Commission and the fisheries sector can exchange information and cooperate more closely on the development of the CFP. According to Art. 2 of

[182] Legislation establishing and modifying the ACFA Committee: Décision de la Commission, (CEE) 128/71; portant création d' un comité consultative de la pêche, OJ 1971, No. L68/18; Commission Decision (EC) 478/99 renewing the Advisory Committee on Fisheries and Aquaculture; OJ 1999, No. L187/70; Commission Decision (EC) 864/2004 amending Commission Decision 1999/478/EC of renewing the Advisory Committee on Fisheries and Aquaculture, OJ 2004 No. L370/91.

[183] Art. 31 of Regulation 2371/02; Council Decision (EC) 585/04 establishing Regional Advisory Councils under the Common fisheries Policy, OJ 2004, No. L256/17.

[184] Communication Com (2002) 135 final Green Paper on the Future of the Common Fisheries Policy Vol. 1, p. 28.

[185] Art. 31 (4) of Regulation 2371/02, Art. 2 Commission Decision 478/99, Art. 3 (3) of Council Decision 585/04.

[186] Whether stakeholder involvement has actually much improved through the 2002 CFP reform, remains disputed; see Hatchard, J. and Gray. T., 'The 2002 Reform of the Common Fisheries Policy's System of Governance – Rhetoric or Reality? (2003) 27 *Marine Policy*, pp. 545-554; J.B. Ingerowski and M Salomon, Ein kritischer Blick auf die aktuellen Entwicklungen in der Gemeinsamen Fischereipolitik unter Einbeziehung der neu geschaffenen regionalen Fischereibeiräte' (2006) *Natur und Recht*, pp. 539-542.

[187] Arts. 1 (2), 3 (1-2), 7 of Commission Decision 478/99.

[188] Reason (3) of Commission Decision (EC) 478/99.

[189] Reason (27) Council Regulation 2371/02.

the Council Decision (EC) 585/2004,[190] which establishes Regional Advisory Councils under the Common Fisheries Policy, such Councils have been formed for the Baltic Sea, the Mediterranean, the North Sea, North-Western Waters, South-Western Waters, pelagic stocks and the 'high seas/long distant fleet'.[191] RACs consist of a general assembly and an executive committee.[192] Both will be composed of representatives whose interests are affected by the CFP.[193] Art. 1(2) of Council Decision 585/04 defines 'representatives' to include shipowners, small-scale fishermen, employed fishermen, producer organisations, as well as, *inter alia*, processors, traders, other market organisations and environmental organisations, etc.[194] The general assembly appoints an executive committee of up to 24 members.[195] The latter shall draw up annual reports, annual strategic plans, as well as recommendations.[196] Art. 5(3) of the Decision requires that two-thirds of the seats in the general assembly and the executive committee of the Regional Advisory Councils be allotted to representatives of the fisheries sector.

5.3 Assessment and critique

At various points in the reform process, the Commission has recognised the importance of improving governance structures by including interested parties in the policy-making process.[197] The current scope of participation rights is, however, quite restricted, giving ample grounds for complaint. First, given that the current state of fish stocks mainly results from the activities of the fishing industry (including lobbying), it proves mistaken that industry representative obviously dominate both ACFA and RACs. A second criticism

---

[190] Council Decision (EC) 585/04 establishing Regional Advisory Councils under the Common Fisheries Policy, OJ 2004 No. L256/17; as amended by Council Decision (EC) 409/07, 2007 No. L155/68.

[191] Commission Decision (EC) 222/07 declaring operational the Regional Advisory Council for the south-western waters under the common fisheries policy, OJ 2007 No. L95/52; Commission Decision (EC) 206/07 declaring operational the Regional Advisory Council for the High Seas/Long Distance Fleet under the common fisheries policy, OJ 2007 No. L91/52; Commission Decision (EC) 191/06 declaring operational the Regional Advisory Council for the Baltic Sea under the common fisheries policy, OJ 2006 No. L66/50; Commission Decision (EC) 668/05 declaring operational the Regional Advisory Council for the north-western waters under the common fisheries policy, OJ 2005 No. L249/18; Commission Decision (EC) 606/05 declaring operational the Regional Advisory Council for pelagic stocks under the common fisheries policy, OJ 2005 No. L206/21; Commission Decision (EC) 774/04 declaring operational the Regional Advisory Council for the North Sea under the common fisheries policy, OJ 2004 No. L342/28.

[192] Art. 4(1) of Decision 585/04.

[193] Art. 31(2) of Regulation 2371/02 and Art. 5(1) of Decision 585/04.

[194] For example, the North Sea RAC includes, *i.a.*, representative from fishermen, producer, port, environmental and trader associations and organisations, see online at: http://www.nsrac.org/.

[195] Art. 4(3) of Decision 585/04.

[196] Arts. 4(2), 7(3), 10 of Decision 585/04.

[197] Road Map, pp. 23-24. and Green Paper, pp. 28-30;

regarding the composition of the RACs is that scientists are not granted any voting rights.[198] Third, participation is limited to the pre-decision phase, meaning that committees are only consulted by the Commission.[199] Fourth, when setting TACs annually, Stakeholders have less than one month to analyse and comment on the Commission's proposal to the Council.

It must, however, be kept in mind that participation and lobbying also take place at the national and the parliamentary levels. In fact, all-important CFP issues are decided by the Council in accordance with Art. 37(2), third paragraph of the Treaty. The policies of Member States are, of course, largely influenced by regional political actors and interest groups.[200] As stated above, interest groups also help to shape the CFP through the Fisheries Committee of the European Parliament. Most members of this Committee come from areas dependent on fisheries and act, according to *Lequesne*, 'as intermediaries for their fishing industry's territorialized interests in the Community sphere'.[201]

[198] J.B. Ingerowski and M Salomon, ,Ein kritischer Blick auf die aktuellen Entwicklungen in der Gemeinsamen Fischereipolitik unter Einbeziehung der neu geschaffenen regionalen Fischereibeiräte', pp. 540-541.

[199] Hatchard, J. and Gray. T., 'The 2002 Reform of the Common Fisheries Policy's System of Governance – Rhetoric or Reality?, pp. 546-550.

[200] C. Lequesne, *The politics offisheries in the European Union*, (Manchester: 2004), pp. 32-37.

[201] *Ibid.*, p. 40.

CHAPTER 3

Fisheries Management under the CFP

1 Fisheries management instruments

The Community fisheries management regime aims at guaranteeing sustainable fishing. Since the reform of the CFP in 2002, environmental objectives are also increasingly taken into account. The Community's management regime, which includes control and enforcement measures, will be described and assessed in the following chapter.

1.1 Pertinent terms and concepts under the Community's management regime

Fisheries management has been described as:

> [the] integrated process of information gathering, analysis, planning, decision making, allocation of resources and formulation and enforcement of fishery regulations by which the fisheries management authority controls the present and future behaviours of the interested parties in the fishery, in order to ensure the continued productivity of the living resources.[1]

Community fisheries law draws a distinction between management and conservation.[2] However, these terms are often used interchangeably and neither term is explicitly defined in the Community's fisheries legislation.[3] In general, it could be assumed that while conservation includes undertakings directed at the sustained use of fisheries resources, management refers to the process of adopting and implementing conservation measures, such as TACs and effort limitations. To successfully manage resources, fisheries policies must pay due regard to several external factors such as, for example, the socioeconomic situation of fishers, the efficient operation of the fleet, consumer demands, etc. Measures targeting these external factors are not categorised as fisheries management *stricto sensu*.

ICES defines 'fishery' as a group of fishing vessels targeting the same (assemblage of) species and/or stocks, using similar gear, during the same period of the year and within the same area.[4] The CFP's management regime does little to regulate individual fisheries. For example, the 'Baltic Cod Fishery' includes fishers who fish for the two major cod stocks existing in the Baltic Sea;[5] however, it is not a legal term under Community fisheries law and there are no conservation measures which explicitly limit fishing for the 'Baltic Cod Fishery'. In general, conservation measures concern fish stocks or fishing areas.

1 Available online at www.fao.org/fi/glossary/.

2 Art. 1(1) and (2) of Regulation 2371/02.

3 This is in accordance with the use of the terms at international law; see Wolff, *Fisheries and the Environment*, p. 54-55; see also Art. 1(1)(b) of the Straddling Stocks Agreement.

4 ICES, *Report of the Baltic Fisheries Assessment Working Group 2007* (Copenhagen: 2007), p. 6.

5 *Ibid.*, pp. 6-22.

However, when adopting conservation measures, the distinctive features of existing fisheries are taken into consideration.[6]

1.2 Quantitative catch and effort limitations

Setting maximum catch limits for specific stocks (total allowable catches or 'TACs') is the main management tool established in Community law. Since the 2002 Reform, however, effort limitations play a growing role. Arts. 20(1) and 4(2)(d), (f) of Regulation 2371/2002 provide that the Council, acting by qualified majority, will decide on 'catch and/or fishing effort limits', the conditions associated with those limits, as well as the allocation of fishing opportunities among Member States.

1.2.1 Total allowable catch and quotas

According to Art. 3(m) of Regulation 2371/02, 'catch limit' means a quantitative limit on landings of a stock or group of stocks over a given period.

1.2.1.1 The creation of the Community TAC system

In the course of the adoption of a common fisheries management regime in 1983, the Community chose maximum catch limits as the main conservation approach. Supplemental instruments also included technical measures and the setting of minimum sizes for commercially exploited stocks.[7] The decision to adopt TACs instead of limiting fishing by way of effort regulations was due to the fact that policy makers and administrators working on behalf of the Community and Member States were accustomed to TACs as a result of their prior work in the North East Atlantic Fisheries Commission ('NEAFC') and North West Atlantic Fisheries Organisation ('NAFO') during the

---

6 This is particularly so under the effort regulation, see Art. 2(1) and Annex I and II of Council Regulation (EC) 1415/04 fixing the maximum annual fishing effort for certain fishing areas and fisheries, OJ 2004 No. L258/1.

7 See Art. 3 of Regulations 170/83 and 171/83; a control regulation had already been adopted in 1982, i.e. Regulation 2057/82. Art. 5 of the first structural Regulation 2141/70 already included the possibility for the Council to adopt restrictions relating to the catching of certain species. However, the provision did not refer to TACs. Prior to 1983, merely a ban on the landing of herring for purposes other than human consumption had been adopted, OJ 1977 No. L48/28. Furthermore, non-binding TACs and quotas were proposed by the Commission to the Member States from 1978 to 1983, see, for example, Council Regulation (EEC) 754/80 concerning, for certain fish stocks occurring in the Community fishing zone, the fixing for 1980 of the total allowable catches, the share available for the Community and the means of making the catches, OJ 1980 No. L84/36.

1960s.[8] Besides, in view of the struggles linked to the distribution of fish stocks amongst Member States, TACs are comparatively straightforward to negotiate, simple to define, and TAC-quotas are easy to predict for the fishing industries.[9] The limitation of fishing opportunities through catch limits is a question of the division of available resources. After difficult negotiations, the Member States agreed on an allocation formula according to which each Member State is guaranteed a specified percentage of the TAC for each type of stock. This formula has remained the distribution key according to which the Community has allotted fishing opportunities among Member States.[10] The distribution ratio has been based on different considerations: (a) historic catches by Member States based on the reference period from 1973 to 1978, (b) preferences favouring Ireland, the UK and (at that time) Greenland, and (c) losses of fishing opportunities suffered particularly by German and UK fishers in the course of the creation of the EEZ.[11] In 1983, the Community decided to fix the allocation formula according to the principle of relative stability.[12] Every year since then, each Member State has received the same share of stock specific TACs, *i.e.*, the same shares relative to the fluctuating levels of the TAC.[13] It is only when new Member States gain membership to the Community that the existing allocation formula is carefully amended.[14] Given the difficulties associated with such distributive negotiations (mainly due to commercial and planning interests of the fishing industry

8 NAFO was formerly referred to as International Commission for the Northwest Atlantic Fisheries ('ICNAF'). S.S. Gezelius, 'The Arrival of Modern Fisheries management in the North Atlantic: A Historical Overview', in Gezelius and Raakjær (eds.), Making Fisheries Management Work (Berlin: 2008), pp. 27-40; T.J. Hegland and J. Raakjær, 'Recovery Plans and the Balancing of Fishing of Fishing Capacity and Fishing Possibilities – path dependence in the Common Fisheries Policy', in Gezelius (ed.), *Making Fisheries Management Work – Implementation of Politics and Politics of Implementation in Resources Conservation*, (Berlin: 2008), p. 138.

9 S.S. Gezelius, 'The Problem of Implementing Policies for Sustainable Fishing', in Gezelius and Raakjær (eds.), Making Fisheries Management Work (Berlin: 2008), p. 22. pp. 9-10.

10 The principle of relative stability was included in Art. 4 of the first basic Regulation 170/1983 (see also Reasons (4) to (7) of that Regulation) and maintained under Art. 8 (4) (ii) of Regulations 3760/1992 (see also Reasons (12) and (14) of that Regulation). It still applies today; see Art. 20 (1) of Regulation 2371/02.

11 Preferences favouring Ireland and the UK were granted to these areas because they were heavily dependant on fishing. The content of these preferences was declared at a Council meeting at The Hague and therefore generally referred as the 'Hague Preferences'; see Holden, *The Common Fisheries Policy*, pp. 43-44; Leigh, *European integration and the common fisheries policy*, pp. 90 et seq.; Fischer, 'Fischereipolitik', p. 12.

12 See Art. 4 of Regulation 170/83.

13 Case 46/86, Romkes v Officier van Justitie for the District Zwolle [1987] ECR 2671, para. 17.

14 This gave rise to disputes before the ECJ; see, for example, Case 73/90, Kingdom of Spain v Council of the European Communities [1992] ECR I-5191 and Case C-206/97 Kingdom of Sweden v Council of the European Union, [1999] ECR I-3885.

and Member States), major changes to the current system are considered to be unlikely without a substantial change to the CFP.[15]

1.2.1.2 The annual setting of TACs

The Council sets maximum catch limits annually for specific stocks based on the Commission's proposals.[16] TACs apply to stocks located in Community waters and, for stocks exploited by Community vessels on the high seas and in third country waters (where catch limits are required). Around June of each year, the Commission starts to draw up its proposals. The first step of this process involves informal consultations with stakeholders through meetings with regional committees, workshops, internet-communications, and formally in Regional Advisory Councils (RACs). At the end of October, the Technical Committee for Fisheries (STECF) receives data from different institutions, such as the ICES, international fisheries commissions (IFOs), Member States and third countries.[17] Based on the STECF recommendations and the information complied from informal and formal consultations, the Commission prepares a proposal for the Council. Before sending the proposals to the Council, the Commission sometimes consults its own environmental, social or regional departments. Following this, Council working groups, composed of national officials and experts as well as at least one member of the Commission, examine the proposal and send it to the European Parliament's Fisheries, Economic and Social and Regions Committees for consultation. After receiving Parliament's non-binding recommendations,[18] the Council, generally at the end of each year, sets TACs for the upcoming year for the commercially most important stocks.[19] If the Council is not able to agree with the Commission's suggested course of

---

15 This distribution pattern aims at creating predictability in the fisheries sector; see Reasons (16) and (17) of Regulation 2371/2002; see also van Rijn, 'Fischereipolitik', p. 1257; when new fishing opportunities are established, for example in the waters of third countries, a different allocation pattern might be applied: The 'Council shall decide on the allocation for those opportunities, taking into account the interests of the Member States'; see Art. 20(2) Regulation 2371/2002.

16 See, for example, Council Regulation (EC) 40/08 fixing for 2008 the fishing opportunities and associated conditions for certain fish stocks and groups of fish stocks, applicable in Community waters and, for Community vessels, in waters where catch limitations are required, OJ 2008 No. L19/1; see also Council Regulation (EC) 41/07 fixing for 2007 the fishing opportunities and associated conditions for certain fish stocks and groups of fish stocks, applicable in Community waters and, for Community vessels, in waters where catch limitations are required, OJ 2007 No. L15/1; Council Regulation (EC) 1941/06 fixing the fishing opportunities and associated conditions for certain fish stocks and groups of fish stocks applicable in the Baltic Sea for 2007, OJ 2006 No. L367/1.

17 Arrangement in the form of an exchange of letters between the European Economic Community and the International Council for the Exploration of the Sea, OJ 1987 No. L149/14.

18 See Art. 37(2), paragraph 3 of the Treaty.

19 See Art. 3(m) of Regulation 2371/02; see Council of the European Union, Press Release 15479/05 (Presse 349), regarding 2702nd Council Meeting, Brussels, 20-22 December 2005.

action, the latter can institute provisional TACs, which are applicable only until the Council is able to decide on new TACs.[20] TAC regulations are sometimes modified over the course of the year.

With regards to the exploitation of fisheries resources on the high seas, TACs are established with reference to the relevant IFOs such as, for example, NEAFC, NAFO or ICCAT, etc.[21] Many IFOs adopt binding TACs and other management measures; however, the Community maintains the right to opt out of these decisions within a certain time period.[22]

1.2.1.3 Quotas, quota flexibility and quota hopping

The resources listed in each TAC regulation are divided according to the principle of *relative stability*,[23] according to which each Member State receives a given percentage of the stock in the TAC. A set amount of quota is assigned to specific marine areas in Community waters ('ICES-areas').[24] Fishing up the quotas may only take place in the ICES-areas listed in the TAC regulations. Quota is also referred to as 'fishing opportunities', meaning 'a quantified legal entitlement to fish, expressed in terms of catches and/or effort'.[25] Member States are obliged to implement quotas properly.[26] Where their fishers overfish the allocated quota, it is the Member States who have transgressed Community law by not preventing overfishing by vessels flying their flags.[27]

After having received the annual quota, Member States face the problem of fishermen competing with each other to exhaust the annual quota first. Thus, the national quota must be divided and allocated to individual fishers. Member States can apply their own method for distributing TAC quota,[28] deciding, for

---

20 Art. 7 of Regulation 2371/02. Paul Nemitz, 'Fischereipolitik – Nach Art. 34 EGV', p. 577. Since the CFP has become an exclusive Community competence, disagreements between the Commission and the Council have only arisen on two occasions; see Christoph Becker and Karl J. Spurzem, 'Brüsseler Rituale' (2005) 51 *Mare*, available online at: http://www.mare.de/mare/.

21 See above, Chapter 2.

22 See chapter on international law.

23 On the legality of the principle of relative stability, see Case 46/86, Romkes v. Officier van Justitie, [1987] ECR, paras. 12 et seq.

24 ICES-areas can be viewed at www.ices.dk.

25 Arts. 3(q) and 20(1) of Regulation 2371/02. The term quota has been defined in Art. 3(b) of Council Regulation (EC) 40/08: 'Quota means a proportion of the TAC allocated to the Community, Member States or third countries'.

26 The Commission has aright to close a fishery on its own initiative, see Art. 21(3) of Regulation 2847/93 and Art. 26(4) of Regulation 2371/02. See below Chapter 3.2.2.2.

27 Case C-62/89, Commission v. France [1990] ECR, paras. 8 et seq. Member States are obliged to take penal or administrative actions against the violator, see Case C-52/95, Commission v. France [1995] ECR, para. 3 (Summary).

28 Art. 20(3) of Regulation 2371/02.

example, to allocate it to producer organisations or individual fishers.[29] The grant of quota may, for example, be contingent upon the fisher meeting certain licensing requirements. The Netherlands, again, has devised a system of individual tradeable quotas (ITQs), which allows fishermen to exploit, market or transfer their specific quotas.[30] Whatever system is adopted, Member States must notify the Commission of its choice. Although Member States are entitled to freely choose a national quota allocation system, the system itself cannot conflict with Community law.

Greater flexibility in the quota system is built into the system by providing for a quota exchange among Member States.[31] Around September of each year, the Commission submits a report on the current state of the quota to facilitate the exchange of surplus quota between Member States.[32] The term *exchange* implies that Member States do not sell and buy quota.[33] Furthermore, under certain circumstances, Member States can carry over unused quota to the forthcoming year.[34] If a Member State exceeds its annual quota, the Commission may deduct future fishing opportunities from the violating Member State, or, under specific circumstances, reallocate it to the Member State that suffered the violation.[35]

In fact, flexibility is also achieved through so-called 'quota-hopping', where the fishermen from one Member State relocate to or register their vessels in another Member State to gain access to its quota. This practice has opened the conflict between the right of establishment and free movement of workers, on the one hand, and the principle of relative stability on the other. Member

---

[29] Churchill, *EEC Fisheries Law*, p. 118.

[30] A. Berg, *Implementing and Enforcing European fisheries Law*, (The Hague: 1999), p. 124; W. Smit, 'Common Fishery Policy and National Fisheries Management' (1997) 12 *Marine Resource Economics*, pp. 355-359. Transferability of and market in fishing rights exist *de facto* in other Member States, too; see Fernando Gonzáles Laxe, 'Transferabilty of fishing rights: The Spanish case' (2006) 30 *Marine Policy*, pp. 379-388; On the United Kingdom, see OECD, *Using Market Mechanisms to Manage Fisheries* (Paris: 2006), pp. 279-287; Commission Staff Working Document SEC(2007) 247 accompanying the Communication from the Commission to the Council and the European Parliament on Right-based management tools in fisheries, p. 12. See Commission Communication Com (2007) 73 final on rights-based management tools in fisheries, pp. 3-4.

[31] Art. 20(5) of Regulation 2371/02.

[32] Nonnenmacher, *Die gemeinsame Fischereipolitik*, p. 203.

[33] Churchill, *EEC Fisheries Law*, pp. 117-118. Given the scarcity of fisheries resources, it is unlikely that exchange rates are significant. These numbers are not published by the Commission, see Proelß, *Meeresschutz*, p. 382.

[34] Art. 4 of Council Regulation (EC) No. 847/96 introducing additional conditions for year-to-year management of TACs and quotas, OJ 1996 No. L115/3.

[35] See Art. 23(4) of Regulation 2371/02; see, for example, Commission Regulation (EC) 147/07 adapting certain fish quotas from 2007 to 2012 pursuant to Article 23(4) of Council Regulation (EC) No. 2371/2002 on the conservation and sustainable exploitation of fisheries resources under the Common Fisheries Policy, OJ 2007 No. L46/10.

States have tried to prevent quota hopping by reserving national quota for their own citizens. Belgium, France, Greece and the United Kingdom, for example, imposed new laws that required that a certain proportion of vessels' crews be citizens or residents within their country.[36] Quota hopping then became an issue to be determined by the ECJ. The ECJ specifically ruled on the legality of the new laws imposing nationality conditions on crews, finding Member States may only reserve certain important positions for its own nationals (*e.g.*, master and first mate). However, requiring vessels` crews to be citizens or residents within their country was found to be in violation of Community law, *i.e.*, the principle of the free movement of workers.[37] Another matter of dispute concerned a United Kingdom ministry decision requiring vessels registered within the United Kingdom to have a *real economic link* to the country. In this case, the Court stated that, in principle, registering vessels in other Member State is possible pursuant to Art. 43 of the Treaty, which guarantees the freedom of establishment. However, it did allow Member States to require a *real economic link* between itself and the quota granted to certain vessels under specific conditions.[38] To prove the real economic link, Member States are allowed to require other Member States vessels to land a proportion of their catches in their ports or to be present periodically, 'provided that the frequency with which the vessel is required to be present in those ports does not impose, directly or indirectly, an obligation to land the vessel's catches in national ports or hinder normal fishing operations'.[39] It also decided that Member States could not force other Member States vessels to depart from there in all cases.[40]

The ECJ's rulings in these cases gave the freedom of establishment and the free movement of workers priority over the principle of relative stability. Subject to few (rather vague) restrictions, which can be adopted by Member States, quota hopping as a practice can continue. It is thus likely that quota hopping will slowly erode the relative strength of the principle of relative stability.[41]

---

36 Case 167/73, Commission v French Republic [1974] ECR 359; Case C-37/93, Commission v Belgian Kingdom [1993] ECR 6295; Case C-290/94, Commission v Greek Republic [1996] ECR 3285; Case C-3/87, The Queen v Ministry of Agriculture, Fisheries and Food, ex parte Agegate Ltd [1989] 4495.

37 Art. 39 of the Treaty.

38 Case 216/87, The Queen v. Ministry of Agriculture, Fisheries and Food, ex parte: Jaderow Ldt. et seq. [1989] ECR I-4544.

39 *Ibid.* pp. 4547-4548, 4550-4551.

40 *Ibid*, pp. 4547, 4550-4551.

41 M. Morin, 'The fisheries resources in the European Union. The distribution of TACs: principle of relative stability and quota-hopping' (2000) 24 *Marine Policy*, pp. 266-272.

1.2.1.4 Legal commitment to sustainability, precaution and ecosystems

When setting TACs (and other management measures),[42] the Council has to balance conservation and socioeconomic considerations. Art. 2(1) of Regulation 2371/02 makes a clear commitment to sustainable fishing, requiring that the Council ensures sustainable exploitation, applies the precautionary principle, and to minimise the impact of fishing on marine ecosystems.

The requirement of *sustainability*, set out in Regulation 2371/02, is concretised in Art. 2(1), which states that the CFP 'shall ensure exploitation of living aquatic resources that provides sustainable economic, environmental and social conditions'. One interpretation of this article is that economic and social priorities can override environmental ones; and, in practice, the Council regularly downgrades environmental concerns when adopting TACs. This is not to suggest that the Council's practice of setting TACs at unsustainable levels is incompatible with the legal requirements of Regulation 2371/02. However, it must be noted that the economic and social state of the fishing industry is contingent on the health of fish stocks. Therefore, 'sustainable economic, social and environmental conditions' can only be guaranteed where stocks are not exploited at levels which ultimately lead to their collapse.[43] This interpretation of Art. 2(1) is supported by Art. 3(e) of Regulation 2371/02, which defines 'sustainable exploitation' as 'exploitation of a stock in such a way that the *future exploitation of the stock will not be prejudiced* and that it does not have a negative impact on the marine ecosystem'.[44] Sustainable in this case means that the exploitation of stocks at any given time must not render future takes impossible and that it cannot negatively impact the marine ecosystem. From this perspective, the Council's discretion to balance economic, environmental and social objectives must be regarded as limited, preventing it from adopting one-dimensional measures which favour economic interests over conservation objectives and threaten the long-term survival of fish stocks. Fisheries scientists agree that in order to *not prejudice the future exploitation of stocks*, TACs must be set at levels that are within safe biological limits (*i.e.*, catch rates must not exceed the critical point at which stocks are threatened by a substantial

---

[42] The objectives of Art. 2 of Regulation 2371/023 apply to all management actions taken under CFP.

[43] See also G. Winter, 'A Fundament and Two Pillars', in Bugge and Voigt (ed.), *Sustainable Development in International and National Law*, (Groningen: 2008), pp. 24-45; *Ibid.*, *Towards a legal clinic for fisheries management*, p. 38; see on the content of the principle of sustainable development in Community's primary law, *Ibid.* 'Umweltrechtliche Prinzipien des Gemeinschaftsrechts' (2003) Sonderheft *Zeitschrift für Umweltrecht*, pp. 137-145-at 144. See also arguments of Marion Markowski, *Allocation and management of fisheries resources: an in-depth analysis of instruments in comparative perspective*, (Bremen; forthcoming 2008), p. 11. The Commission also recognises the benefits of setting fishing rates at MSY levels, see Commission Communication Com(2006)360 final on implementing sustainability in EU fisheries through maximum sustainable yield.

[44] Italics have been inserted by the author.

decline or collapse, the so-called F*lim* reference level).[45] Scientist also agree that exploiting fish stocks *continuously* beyond the MSY-level, *i.e.*, the largest harvest that can be taken from a fish stock year after year while still maintaining the average stock size, will eventually put stocks at this critical reference point.[46] As a result, the sustainability criteria in Art. 2 of Regulation 2371/02 should be interpreted as requiring the Council to abide by the following minimum management standards: First, adopting TACs, the Council, and in some cases the Commission, shall not exceed the MSY level. However, where there are compelling socio-economic grounds, Community organs are granted discretion to weigh economic and social objectives higher than environmental factors. In such exceptional cases, catch rates may exceed the MSY-level. Ultimately, however, it must be regarded as a clear violation of Art. 2 of Regulation 2371/02 if Council and the Commission grant fishing opportunities which cause stocks to fall below safe biological limits (F*lim*) or which maintain stocks within these limits.[47]

The meaning of the phrase 'exploitation under sustainable economic, environmental and social conditions' found in Art. 2(1), para. 1 is, again, qualified by Art. 2(1), para. 2, which states that 'the Community shall apply the precautionary approach in taking measures designed to protect and conserve living aquatic resources [...]'. Art. 3(i) of Regulation 2371/02 states:

> [The] precautionary approach to fisheries management means that the absence of adequate scientific information should not be used as a reason for postponing or failing to take management measures to conserve target species, associated or dependant species and non-target species and their environment.

The approach set out under Art. 2(1) and 3(i) requires the Community to take management measures *to conserve stocks* even in the absence of adequate scientific information. The meaning of conservation under the basic regulation is, as described above, determined by the sustainability criteria. The latter requires that catch restrictions should, in principle, prohibit exploitation beyond MSY-levels; it is only where there are compelling socio-economic reasons that MSY-levels may be exceeded. However, catch levels must be maintained *at a minimum*

45 G. Hubold, 'Wege zu einer nachhaltigen Fischerei' (2003) 14 *Zeitschrift für Umweltrecht*, 5, p. 338-339; Commission Communication (2000) 803 final on the application of the precautionary principle and the multiannual arrangements for setting TACs, pp. 4-5; see also ICES, Report of the ICES Advisory Committee on Fishery Management, Advisory Committee on the Marine Environment and Advisory Committee on Ecosystems – Book 1, (Copenhagen: 2007), available online at www.ices.dk; Note that the term 'safe biological limit' is defined differently in Art. 3 (l) of Regulation 2371/02.

46 Yumiko Kura, Carmen Revenga, Eriko Hoshino and Greg Mock, *Fishing for Answers – Making Sense of the Global Fish Crisis* (Washington D.C.: 2004), p. 91; Iudicello *et al.*, *Fish, Markets, and Fishermen – The Economics of Overfishing*, (London: 1999), pp. 45-47.

47 This would be in accordance with the UNCLOS requirements as interpreted by: Rat von Sachverständigen für Umweltfragen (SRU), *Meeresschutz für Nord- und Ostsee – Sondergutachten*, p. 126.

within safe biological limits. These two reference points, MSY-levels in normal circumstances and safe biological limits in exceptional ones, also serve as the reference points for the application of the precautionary approach. As a result, a *safety margin* or *precautionary buffer* that takes into account the unpredictable nature of stock levels and uncertainties in assessment must be factored into the sustainable exploitation rate (which respects the MSY- and *Flim*-level).[48] This applies in particular to underassessed stocks.[49] Under the CFP, TACs for such species are based on 'intelligent and educated guesswork'.[50] In these cases, safety margins must be particularly wide, which means that exploitation rates may not exceed the limit that scientists consider to be *clearly sufficient* to protect stocks from falling below the MSY- and, under exceptional circumstances, *Flim*-level.

Finally, Regulation 2371/02 requires the Community to aim at a progressive implementation of an ecosystem-based approach to fisheries.[51] The wording of its provisions, however, does not establish a clear legal obligation. The most important features of eco-system-based management are listed in several Commission Communications on Fisheries management and nature conservation in the marine environment, as well as in the Biodiversity Action Plan.[52] In these documents, the Commission lists a number of important measures and objectives such as, *inter alia*, protecting habitats, other species (particularly by reducing by-catch and discards), animals, coastal areas particularly by space-time limitations, etc.[53] With regards to quantitative catch limitations, the Commission has recognised the importance of reducing fishing pressure and the need for improved scientific research. Where information is lacking, efforts

[48] Arguing similarly: Wolff, *Fisheries and the Environment*, p. 152; Report of the ICES Advisory Committee on Fishery Management, Advisory Committee on the Marine Environment and Advisory Committee on Ecosystems – Book 1, (Copenhagen: 2007), available online at www.ices.dk., p. 2; See also G. Winter, *Towards a legal clinic for fisheries management*, pp. 38-39.

[49] Communication Com(2007) 295 final from the Commission to the Council on Fishing Opportunities for 2008 – Policy Statement from the European Commission, p. 5.

[50] , see A. Karagiannakos, 'Total Allowable Catch (TAC) and quota management system in the European Union' (1996) 20 *Marine Policy*, p. 244.

[51] The protection of ecosystems is considered in Arts. 2(1), second paragraph, 4(1) and (2)(g)(iv), 5(2), 6(2), 7(1), 8(1) of Regulation 2371/02.

[52] Commission Communication Com(1999) 363 final on Fisheries Management and Nature Conservation in the Marine Environment, pp. 9-16; Commission Communication Com(2001) 162 final Biodiversity Action Plan for Fisheries, pp. 11-29; see also Commission Communication Com (2001), 143 final Elements of a Strategy for the Integration of Environmental Protection Requirement in the Common Fisheries Policy, p. 9 and p. 22, Commission Communication Com(2002) 186 final Action Plan to integrate environmental protection requirements into the Common Fisheries Policy.

[53] Other measures mentioned are the improvement of the selectivity of gear, developing new technical conservation measures to reduce fishing impact, vocational training, information initiatives and consultation activities, improve the level of coherence between CFP and environmental measures, Commission Communication Com(2001), 143 final, pp. 13-14.

must be made to improve on current levels of understanding.[54] The implementation process must be monitored and revised based on a system of indicators.[55]

To sum up, the Council's discretion concerning the balancing of social, environmental and conservation objectives when setting TACs (and adopting other management instruments) is fettered by the sustainability objective and the precautionary approach; it must also employ the ecosystem approach in Art. 2(1) of Regulation 2371/02. As will be shown below, the precautionary approach has become particularly relevant in the context of the recovery and management plans, two instruments introduced during the 2002 Reform.

1.2.1.5 Critique and perspectives

There has been much criticism of the use of TACs as the central management instrument in the CFP. First, setting TACs at proper levels is difficult and cost intensive. Given the general unpredictability of the development of fish populations in combination with uncertainties in scientific data, TACs are based on broad assumptions.[56] According to the Commission,[57] in 2007, only 29 of 126 Community TACs were based on 'full assessment and forecast'. In contrast, 62 TACs were based on 'relevant quantitative advice'; and 35 TACs were not based on scientific advice at all. Second, TAC-quotas are often considered as unsuitable for application to multi-species fisheries. Many of the commercially important stocks harvested in Community fisheries are multi-species fisheries.[58] Where TAC-quotas are applied to such fisheries, fishers are inclined to continue fishing until all of their quotas are used up, continuing to fish those stocks for which the quotas are already exhausted. Besides, they often prefer not to keep low value fish species on board (upgrading). The problem is aggravated by the fact that Community law still requires catches of excess of quotas (or juvenile undersized fish) to be discarded. This combination of interests and statutes leads to huge amounts of wasted resources.[59] Ultimately, this creates a major problem with regards to the implementation of an ecosystem approach.[60] Third, as explained above, setting TACs and national quotas does not itself eliminate the incentive for fisheries to 'race for fish'. Well-defined

---

54 *Ibid.*, pp. 13-14.

55 Commission Communication Com (2001) 143, pp. 20-21.

56 A. Karagiannakos, 'Total Allowable Catch (TAC) and quota management system in the European Union', p. 244.

57 Communication Com(2007) 295 final from the Commission to the Council on Fishing Opportunities for 2008 – Policy Statement from the European Commission, p. 5.

58 D. Symes, 'The European Community's Fisheries Policy' (1997) 35 *Ocean & Coastal Management*, p. 147.

59 Commission Communication Com(2007) 136 final – A policy to reduce unwanted by-catches and eliminate discards in European fisheries; Communication Com(2002) 656 final from the Commission to the Council and the European Parliament – On a Community Action Plan to reduce discards of fish, pp. 6-7.

60 WWF, *WWF Mid-Term Review of the EU Common Fisheries Policy*, pp. 39-41.

and enforced individual (or collective) quotas are necessary to accomplish this. However, under the CFP, the implementation and enforcement of individual quotas is primarily the responsibility of Member States,[61] which do not enforce individual fishing quotas in a uniform and strict manner. Against this background, fishers may get the impression that exploiting the quota before the Commission declares it to be exhausted is the right strategy.[62] Where the race to fish continues, it is likely to cause the oversupply of fish markets early in the fishing season, which results` in an undersupply later on in the year. Another danger associated with the 'race to fish' is that fishers are driven to fish under all conditions, leading to an increase in accidents.[63] Fourth, another critique of the CFP management system concerns the fact that the Council often simply ignores scientific recommendations. This practice, however, conflicts with Art. 4(1) of Regulation 2371/02, which requires the Council to take 'into account available scientific, technical and economical advice [...]'. This is illustrated in the following examples. In response to ICES recommendation, the Commission proposed and the Council adopted the following catch reductions in 2003 for certain species.[64] In this respect, it should be noted that ICES proposes moratoria only where stocks are well below their safe biological limits.[65]

Table: 3.1. TAC setting

	ICES-recommendation	Commission's proposal	Council's TAC
Cod	Moratorium	- 66%	- 45%
Haddock		- 70%	- 50%
Whiting		- 76%	- 60%
Plaice		- 17%	- 5%

Despite these recommendations, the average deviation of Council TAC from scientific recommendations between 2003 and 2007 was as follows:[66]

61 Report from the Commission to the Council and the European Parliament on the monitoring of the Member States' implementation of the Common Fisheries Policy, Com (2007) 167 final of 10 April 2007.

62 A similar argument was put forward in the following paper: European Commission, 'A level playing field for better enforcement of CFP rules' (2003), 19 *Fishing in Europe*, pp. 3-6; see also Court of Auditors, Special Report 7/2007, para. 55.

63 Yumiko Kura *et al.*, *Fishing for Answers – Making Sense of the Global Fish Crisis*, p. 91.

64 The table is taken from Rat von Sachverständigen für Umweltfragen (SRU), *Meeresschutz für Nord- und Ostsee – Sondergutachten*, p. 126. See also Court of Auditors, Special Report 7/2007, para. 111.

65 WWF, *WWF Mid-Term Review of the EU Common Fisheries Policy*, p. 12, Olav Schramm Stokke and Clare Coffey, 'Precaution, ICES and the common fisheries policy: a study of regime interplay' (2004) 28 *Marine Policy*, p. 119.

66 Communication Com(2007) 295 final from the Commission to the Council on Fishing Opportunities for 2008 – Policy Statement from the European Commission, p. 5

Table: 3.2. Average deviation of Council TAC from scientific recommendations between 2003 and 2007

	2003	2004	2005	2006	2007
Average deviation	42%	48%	57%	47%	43%

Council decisions are often politically motivated, as no fisheries minister wants to tell its national electorate that there will be losses due to quota reductions. This is the main reason why the Council (in most cases with the Commission's assent) regularly sets catch limits at levels higher than recommended by scientists.[67] This behaviour, *inter alia*, has lead to the fact that in 2007, the state of the stocks was as follows: 20 stocks received a zero-TAC advice from ICES (moratorium); 26 stocks were outside safe biological limits; and only 12 stocks were found to be within safe biological limits.[68] Fifth, with a view to the TAC's interconnectedness with the principle of relative stability, *Symes* states that it acts as a force which maintains the *status quo* of CFP management system.[69] Given that the distribution provided by the principle is strongly linked to the use of TACs[70] (including industry expectations of specific TAC levels), it has impeded the introduction of more effective management instruments such as effort controls or a rights-based management system.

---

67 See, for example, Communication Com(2007) 295 final from the Commission to the Council on Fishing Opportunities for 2008 – Policy Statement from the European Commission, p. 5; ICES, *Environmental Status of the European Seas*, pp. 37-42; Communication Com(2002) 135 final Green Paper on the Future of the Common Fisheries Policy Vol. 1, pp. 6-8; Holden, *The Common Fisheries Policy*, p. 57-60; A. Karagiannakos, 'Total Allowable Catch (TAC) and quota management system in the European Union', p. 244; J.-P. Boude, J. Bonceur and D. Bailly, 'Regulating the access to fisheries: learning from the European experiences' (2001) 25 *Marine Policy*, p. 315; D. Symes, The European Community's Fisheries Policy', pp. 145-147; Court of Auditors, Special Report 7/2007, para. 111.

68 *Ibid.*, p. 5.

69 David Symes, 'Part 2 – Institutional Issues', p. 72; *Majone*, *Payne* and *Symes* suggest that the CFP has been shaped more by the desire to promote European integration than to manage fisheries resources properly, Giandomenico Majone, *Dilemmas of European integration – The ambiguities and pitfalls of integration by stealth*, (Oxford: 2005), pp. 111-114; D.C. Payne, Policy-Making in Nested Institutions: Explaining the Conservation Failure of the EU's Common Fisheries Policy' (2000) 38 *Journal of Common Market Studies*, pp. 303-324, David Symes, 'Part 2 – Institutional Issues' in Sissiwine and Symes (ed.), *Reflections on the Common Fisheries Policy*, online at www.greenpeace.org-/raw/content/denmark/press/, pp. 71-73. Payne and Majone also deem the *principle of free access* to be responsible for CFP failures. They consider it to perpetuate the international commons problem. However, this statement seems to be based on the wrong assumption that the principle of free access allows fishers to fish freely in offshore water of every Member States. In fact, quotas are assigned to specific ICES areas and there is a 12 nm exception to the free access principle, see Art. 17(2) of Regulation 2371/02.

70 The use of TACs allows a relatively easy application of the principle of relative stability compared to other management measures, see Payne, Policy-Making in Nested Institutions: Explaining the Conservation Failure of the EU's Common Fisheries Policy', p. 314.

Existing Community law thus requires that the Community change its current management practice. As a first step, the Council must decide whether it wants to continue using TACs as its main management tool. However, given the difficulty in achieving consensus on the distribution of resources amongst the Member States as well as the fact that non-discrimination can be implemented comparably easily using TACs, it is likely that the Community will continue with this course.[71] Due to this reality, the Community should consider ways to reduce the negative factors at work under a TAC and quota system. In this regard, it will be necessary to improve scientific assessment and strengthen its control and enforcement system.[72] However, a better course of action would be to modify the TAC system itself. To this end and despite the costs and extensive labour efforts, it should continue to move away from the setting of annual TACs towards a multi-annual approach.[73] TACs must also consider the multi-species nature of many Community fisheries, for example, by adopting multi-species TACs.[74] Discards should be forbidden, *i.e.*, fishers should be obliged to land all fish they catch. To not pressure fishers economically, they would receive a small proportion of the proceeds of the sale of these unlawfully caught fish to cover their costs.[75] The Community may also try to increase economic efficiency and transparency of the TAC system by allocating clearly defined individual tradable fishing rights to fishers, communities, or cooperatives, etc.[76] However, the most important thing to change would be the Council's practice of fixing TACs. This would require a change in the existing fisheries legislation. The new rules must ensure that keeping stocks at sustainable levels takes priority over short-term economic interests. It is clear that the Council's wide margin of discretion must be restricted. To this end, the principles in Art. 2 of Regulation

---

71 D.C. Payne, 'Policy-Making in Nested Institutions: Explaining the Conservation Failure of the EU's Common Fisheries Policy' (2000) 38 *Journal of Common Market Studies*, pp. 315-316; see also Communication Com(2002) 135 final Green Paper on the Future of the Common Fisheries Policy Vol. 1, pp. 23-24.

72 See chapters below.

73 See critiques on the current practice of the Council with regards to the multi-annual approach paragraph on recovery and management plans below. See also Communication Com(2002) 135 final Green Paper on the Future of the Common Fisheries Policy Vol. 2, pp. 6-8; Commission Communication to the Council and the European Parliament Com(2000) 803 final on the application of the precautionary principle and multi-annual arrangements for setting TACs.

74 Communication Com(2002) 656 final from the Commission to the Council and the European Parliament – On a Community Action Plan to reduce discards of fish, pp. 9-10.

75 Such a system exists in Norway, see Court of Auditors, Special Report No. 7/2007 on the control, inspection and sanction systems relating to the rules on conservation of Community fisheries resources, OJ 2007 No. C317/1, para. 26.

76 The Commission started an initiative in 2007 on rights-based management, see Commission Staff Working Document SEC(2007) 247 accompanying the Communication from the Commission to the Council and the European Parliament on Right-based management tools in fisheries, p. 12. See Commission Communication Com(2007) 73 final on rights-based management tools in fisheries, pp. 3-4.

2371/02 should be interpreted as proposed here, or legislation simply revised.[77] Moreover, given the specific competitive situation within the Council, a clear commitment to sustainability by a majority of Member States would be required. The ECJ must also take a strong stance by declaring certain TACs found in tertiary regulations as incompatible with higher-ranking Community principles (*e.g.*, Arts. 2 or 4 of Regulation 2371/02).[78] Another suggestion would be to shift the power of setting TACs to the Commission or a somewhat independent expert Agency to depoliticise the process (diagonal competence shift).[79]

1.2.2 Effort limitations

Generally speaking, regulating fishing effort can involve limiting all factors that enable fishing vessels to exploit fisheries resources (input-regulations). Such measures may include the limitation or reduction of the number of fishing vessels, together with decreases in size and engine power, a change of fishing gear, and reduced catching times or areas.[80] Community legislation refers to fishing effort more specifically as the 'product of the capacity and the activity of a fishing vessel; for a group of vessels it is the sum of the fishing effort of all vessels in the group'.[81] Compared to other management instruments, effort limitations are generally deemed to be 'simpler to enforce'. The Community considers effort reductions as a suitable instrument to overcome the problem of overcapacities as well as a useful supplement to TACs and technical regulations. Its current effort system will be described as follows.[82]

1.2.2.1 The development of the effort regime

Already in 1983, the Community introduced basic effort limitations for the marine areas around the Shetland and Orkney islands, *i.e.*, in the

---

77 Objectives under Canadian, US and Australian fisheries laws provide less discretion with regards to the setting of exploitation levels and place a greater emphasis on conservation over exploitation, see Michael Sissiwine, 'Part 1 – Challenges, Performance and the Future' in Sissiwine and Symes (ed.), *Reflections on the Common Fisheries Policy*, available online at www.greenpeace.org/raw/content/denmark/press/, pp. 15-17.

78 Under Community law, a basic regulation is higher ranking than a subsequent implementation regulation, see Case 38/70, Deutsche Tradax GmbH v. Einfuhr- und Voratsstelle für Getreide und Futtermittel, [1971] ECR 145, para. 3 (Summary).

79 Rat von Sachverständigen für Umweltfragen (SRU), *Meeresschutz für Nord- und Ostsee – Sondergutachten*, p. 127; more general, see G. Winter, *Towards a legal clinic for fisheries management*, pp. 39-40.

80 Holden, *The Common Fisheries Policy*, p. 196; Yumiko Kura *et al. Fishing for Answers – Making Sense of the Global Fish Crisis*, pp. 90-91.

81 Art. 3(h) of Regulation 2371/02 and Art. 2(b) of Regulation 1954/03.

82 The fleet reduction schemes are described in the chapter on promotion.

so-called 'Shetland Box'.[83] Besides, in the course of Spain's and Portugal's accession in 1985, the number of Spanish and Portuguese vessels allowed to fish in the existing Community waters was limited.[84] Then, in the early 1990s, policy makers became increasingly aware that resources were declining and saw a need for improvements to the existing TAC system. Many reasons for these failures were identified, such as weak enforcement and insufficient reductions to the over-capitalized fleet.[85] In response, the basic effort measures were advanced regarding specific marine areas, including the now so-called 'Western Waters', the 'Irish Box' and the Baltic Sea.[86] 'Western Waters' and 'Irish Box' are not legal terms, but correspond to several large marine areas in the Northeast Atlantic and waters close to Ireland.[87] The Community regulated and limited the use and access of specific gear and vessel types.[88] To effectively implement fishing effort, it established a fleet register, minimum licensing requirements, logbook requirements and new provisions on special fishing permits.[89] The control regulation was amended to correspond with the new requirements.[90]

---

[83] Art. 7 and Annex II of Regulation 170/83 and Commission Regulation (EEC) 2166/83 establishing a license system for certain fisheries in an area north of Scotland (Shetland area), OJ 1983 No. L206/71.

[84] See Arts. 156 to 160 of the 1985 Act of Accession. The Restrictions for Portuguese vessels were adopted by the Council annually, see Churchill, *EEC Fisheries Law*, p. 139.

[85] Commission SEC(1991) 2288 final Report from the Commission to the Council and the European Parliament on the Common Fisheries Policy.

[86] Arts. 3(f) and 4(2)(d),(e), 7, 8(1) of Regulation 3760/92; Council Regulation (EC) No. 685/95 on the management of the fishing effort relating to certain Community fishing areas and resources, OJ 1995, No. L71/5; Council Regulation (EC) No. 2027/95 establishing a system for the management of fishing effort relating to certain Community fishing areas and resources, OJ 1995 No. L199/1; Council Regulation (EC) No. 779/97 introducing arrangements for the managing of fishing effort in the Baltic Sea, OJ 1997 No. L113/1. See also Annex II of Regulation 51/2006.

[87] As defined in Council Regulation 3880/91 on the submission of nominal catch statistics by Member States fishing in the Northeast Atlantic, OJ 1991 No. L365/1; see also Commission Regulation (EC) 2945/95 amending Regulation (EEC) 2807/83 laying down detailed rules for recording information on the Member States' catches of fish, OJ 1995 No. L308/18.

[88] Council Regulation (EC) No. 2027/95 establishing a system for the management of fishing effort relating to certain Community fishing areas and resources, OJ 1995 No. L199.

[89] Commission Regulation (EC) 2090/98 concerning the fishing vessel register of the Community, OJ 1998 No. L266/27 and Commission Regulation (EC) 2092/98 concerning the declaration of fishing effort relating to certain Community fishing areas and resources, OJ 1999 No. L118/66; Council Regulation (EC) 3690/93 establishing a Community system laying down rules for the minimum information to be contained in fishing licences, OJ 1993 No. L341/93; Council Regulation (EC) 1627/94 laying down general provisions concerning special fishing permits, OJ 1994 No. L171/7. Commission Regulation (EC) 2091/98 concerning the segmentation of the Community fishing fleet and fishing effort in relation to the multiannual guidance programmes, OJ 1998 No. L266/36.

[90] Council Regulation (EEC) 2847/93 establishing a control system applicable to the common fisheries policy, OJ 1993 No. L261/1; as amended by Council Regulations 2205/97, OJ 1997 No. L304/1; 2635/97, OJ 1997 No. L356/14.

However, in 2001, the Commission observed that its attempt to improve fisheries management through effort limitations was largely a failure. According to the Commission, this was mainly due to the Member States' strong opposition to the joint use of TACs and effort limitations ('duplicate measures') as well as a 'lack of specific scientific advice establishing the proper quantitative basis for effort reductions'.[91]

The Commission also emphasised that the Community's fleet was much too large with respect to the available resources. Overcapacity had led to overfishing, and neither TACs, fleet reductions nor technical measures were regarded as sufficient to guarantee effective conservation.[92] Given that attempts to reduce overcapacities had largely failed in the past, the Community decided to refrain from adopting mandatory capacity reductions itself; and in 2002, it handed back the responsibility to reduce the fleet to Member States. From its side, it chose to handle overcapacities by regulating fishing effort limitations, restricting both fishing activities and fishing power.

As a result, the Council later adopted a new general system for the management of the fishing effort in the Western Waters, the Irish Box and the Baltic Sea.[93] The basic regulation 2371/02 also refers to effort limitations as the central management tool in the context of recovery plans.[94] The new regime has introduced certain new rules, which extend the application of the effort regime to new species and fisheries. The effort regime for the Baltic Sea has been replaced by a recovery plan and specific provisions on fishing in the Gulf of Riga.[95]

1.2.2.2 The current effort regime

With the adoption of basic regulation 2371/02, the Council opted for a Community specific definition of fishing effort. Art. 3(h) of Regulation 2371/02 and Art. 2(b) of Regulation 1954/03 refer to fishing effort as the

91 Communication Com(2002) 135 final Green Paper on the Future of the Common Fisheries Policy Vol. 2, p. 6; For other reasons, see Holden, *The Common Fisheries Policy*, pp. 196-198.

92 Commission Communication Com(2001) 135 final Green Paper on the Future of the Common Fisheries Policy, vol. I, pp. 8-11; vol. II, pp. 5-9 and 18-19.

93 Council Regulation 1954/2003 on the management of the fishing effort relating to certain Community fishing areas and resources and modifying Regulation (EC) No. 2847/93 and repealing Regulations (EC) 685/95 and (EC) No. 2027/93, OJ 2003 L289/1; Council Regulation (EC) 1098/07 establishing a multiannual plan for the cod in the Baltic Sea and the fisheries exploiting those stocks, amending Regulation (EEC) No. 2847/93 and repealing Regulation (EC) No. 779/97, OJ 2007 No. L248/1.

94 See Art. 5(4), paragraph 2 of Regulation 2371/02 as compared to management plans in Art. 6(4) of Regulation 2371/02.

95 Council Regulation (EC)1098/07 establishing a multiannual plan for the cod stocks in the Baltic Sea and the fisheries exploiting those stocks, amending Regulation (EEC) 2847/93 and repealing Regulation (EC) 779/97, OJ 2007 No. L248/1. See also Arts. 20 to 22 of Council Regulation (EC) 2187/05 for the conservation of fishery resources through technical measures in the Baltic Sea, the Belts and the Sound, amending Regulation (EC) 1434/98 and repealing Regulation (EC) 88/1998, OJ 2005 No. L349/1.

'product of the capacity and the activity of a fishing vessel; for a group of vessels it is the sum of the fishing effort of all vessels in the group'. While capacity is expressed in tonnage or engine power, fishing activity is expressed in days spent at sea. This definition does not include input factors like gear, the volume of fishholds, freezing capacity of vessels or catching areas.[96]

Regulation 1954/03 provides a general system for the management of fishing effort in the Western Waters and Irish Box (ICES areas V, VI, VII, VIII, IX and X, and CECAF division 34.1.1, 34.1.2. and 34.2.0). It requires Member States to assess different fisheries in each of these areas between 1998 and 2002.[97] Then, Member States must ensure that fishing effort – for all vessels 15 metres or less in length – is kept at the average fishing effort level which existed in this period.[98] With regards to the catching of demersal species, certain molluscs and crustaceans as well as fishing in the areas around Ireland, the Council has fixed the maximum annual fishing effort for each Member State based on the information provided by Member States.[99] The Council sets maximum levels of fishing effort for groups of species, fishing areas and fishery, and by Member State. For example, demersal fisheries carried out by German vessels in ICES area VII are limited to 233,560 (*i.e.*, the product of kw (x) fishing days).[100] Fishing effort levels may be modified by the Commission upon the request of a Member State where such an adaptation enables them to use up their TACs.[101] Further amendments are to be adopted, where the Council adopts recovery plans.[102]

For the waters up to 100 nm from the baselines of the Azores, Madeira and the Canary Islands, the Member States in this area may reserve fishing to vessels registered in the ports of these islands.[103]

Art. 7 requires Member States to establish a list of vessels flying their flag, which participate in fisheries mentioned in the effort regulation. They must take the necessary measures to regulate a fishery's effort. This requires that Member States monitor fishing effort.[104] They are also obliged to issue special fishing permits to vessels which engage in fisheries mentioned in the effort regulation.[105] Art. 8(3) refers to Council Regulation 1627/94 which establishes

---

96 See Commission Communication Com(2007) 39 final on improving fishing capacity and effort indicators under the common fisheries policy.

97 Arts. 3, 4 and 6 of Regulation 1954/03.

98 Arts. 3(4) and 4 (1)-(3) of Regulation 1954/03; these provisions include detailed exceptions.

99 Arts. 10 and 11 of Regulation 1954/03; see Council Regulation (EC) 1415/04 fixing the maximum annual fishing effort for certain fishing areas and fisheries, OJ 2004 No. L258/1.

100 Annex I of Council Regulation (EC) 1415/04.

101 Art. 12 of Regulation 1954/03.

102 Art. 3(3) of Regulation 1954/03.

103 Art. 5 of Regulation 1954/03.

104 Art. 8 of Regulation 1954/03.

105 See paragraph on the licensing system in this Chapter.

general provisions on these special fishing permits.[106] Art. 2(1)(a) of that regulation defines a special fishing permit as

> a prior fishing authorisation issued to a Community fishing vessel to supplement its fishing license, thereby enabling it to carry out fishing activities during a specified period, in a given area, for a given fishery, in accordance with the measures adopted by the Council.

1.2.2.3 Critique of the current effort regime

Scientifically, it is very difficult to precisely establish the fishing effort of a fishing vessel and to convert it to a specific rate of fishing mortality.[107] In comparison to TACs, effort limitations require 'equal, if not more data' and they are even more time consuming.[108] For example, every time a vessel's engine or tonnage is modified, the effort limit must be changed.[109] In addition, just like TACs, effort limitations rely on expensive biological information on stocks, which is difficult and costly to provide. Moreover, like TACs, effort limitations do not solve all problems that arise when the Council limits fishing for individual stocks in mixed fisheries (particularly for demersal).[110] Another problem is that it is difficult to measure engine power, particularly where engines are already installed; and these figures are easy to manipulate.[111] Furthermore, effort limitations (like TACs) are devised in the same deficient political process in the Council. The best thing one may say about the limiting of fishing effort as a management instrument is that, even though effort regulations are susceptible to manipulation, compared to TACs, they are generally perceived to be easier to control and enforce, particularly through the use of satellite monitoring systems.[112] However, the Commission stated in 2007 that the control of fishing effort is implemented only reluctantly by Member States. They do not use their satellite monitoring systems effectively to monitor fishing effort.[113] The Commission concludes that 'there is no evidence that the reduction in fishing effort has

---

106 OJ 1994, No. L171/7.

107 Holden, *The Common Fisheries Policy*, p. 196.

108 *Ibid.* p. 196; see also Long and Curran, *Enforcing the Common fisheries Policy*, pp. 142-144.

109 *Ibid*, p. 196.

110 *Ibid*, p. 197.

111 Long and Curran, *Enforcing the Common fisheries Policy*, p. 178; see also Commission Communication COM(2007) 39 final on improving fishing capacity and effort indicators under the Common Fisheries Policy.

112 Holden, *The Common Fisheries Policy*, p. 198, see also Commission Communication Com(2007) – Fishing Opportunities for 2008 – Policy Statement from the European Commission, p. 6.

113 Commission Report Com(2007) 167 final on the monitoring of the Member States implementation of the Common Fisheries Policy 2003-2005, pp. 9-10.

compensated for over-capacity in the fleet, even taking into account the effect of decommissioning schemes'.[114]

1.3 Licensing system

The FAO generally defines licenses as 'documents giving the producer the right to operate in a fishery according to the terms established by the regulating authority.'[115] Under Community law, a fishing licence 'confers on its holder the right, as limited by national rules, to use a certain fishing capacity for commercial exploitation of living resources'.[116]

Both definitions imply that the granting of fishing licenses can be conditioned on very different requirements. To price the use of a common resource, governments can either allocate licenses for free or by means of a bidding process, and may simply sell them or demand recurring royalties. To limit the overall fishing effort, the number of licenses can be restricted. In addition, to create efficiency in the sector licenses may be transferable or tradeable. States may also require that fishers prove their qualifications as fishermen or can make the grant contingent upon certain environmental gear specifications. Management requirements can also be attached to licenses such as information obligations, gear, area and quantitative restrictions. Licenses may ultimately be withdrawn where fishers do not comply with the requirements at the time of granting or those attached to them.

The establishment of a Community-wide licensing system under the CFP was widely discussed in the 1980s.[117] However, a comprehensive licensing system has never been adopted at the Community level. Instead, the Community has only required that Member States meet certain minimum requirements with regard to licenses. These licensing requirements are currently found in Council Regulation (EC) 700/2006 and Commission Regulation (EC) 1281/2005.[118]

To legally gain access to Community waters, every fishing vessel must carry on board a license and, where provided for, its authorisation for fishing.[119] Art.

[114] *Ibid.*, pp. 9-10.

[115] See online at: www.fao.org/fi/gloassry.

[116] Art. 2 of Regulation 1281/05.

[117] An intense discussion took place in the European Parliament in 1987, see OJ 1987 No. C 76/174.

[118] Council Regulation (EC) 700/06 repealing Regulation (EC) 3690/93 establishing a Community system laying down rules for the minimum information to be contained in fishing licences, OJ 2006 No. L122/1; this regulation gives effect to Commission Regulation (EC) 1281/05 on the management of fishing licences and minimal information to be contained therein, OJ 2005 No. L203/3; Council Regulation (EC) No 700/2006 of 25 April 2006 repealing Regulation (EC) No 3690/93 establishing a Community system laying down rules for the minimum information to be contained in fishing licences.

[119] Art. 22(1)(a) of Regulation 2371/02. Member States shall ensure, no later than 12 months from the date of application of Regulation 1281/05, that all of their licences comply with the law, see Art. 8 of Regulation 1281/05.

2 of Regulation 1281/05 states that a fishing licence 'confers on its holder the right, as limited by national rules, to use a certain fishing capacity for commercial exploitation of living resources'. Member States bear the responsibility of issuing, managing and withdrawing licences.[120] Fishing licences must contain certain information about the vessel, the licence holder and other particulars related to fishing capacity such as engine power, tonnage, length and gear.[121] Member States are required to update this information on a regular basis and ensure that the information is consistent with what is stated in the Community fishing fleet register.[122] Member States must temporarily suspend the licence of vessels which are subject to temporary immobilisation decided by that Member State.[123] Licences must be withdrawn permanently where a vessel is affected by capacity adjustment measures under Art. 11 (3) of Regulation 2371/02, *i.e.*, where capacity withdrawals have been supported by public aid.[124] With regards to fleet management, the total capacity granted in licences issued by a Member State may not be higher than the capacity adopted under the capacity management rules.[125]

The general licensing system is supplemented by a special licences system. Fishers who want to fish, for example, in Western Waters, the Irish Box or in third country waters, are required to obtain these special licences.[126]

Germany implements Community law licensing requirements through its national 'Fisheries Code' ('Seefischereigesetz'). Where Community law restricts fishing activities, German fishers are required under § 3 Seefischereigesetz to obtain a catch allowance ('Fangerlaubnis'). The German government is obliged to grant allowances to the extent fishing opportunities are made available under Community law. Allowances are not granted where applicants have missed the application deadlines, have significantly overfished the most recent quotas assigned to them, or where they are deemed not to apply for themselves. Allowances are only granted to fishers who fish from vessels that

---

[120] In accordance with the provisions of Regulation 1281/05, see Art. 4 of Regulation 1281/05.

[121] see Art. 5(1) and Annex of Regulation 1281/05.

[122] Art. 5(2)(a) and (3) of Regulation 1281/05; the Community's fleet register is accessible at: http://ec.europa.eu/fisheries/fleet/index.cfm.

[123] Art. 6(1) of Regulation 1281/05.

[124] Art. 6(2) of Regulation 1281/05.

[125] See Art. 7 of Regulation 1281/05; Arts. 12 and 13 of Regulation 2371/02 and Council Regulation (EC) 639/2004 on the management of fishing fleets registered in the Community's outermost regions, OJ 2004 No. L102/9, Commission Regulation (EC) 2104/04 laying down detailed implementing rules for Council Regulation (EC) 639/2004, OJ 2004 No. L365/19; Commission Regulation (EC) 1438/2003 laying down implementing rules on the Community Fleet Policy as defined in Chapter III of Council Regulation 2371/2003. OJ 2003 No. L204/21.

[126] Council Regulation (EC) 1627/94 laying down general provisions concerning special fishing permits OJ 1994, No. L171/7. Council Regulation (EC) 3317/94 laying down general provisions concerning the authorisation of fishing in the waters of a third country under a fisheries agreement, OJ 1994 No. L350/13.

have been in operation since 1986, fishers who received a licence in 1987, or which have been financed by the German government or its Germany's federal states.[127] Additional constraints can be imposed on allowances where it is in the general interest of fisheries or where required by Community law. The competent authority to grant allowances is the 'Bundesanstalt für Landwirtschaft und Ernährung.'[128]

To be allowed to fish in third country waters, fishers apply to their national government for special permits, and the Member State then forwards the application to the Commission. The Commission examines whether the application is in conformity with the relevant access agreement, third country fisheries rules and Community law, and also determines whether there is sufficient quota available.[129] It then either forwards the application to the third country or rejects it outright. Where the third country is willing to grant a fishing licence, the Commission notifies the relevant Member State, which, in turn, issues the permit to the applicant.[130] If the third country withdraws the licence, the Commission informs the Member State, which then withdraws the permit.[131]

1.4 Technical measures

Technical measures supplement TACs and effort limitations. Technical measures particularly aim at protecting juvenile fish, non-target species, and the marine environment.[132]

1.4.1 The current technical measures

Art. 4 (2)(g) of Regulation 2371/2002 lists technical measures adopted under the CFP:

> '(i) measures regarding the structure of fishing gear, the number and size of fishing gear on board, their methods of use and the composition of catches that may be retained when fishing with such gear[133];
> (ii) zones and/or periods in which fishing activities are prohibited or restricted including for the protection of spawning and nursery areas;
> (iii) minimum size of individuals that may be retained on board and/or landed;

---

[127] Several exceptions to this principle requirement are listed in Art. 3(1) of the Seefischereigesetz.

[128] § 3(3) Seefischereigesetz. See online at: www.ble.de.

[129] Art. 5 of Regulation 3317/94.

[130] Art. 4(2) of Regulation 3317/94.

[131] Art. 6 of Regulation 3317/94.

[132] See, for example, Commission Communication Com(95) 669 final on Implementation of Technical Measures in the Common Fisheries Policy, pp. 1-2.

[133] See also Commission Regulation (EC) 129/03 laying down detailed rules for determining the mesh size and thickness of twine of fishing nets, OJ 2003 No. L22/5.

(iv) specific measures to reduce the impact of fishing activities on marine ecosystems and non-target species.'

The Community has adopted different technical measures for different marine areas. Fishing in the North Sea and Atlantic is governed by Council Regulation (EC) 850/1998,[134] in the Mediterranean by Council Regulation 1967/06,[135] in the Baltic Sea by Regulation 2187/05,[136] and in Antarctic waters by Regulation (EC) 600/04.[137] General provisions on drift nets are laid down in Regulation (EC) 894/97;[138] and measures adopted under IFO-regimes such as, for example, NAFO are included in the annual regulations establishing fishing opportunities. All of these regulations, to a certain extent, make use of the above-mentioned technical measures. Given the fast pace of developments in fisheries management, technical regulations are regularly amended. For example, from 1998 to January 2006, Regulation 850/1998 had been amended more than 95 times.

Given that the North Sea and Atlantic are the Community's most important fishing grounds in terms of catch quantities, Regulation 850/1998 will be described as an example in more detail.[139] Arts. 4 to 10 of the technical regulations delineate the rules on towed gears, imposing a ban on the use of different kinds of nets and mesh sizes in specific areas, and the composition of the catch in relation to target and non-target species. Similar provisions on fixed gears are found in Art. 11 to 16. Arts. 17 to 19 outline the rules on the treatment of undersized marine organisms. Art. 19(1) reads: 'undersized marine organisms shall not be retained on board or be transhipped, landed, transported, stored, sold, displayed or offered for sale, but shall be returned immediately to the sea'.[140] Art. 18 and Annex XII define the minimum size of the marine organisms that can be caught. Arts. 20 to 29 lay down special provisions on fishing for specific

[134] Council Regulation (EC) 850/1998 for the conservation of fishery resources through technical measures for the protection of juveniles of marine organisms, OJ 1998, No. L125/1. This Regulation had been amended about 70 times before 2006. Few provision of Regulation 894/97 laying down certain technical measures for the conservation of fishery resources, OJ 1998 No. L132/1, are still applicable.

[135] Council Regulation (EC) 1967/06 concerning management measures for the sustainable exploitation of fishery resources in the Mediterranean Sea, amending Regulation (EEC) 2847/93 and repealing Regulation (EC) 1626/94, OJ 2006 No. L409/11.

[136] Council Regulation (EC) 2187/05 for the conservation of fishery resources through technical measures in the Baltic Sea, the Belts and the Sound, amending Regulation (EC) 1434/98 and repealing Regulation (EC) 88/1998, OJ 2005 No. L349/1.

[137] Council Regulation (EC) 600/04 laying down certain technical measures applicable to fishing activities in the area covered by the Convention on the conservation of Antarctic marine living resources, OJ 2004 No. L97/1.

[138] Council Regulation (EC) 894/97 laying down certain technical measures for the conservation of fisheries resources, OJ 1997 No. L.132/1.

[139] The technical regulation for the Mediterranean Sea will be described below.

[140] Reason (6) of Regulation 850/1998 reads that 'the practice of returning marine organisms to the sea should be reduced as much as possible'. Art. 19(1) does not apply to specified exception under Art. 19(2).

marine organisms like herring, sprat, mackerel, tuna, etc. The provisions establish *no-take zones* and *no-take periods* restricting the catch of the respective species. Such measures aim at protecting both target and non-target species.

Arts. 30 to 39 concern the protection of the physical marine environment – particularly the seafloor. For example, the use of trawl nets, which are kept on the seafloor with heavy steel weights, has a major impact on the seabed. This is particularly the case with regard to the use of beam and otter trawls. The weights plough over the seabed, leaving largely destroyed habitats and ecosystems in their wake. These disturbed habitats are particularly vulnerable to invasive species and ecosystem degradation in the long-term. Community law does not generally ban trawling but it is restricted in those areas which are most vulnerable. Subject to some exceptions, beam trawling is forbidden in the 12-mile zone around the UK and Ireland.[141] In addition, the use of beam trawls is entirely forbidden in the Kattegat, and regular trawling is restricted in the Skagerrak.[142] Similar bans have been included in technical measures that apply to other areas of Community waters.[143] Other destructive fishing practices such as the use of explosive, poisonous or stupefying substances and electric current are entirely banned.[144]

Arts. 42 to 44 institute technical and other measures relating to scientific research. Art. 45(1) provides that where the conservation of stocks calls for immediate action, after consulting the STECF, the Commission may adopt any measure necessary.[145]

In addition, according to Art. 45(2), Member States may also take non-discriminatory conservation measures within waters under their jurisdiction. According to Art. 45(2), this can be done in cases where 'any delay would result in damage which would be difficult to repair'. The Commission must receive notification of the adoption of such measure, which may require its cancellation or amendment. The Commission's decision can be referred to the Council within ten working days of receiving the notification of the Commission. The Council may then, by qualified majority, adopt a derogating decision within one month. Member States may, according to Art. 46, also take technical measures that strictly concern local stocks of sole interest to national fishermen. Such measures may either supplement Community legislation, or go beyond its minimum requirements, provided that they apply solely to the fishermen of the Member States and that they are compatible with Community law and in conformity with the CFP.

---

141 Art. 34 of Regulation 850/1998.

142 Arts. 37 and 39 of Regulation 850/1998.

143 See, for example, Art. 4 of Regulation 1967/2006 on trawling in the Mediterranean and Art. 22 of Regulation 2817/05 for the Baltic Sea.

144 Art. 31 of Regulation 850/1998.

145 Art. 45(1) of Regulation 850/1998 refers to Art. 48 of Regulation 850/1998, which again refers to Art. 18 of Regulation 3760/92.

1.4.2 Critique and perspectives of the current technical measures

The Community's technical measures have been criticised on various grounds. The Commission, for example, indicated in 2001 that technical regulations have become increasingly complex and difficult to apply,[146] and that they do not sufficiently address environmental concerns.[147]

The obligation to return undersized species to the sea in Arts. 19 of Regulation 850/1998 is particularly problematic.[148] Given the huge amount of by-catch and that only approximately 6% of the fish returned to sea survive, Art. 19 creates an enormous amount food waste.[149]

There is also extensive criticism of the rules on mesh sizes. The Community regulations on mesh sizes are regarded as not ambitious enough to protect juvenile fish. Sissiwine, for example, explains that while the Community requires 120 mm for cod trawling, the US requires a mesh size ranging from 165 to 179 mm, depending on the location and species.[150] Some Member States have also been reluctant to properly implement mesh size regulations.[151] Furthermore, in practice, it is difficult for fisheries inspectors to measure the exact size of the mesh. This is due to the fact that the gauges used for measurement of mesh sizes are not very precise control instruments.[152] Fishers may also easily undermine mesh size regulation by intentionally placing heavy objects in the nets to diminish mesh sizes.[153]

Another problem is that, in some cases, fishers are allowed to carry gear on board which they are not allowed to use. According to the conditions set out in the control regulation,[154] this gear must be lashed and stowed.[155] Given the problems with enforcing the CFP, fishers may be tempted to use that gear.[156] The Commission stated in 2007 that especially in southern Europe banned gears

[146] Commission Communication Com(2002) 135 final Green Paper on the Future of the Common Fisheries Policy Vol. 2, p. 10.

[147] Commission Communication Com(2002) 135 final Green Paper on the Future of the Common Fisheries Policy Vol. 1, p. 23.

[148] J.B. Ingerowski and M Salomon, ,Ein kritischer Blick auf die aktuellen Entwicklungen in der Gemeinsamen Fischereipolitik unter Einbeziehung der neu geschaffenen regionalen Fischereibeiräte', p. 539.

[149] *Ibid.*, p. 546 and 539.

[150] Michael Sissiwine, 'Part 1 – Challenges, Performance and the Future', pp. 26-27; on the history of the setting of mesh sizes and gear measures, see Holden, *The Common Fisheries Policy*, pp. 72-84.

[151] See, for example, Case C-64/88, Commission v French Republic, [1991] ECR I2727.

[152] Long and Curran, *The Common Fisheries Policy*, p. 170. This has given rise to disputes before the ECJ, see Case C- 348/88, Criminal proceedings against Hakvoort [1990] ECR I-1647.

[153] Long and Curran, *The Common Fisheries Policy*, pp. 165-166.

[154] Art. 20 of Regulation 2847/93.

[155] Art. 4(2)(d) of Regulation 850/98.

[156] This used to be a serious problem throughout the 1980s and early 1990s, see Holden, *The Common Fisheries Policy*, pp. 72-73; Long and Curran, *Enforcing the Common Fisheries Policy*, pp. 165-166.

such as driftnets are still being used.[157] Fishers can also easily disengage illegal gear with the arrival of control vessels. [158]

Finally, scientists argue that more no-take zones should be established to protect nursery grounds and sensitive marine areas against fishing activities. *Froese* and *Pauly*, for example, have proposed that almost 40% of the North Sea area be closed for fishing.[159]

The Community's technical measures regime must be improved. First, mesh sizes should be enlarged to sizes that allow exploited species to reach spawning age.[160] Second, the selectivity of gear should continuously be improved in accordance with current scientific advances. Third, given the difficulties associated with enforcing technical measures, the Council should consider adopting so-called 'one net rules' for specific fisheries.[161] This would restrict fishers to carrying only one or two nets when fishing in specific areas for specific stocks. Fourth, should fishers discharge their gear prior or during the course of an inspection, a rebuttable presumption of malpractice should apply in national administrative or judicial proceedings.[162] Fifth, more no-take areas should be established to protect nursery grounds and vulnerable marine areas and ecosystems. Ultimately, however, more far-reaching changes should be considered, including a general ban on discards. All fish caught would have to be landed and deducted from the individual fisher's quota. 'By-catch' could still be used for the production of fishmeal or fishoil. Such a ban would also allow for the exact assessment of overall fish take from the seas.[163] For example, in lieu of imposing technical measures, Norway has chosen to close marine areas to fishing where a certain amount of undersized fish is found within a catch.[164]

---

[157] Commission Report Com(2007) 167 final on the monitoring of the Member States implementation of the Common Fisheries Policy 2003-2005, pp. 9-10.

[158] Long and Curran, *The Common Fisheries Policy*, pp. 165-166.

[159] R. Froese and D. Pauly, 'Dynamik der Überfischung', in Lozán, Rachor, Resie, Sünderman and v. Westernhagen (eds.), *Warnsignale aus Nordsee & Wattenmeer – Eine Aktuelle Umweltbilanz*, (Hamburg: 2003), p. 288, 294.

[160] The Community financed the project 'INCOFISH' (Project Number: INCO 003739), a consumer campaign entitled 'Don't Eat the Babies' which referred to the consumption of undersized fish marketed on German fish markets. Another project, called 'Fish Ruler', aimed at empowering consumers to measure whether they were being sold undersized fish.

[161] The Council may actually oblige fishers on the basis of Art. 20(3) of regulation 2847/93 to not carry nets of different minimum mesh sizes during a single fishing trip. See also Long and Curran, *The Common Fisheries Policy*, pp. 168-169.

[162] *Ibid.*, pp. 168-169.

[163] Ingerowski and Salomon, 'Ein kritischer Blick auf die aktuellen Entwicklungen in der Gemeinsamen Fischereipolitik unter Einbeziehung der neu geschaffenen regionalen Fischereibeiräte', p. 539.

[164] *Ibid.*, p. 170.

1.5 Procedural management instruments

To improve and accelerate the cumbersome decision-making process within the CFP management system, CFP measures provide for different procedural instruments. Arts. 5 and 6 of Regulation 2371/2002 lay down the legal structures for the so-called multi-annual recovery and management plans.[165] Additionally, according to Arts. 7 and 8 of Regulation 2371/2002, under specific conditions the Commission and Member States may adopt emergency measures.

1.5.1 Multi-annual measures

For a long time, the Commission has promoted the adoption of a multi-annual approach to fisheries management. According to the Commission, this kind of approach should foster the use of objective criteria in the Council's political processes. Moreover, it improves long-term planning in the commercial sector by avoiding abrupt changes in the annual setting of TACs. Finally, long-term perspectives make it easier to link fleet policies to management schemes.[166] As of 1992, the Council has already had the opportunity to make use of this approach under the prior basic regulation.[167] However, this type of approach was not adopted due to a lack of sufficient scientific information, industry opposition and the fear that TACs would become too inflexible.[168] The current basic management regulation has introduced two types of long-term plans, which set the framework for the adoption of the multi-annual recovery and management plans.[169] The Council has made use of this recovery plan option since 2003, and after this time has adopted several of such plans (*e.g.*, recovery plans for cod and hake in 2003 and eel and tuna in 2007).[170]

---

165 Reason (7) of Regulation 2371/2002.

166 Commission Communication Com(2000) 803 final on the application of the precautionary principle and multi-annual arrangements for setting TACs, pp. 3, 8-17.

167 See Art. 8(4)(i) of Regulation 3760/92.

168 Commission Communication Com(2002) 135 final Green Paper on the Future of the Common Fisheries Policy Vol. 2, pp. 6-7.

169 Arts. 5 and 6 of Regulation 2371/02.

170 Council Regulation (EC) 423/04 establishing measures for the recovery of cod stocks, OJ 2004 No. L70/8; Council Regulation (EC) 811/04 establishing measures for the recovery of the Northern hake stock, OJ 2004 No. L150/1; Council Regulation (EC) 643/07 amending Regulation 41/07 as concerns the recovery plan for bluefin tuna recommended by the International Commission for the Conservation of Atlantic Tunas, OJ 2007 No. L151/1; Council Regulation (EC) 1100/07 establishing measures for the recovery of stocks of European eel, OJ 2007 L248/17. Currently, recovery plans for 11 stocks have been adopted or proposed, see WWF, *WWF-Mid-Term Review of the Common Fisheries Policy*, p. 25.

1.5.1.1 Recovery plans

Recovery plans are adopted if stocks are found to be outside safe biological limits. The overall objective of the recovery plans is to ensure the return of fish stocks back to safe biological limits.[171] The plans are multi-annual and should indicate the expected time frame for reaching the established targets.[172] They must also include *conservation reference points*,[173] defined in Art. 3(k) as 'values of fish stock population parameters (such as biomass or fishing mortality rate) used in fisheries management, for example with respect to an acceptable level of biological risk or a desired level of yield'. These reference points serve as the basis for assessing the recovery process. Reference points can relate to targets such as population size, long-term yields, fishing mortality rates, and stability of catches. Recovery plans are to be drawn up in accordance with the precautionary approach, taking into account the so-called *limit reference points* recommended by the pertinent scientific bodies. Limit reference points are defined in Art. 3(j) as 'values of fish stock population parameters (such as biomass or fishing mortality rate), which should be avoided because they are associated with unknown population dynamics, stocks collapse or impaired recruitment'. Targets *may* relate to other living aquatic resources and the maintenance or improvement of the conservation status of marine ecosystems.

The Council decides which measures are to be included in the recovery plans. Measures must be proportionate to the objectives, targets and the expected time frame. They must have regard to (a) the conservation status of the stocks, (b) the biological characteristics of the stocks, (c) the characteristics of the fisheries in which the stocks are caught, and (d) the economic impact of the measure on the fisheries concerned. The plans should emphasise the use of fishing effort as a management tool. Art. 5(4) states that recovery plans *may* include any measure referred to in Art. 4(2)(c)-(h) but *shall* include limitations on *fishing effort* unless this is not necessary to achieve the objective of the plan.[174]

1.5.1.2 Management plans

Management plans are adopted in cases where the Council considers it *necessary* to maintain stocks within safe biological limits.[175] Accordingly, the main objective of the management plans is to 'maintain stocks within safe biological limits for fisheries exploiting stocks at/or within safe biological

---

[171] Art. 5(1) and (2) of Regulation 2371/02.

[172] Art. 5(3), third paragraph of Regulation 2371/02.

[173] Arts. 3 and 5 of Council Regulation (EC) 423/04 refer to conservation reference points as target and minimum levels.

[174] Italics inserted by author.

[175] Currently, management plans for 10 stocks have been adopted or proposed, see WWF, *WWF-Mid-Term Review of the Common Fisheries Policy*, p. 25.

limits'.[176] From a technical point of view, the difference between recovery and management plans is that management plans do not primarily refer to effort limitations as the central instrument to limit fishing activities. In other respects, management plans shall also be based on reference points which can also relate to targets such as population size, long-term yields, fishing mortality rates, and stability of catches. Management plans must incorporate the precautionary approach and take account of the limit reference points by relevant scientific bodies. Targets *may* relate to other living aquatic resources and the maintenance or improvement of the conservation status of marine ecosystems.

1.5.1.3 Critique of the multi-annual measures

The idea of introducing multi-annual recovery and management plans is based on three central objectives. First, stocks that are outside save biological limits must return to (or remain in) save biological limits. Second, to attain this goal, *multi-annual plans* are to be adopted. Third, under the recovery plans, TACs are supplemented by effort limitations.[177] The following section examines whether the legal requirements included in Art. 5 together with Council's current practice in this respect effectively meet these objectives.

The first objective set out in Art. 5 is to ensure that overexploited stocks return to (or stay within) safe biological limits. However, this article does not establish a firm obligation, instead leaving the Council with a large degree of discretion. Art. 5(3) merely provides that 'recovery plans shall be drawn up on the basis of the precautionary approach to fisheries management and *take account* of the limit reference points recommended by relevant scientific bodies'.[178] Besides, there are no deadlines for when plans have to be established.[179] Thus, every recovery plan is still adopted against the background of the struggle in which Member States push for high exploitation rates. The Council, in fact, has already dismissed an attempt to depoliticise the process of adopting recovery plans by shifting this capacity to the Commission. During the negotiations on Regulation 2371/02, the Commission suggested that the Council 'merely' determines the strategic elements of the multi-annual plans and that subsequent implementation should be carried out by the Commission under the auspices of a Comitology management committee.[180] The Commission's ideas were not accepted. Instead, the Council has maintained all powers regarding the setting of catch and effort limitations as well as the ability to amend catch rates

---

[176] Art. 6(1) of Regulation 2371/02.

[177] Art. 5(2), (3), third paragraph and Art. 5(4), second paragraph of Regulation 2371/02.

[178] Italics have been inserted by the author.

[179] House of Lords, *The Progress of the Common Fisheries Policy*, p. 14.

[180] Council of the European Union Working Paper 13595 on the Reform of the Common Fisheries Policy, Brussels 31 Oct. 2002, p. 12.

annually, depending on the development of the stocks.[181] Not surprisingly, the Council's recovery plans have already been severely criticised and even led to an action before the ECJ. For instance, scientists have stated that the initial fishing mortality limits adopted under the cod recovery plan have been much too high to put an end to overexploitation and that they are inconsistent with the precautionary approach.[182] Furthermore, in proceedings before the ECJ, the WWF has claimed that the Council had not adhered to its own recovery plan by subsequently adopting fishing rates well beyond limits established in the plans.[183] In addition, the overall number of plans that have been adopted is considered very small in comparison to the overall number of adopted TACs (6 to 130).[184] Thus, one of the CFP's central problems – the Council's unwillingness to set catch rates at sustainable levels – continues under the current multi-annual approach. As a consequence, the results of the multi-annual plans appear to be mixed: while some stocks have recovered to a certain extent, others have not.[185]

The second reason for adopting a multi-annual approach to fisheries management was to provide planning security for the industry as well as structural policies. However, at any time the Council can increase exploitation rates under recovery plans.[186] Doing that, it must take into account the target levels established in the multi-annual plans. However, whenever the Council decides to change exploitation rates, this can erode the (potential) benefits of using a multi-annual approach.

The third goal was to include effort restrictions within recovery plans to increase the effectiveness of existing management tools. However, the Council has inserted a loophole into Art. 5(4), paragraph 2, which gives it the discretion to reject any effort restrictions which are *'not necessary to achieve the objectives of the plan'*. Art. 5(4) thus leaves open the possibility that some Member States in the Council will reopen negotiations on effort limitations even if stocks are not within safe biological limits. However, given the existing problems within the

---

[181] For example, the Council has increased catch rates for cod in Regulation 41/2006 for cod in the zones covered by the cod recovery plan as regulated in Regulation 423/04.

[182] Michael Sissiwine, 'Part I – Challenges, Performance and the Future', p. 26; with regards to a herring recovery plan in the North Sea, see Ices, *Report of the ICES Advisory Committee on Fishery Management, Advisory Committee on the Marine Environment, and Advisory Committee on Ecosystems – Book 6 – North Sea*, p. 26, see also the critique in WWF, *WWF-Mid-Term Review of the Common Fisheries Policy*, pp. 26-32.

[183] See Case T-91/07, WWF-UK v Council, not yet in the ECR.

[184] House of Lords, *The Progress of the Common Fisheries Policy*, p. 18.

[185] House of Lords, *The Progress of the Common Fisheries Policy*, p. 19.

[186] See, for example Arts. 5-7 of Council Regulation (EC) 423/04. According to Art. 6(1) of Regulation 423/04, the Council is to decide each year the total allowable catches for each cod stock under the recovery plan.

CFP management regime, 'effort limitation must be applied in most recovery plans' for them to have any chance of success.[187]

1.5.2 Emergency measures

Two procedural mechanisms have been put in place to supplement the management system when there is an unforeseeable threat to fish resources. Here, the Commission and Member States may decide to take *emergency* management measures, deviating from the regular legislative procedure.[188] The Commission may take emergency measures if there is a serious threat to the conservation of living aquatic resources or the marine ecosystem that results from fishing activities and which requires immediate action. It may do so on its 'own initiative or at the substantiated request of a Member State'.[189] The emergency measure cannot be in force for longer than six months, and can only be extended for an additional six months.[190] If a Member State wants the Commission to take emergency measures, it must forward a formal quest to the Commission, all Member States and the Regional Advisory Councils. Other Member States and the RACs may submit written comments within five working days to the Commission. The Commission must then decide within 15 days after receiving the request whether it will allow the emergency measure. If the emergency measure is adopted, it takes immediate effect. Any objecting Member States may refer the Commission's decision on the emergency measure to the Council.[191] The Council then has one month, acting by qualified majority, to overrule the Commission and take a different decision.

According to Art. 8 of Regulation 2371/2002, Member States are also allowed to take emergency measures in waters falling under their sovereign control. They may do so where there is evidence of a 'serious and unforeseen threat to the conservation of living aquatic resources, or the marine ecosystem resulting from fishing activities'. This competence only applies in cases where 'any undue delay would result in damage that would be difficult to repair'.[192] The measures cannot be longer than three months in duration.

[187] Hegland and Raakjær, 'Recovery Plans and the Balancing of Fishing of Fishing Capacity and Fishing Possibilities, chapter 5.

[188] See Arts. 7 and 8 of Regulation 2371/02; for example, Commission Regulation (EC) 1037/05 establishing emergency measures for the protection and recovery of the anchovy stock in ICES Sub-area VIII, OJ 2005 No. L171/24.

[189] Art. 7(1) of Regulation 2371/02.

[190] Commission Regulation (EC) 1539/05 extending the emergency measures for the protection and recovery of the anchovy stock in ICES subs-area VIII, OJ 2005 No. L247/9.

[191] Art. 7(4) of Regulation 2371/02.

[192] Art. 8(1) of Regulation 2371/02.

1.6 The Mediterranean

Fishing in the Mediterranean is carried out under natural, legal and socioeconomic conditions, which are quite distinct from other Community waters. This has implications for the management of fisheries in this area by the Community.

1.6.1 Mediterranean specificity[193]

To this day, only Egypt, Morocco, Syria and Cyprus have claimed EEZs in the Mediterranean.[194] Only four states, including Spain and Malta, have established exclusive fishing zones (EFZ). While Malta claims a 25-mile exclusive fishing zone, Spain claims a 37-mile wide fisheries protection zone, measured from the outer limit of the territorial sea.[195]

Fishing vessels are comparatively small in the Mediterranean. Most fisheries are classified as artisinal or small-scale fisheries;[196] and the numerous minor landing sites make control and enforcement difficult.

Fishing is often carried out close to the shore, which leads to high competition among fishers and also conflicts with tourism and leisure fisheries.[197]

The Commission views the current scientific advice on Mediterranean fisheries as insufficient. This is mainly due to the lack of an overarching scientific forum in the Mediterranean, such as ICES.[198]

The state of fisheries resources in the Mediterranean is critical. The Commission has stated in 2002 that most of the demersal, small pelagic or highly migratory species are overexploited.[199] The EEA stated in its 2007 report on the state of Europe's environment that the percentage of assessed commercial stocks outside safe biological limits in the Mediterranean Sea in 2005 ranged from 10-20%, with Aegean and Cretan waters being in the worst condition. [200]

[193] Commission Communication Com(2002) 535 to the Council and the European Parliament laying down a Community Action Plan for the conservation and the sustainable exploitation of fisheries resources in the Mediterranean Sea under the Common Fisheries Policy.

[194] France and Spain have proclaimed EEZs, but the legislation claiming these zones does not apply to the Mediterranean. The reasons for this relate to the difficulties in delimitation and the need to preserve the freedom of navigation, naval mobility and access to fisheries; see FAO, *Fisheries Laws and Regulations in the Mediterranean: A Comparative Study – General Fisheries Commission for the Mediterranean*, (Rome: 2005), pp. 34-35.

[195] FAO, *Fisheries Laws and Regulations in the Mediterranean*, pp. 34-35. Other states claiming a fisheries zone are Tunisia and Algeria.

[196] *Ibid.*, p. 5.

[197] *Ibid.*, p. 5.

[198] *Ibid.*, p. 6.

[199] *Ibid.*, p. 7.

[200] European Environmental Agency, *Europe's environment 2007*, p. 223; more numbers are provided in Communication Com(2007) 295 final from the Commission to the Council on Fishing Opportunities

1.6.2 Community measures regulating Mediterranean fisheries

CFP rules, in principle, also apply to fisheries carried out within the Mediterranean.[201] This is particularly true for market, structural and control measures. However, the distinctive characteristics of the Mediterranean require the Community to provide a tailor-made management regime for fisheries operating in this area. This regime is established by Council Regulation (EC) 1967/06 concerning management measures for the sustainable exploitation of fishery resources in the Mediterranean Sea,[202] which primarily refers to input restrictions, *i.e.*, mainly gear and area restrictions. With the exception of bluefin tuna, the Community does not adopt TACs for Mediterranean stocks.[203] The regulation provisions apply to fishing operations carried out by vessels flying the flags of Member States in all Mediterranean waters, as well as to third country vessels fishing in Community waters.[204] It prohibits the catch, retention, transhipment and landing of certain species,[205] and provides a framework for the creation of protected areas where certain types of fishing may be prohibited.[206] Fishing with certain gear in certain areas, such as, for example, above seagrass beds or coralligenous habitats, is explicitly forbidden.[207] Arts. 8 to 14 restrict certain types of gear, such as toxics and explosives, and set down minimum mesh and hook sizes, and other gear requirements. Arts. 15 and 16 provide rules on minimum fish sizes. Art. 17 concerns leisure fisheries, a significant activity in the Mediterranean which should be limited so that it does not interfere with resource management. To this end, Art. 17 sets out specific gear restrictions for leisure fisheries as well as a ban on the marketing of fish caught in leisure fisheries. Arts. 18 and 19 establish a legal framework to adopt Community-level management plans for fisheries requiring increased protection. Such plans may include specific effort and technical restrictions as well as increased control and scientific monitoring requirements. Management plans must also provide for the issuing of special fishing permits. Member States are required to adopt such plans for certain fisheries within their territorial zones. Arts. 20 to 24 lay down

---

for 2008, p. 5.

[201] In principle, the basic management regulation is applicable also to Community Mediterranean fisheries, see Art. 1(1) and 3(a) of Regulation 2371/02.

[202] Council Regulation (EC) 1967/06 concerning management measures for the sustainable exploitation of fishery resources in the Mediterranean Sea, amending Regulation (EEC) 2847/93 and repealing Regulation (EC) 1626/94, OJ 2006 No. L409/11.

[203] Commission Communication Com(2002) 535 final laying down a Community Action Plan for the conservation and sustainable exploitation of fisheries resources in the Mediterranean Sea under the Common Fisheries Policy, p. 9.

[204] See Art. 1 and reason (6) of Regulation 1967/06.

[205] Art. 3 of Regulation 1967/06.

[206] Arts. 6 and 7 of Regulation 1967/06.

[207] Art. 4 of Regulation 1967/06.

provisions on monitoring, including specific requirements for measuring catch compositions, transhipment and port landing locations.

Certain requirements are imposed with regards to fishing in the area covered by the Convention of the General Fisheries Commission for the Mediterranean Area (GFCM). The Commission must be provided with specific data on fishing vessels longer than 15 metres which are permitted to fish in the area, which is then forwarded to the GFCM register. Vessels over 15 metres in length which are not entered in the register are not permitted to fish, retain on board, tranship or land any type of fish or shellfish within the GFCM area.

1.6.3 General Fisheries Commission for the Mediterranean Area

Given the comparably small marine area of Community waters in the Mediterranean, GFCM management is of major relevance to Community fisheries in the Mediterranean. The agreement establishing the GFCM was substantially overhauled in April 2004. The territorial scope of the agreement covers the entire Mediterranean and the Black Sea, as well as waters connecting them.[208] Art. III of the GFCM-Agreement states that the purpose of the GFCM is 'to promote the development, conservation, rational management and best utilisation of living marine resources, as well as the sustainable development of aquaculture'. To this end, the GFCM-Commission, *inter alia*, shall recommend measures regulating fishing methods and gear, minimum sizes, fishing seasons and areas as well as TACs and TAEs.[209] Recommendations have been made, for example, in regards to limitations on the catch of tuna and minimum mesh sizes.[210] Recommendations require a two-thirds majority of the votes cast by the CFCM-Commission members in attendance.[211] Members of the Commission are than required to give effect to these recommendations, unless a member of the GFCM-Commission objects within 120 days from the date of notification of a recommendation. In such a case, the objecting party is not obliged to give effect to the recommendation.[212] The brief existence of the new regime for governing fisheries in international waters in the Mediterranean means it is too early to evaluate the CFCM practices in this respect.

[208] Art. IV of the CFGM-Agreement.

[209] Art. III(1)(b)(i) of the CFCM-Agreement. The GFCM Commission may recommend measures regulating fishing methods and gear, minimum sizes, fishing seasons and areas as well as TACs and TAEs.

[210] Resolutions are available online at : www.fao.org/fishery.

[211] Art. 5(1) of the GFCM-Agreement.

[212] Art. 5(3) of the CFCM-Agreement.

2 Community control and enforcement measures

To be effective, fisheries laws must properly be applied, controlled and enforced.[213] The competence in this regard generally lies with the Member States,[214] although the Community is also involved in control and enforcement. In particular, the Community has adopted a comprehensive control regime which requires Member States to undertake certain control actions. In addition, the Community itself controls the application and control and enforcement of CFP rules by Member States.[215] If Member States do not properly apply, control and enforce CFP rules, the Commission may initiate infringement procedures before the ECJ, as well as take preventive measures, cut financial aid or, in case of quota-overfishing, reduce Member States' quotas. The following section will describe the Community's control regime and control and enforcement powers.

2.1 The Community control regime

The Community has adopted control measures from 1982 on.[216] However, policy makers soon recognised the ineffectiveness of that control regime and overhauled the system in the early 1990s.[217] After a full year of controversial negotiations, a new and more comprehensive control regime was adopted in 1993,[218] *i.e.*, Regulation 2847/93 establishing a control system applicable to the Common Fisheries Policy.[219] This Regulation is still in force (besides Arts. 21-28 of Regulation 2371/02), and applies to 'all fishing activities and associated activities', including the control of management, structural and market

[213] See objectives set out in Art. 21 of Regulation 2371/02; see also Commission Communication Com(2003) 130 final towards uniform and effective implementation of the Common Fisheries Policy; Commission Communication Com(2008) 718 final on the proposal for a Council Regulation establishing a Community control system for ensuring compliance with the rules of the Common Fisheries Policy.

[214] This Community law principle is reiterated in Art. 23(1) of Regulation 2371/02.

[215] Art. 26(1) Regulation 2371/02.

[216] The first control regulation was adopted in 1982 and consolidated and repealed in 1987, see Council Regulation (EEC) 2057/82 establishing certain control measures for fishing by vessels of the Member States, OJ 1982 No. No. L220/1; Council Regulation (EEC) 2241/87 establishing certain control measures for fishing activities, OJ 1987 No. L207/1.

[217] Report by the Commission to the Council on the Enforcement of the CFP, Com(86) 301 final of 09.96.1986; Communication from the Commission to the Council and the European Parliament on the Common Fisheries Policy, SEC (90) 2244 of 6 December 1990; Report 1991 from the Commission to the Council and the European Parliament on the Common Fisheries Policy, SEC (91) 2288 of 18 December 1991; Report on Monitoring Implementation of the Common Fisheries Policy, SEC (92) 394 final of 6 March 1992.

[218] Long and Curran, *Enforcing the Common Fisheries Policy*, p. 79.

[219] OJ 1993, No. L261/1.

measures under the CFP. Associated activities are the transhipment, landing, marketing, transport and storage, and the recording of landings and sales.[220] The regulation applies to Community vessels fishing in the territorial seas and EEZ of all Member States, the high seas and third country waters. With regards to third country waters, the Regulation applies 'subject to the special provisions contained in fisheries agreements [...] or in international conventions'.[221] It also covers third country vessels fishing in Community waters.

Art. 2 provides the general obligation of Member States to ensure compliance with the CFP fisheries regime. To this end, 'each Member State within its territory and within maritime waters subject to its sovereignty or jurisdiction shall monitor, inspect and maintain surveillance of all activities in the fisheries sector'. With regards to vessels fishing outside Community waters, Member States must ensure that their activities are properly monitored and, where required by Community law, subject to inspections and surveillance. Member States are to carry out inspection and monitoring, establishing an inspection system that avoids undue interference with normal fishing activities.[222]

2.1.1 The satellite-based vessel monitoring system (VMS)

Prior to the introduction of modern technologies to control fishing activities, locating fishing vessels was one of the most time consuming and costly problems. The introduction of a technology, such as a satellite monitoring systems, was a major step and has made surveillance 'simpler, safer, more effective and less expensive'.[223] However, given its imprecision, data obtained by VMS alone do not currently provide sufficient evidence to satisfy a finding of culpability in court or administrative proceedings.[224] It is generally employed to augment existing control measures, allowing patrol vessels and inspectors to undertake further investigations.

The VMS was introduced in 1998 for a limited number of vessels.[225] Since 2005, all Community fishing vessels exceeding 15 metres in overall length, except those used exclusively for aquaculture or inland fisheries, must have a satellite tracking device on board when leaving a port.[226] These devices transmit

[220] Art. 2(1) of Regulation 2847/93.

[221] Art. 1(3) of Regulation 2847/93.

[222] Art. 4(1) and (2) of Regulation 2847/93.

[223] Long and Curran, *Enforcing the Common Fisheries Policy*, pp. 314-315.

[224] Long and Curran, *Enforcing the Common Fisheries Policy*, pp. 324-325.

[225] By Council Regulation (EC) 686/97 amending Regulation 2847/93 establishing a control system applicable to the common fisheries policy, OJ 1997 No. L 102/1.

[226] Art. 22(1)(b) of Regulations 2371/02 and Arts. 2 and 4 of Commission Regulation (EC) 2244/03 laying down detailed provisions regarding satellite-based Vessel Monitoring Systems, OJ 2003 No. L230/17. These provisions replace Commission Regulation (EC) 1489/97 laying down detailed rules for the application of Council Regulation (EEC) 2847/93 as regards satellite-based vessel monitoring systems, OJ 1997 No. L202/18. The latter was based on Art. 3(10) of Regulation 2847/93 (Control Regulation). Art. 3

data to so called Fisheries Monitoring Centres (FMCs) operated by Member States. These centres track all Community and third country vessels when they are in the waters under their sovereignty or jurisdiction.[227] Data on the identification of the vessel, its geographical position, the date and time of the fixing of that position and speed and course of the vessel must be transmitted at least once an hour.[228] FMCs specifically monitor the entry and exit of vessels into and from maritime areas with restricted access and fishing, areas covered by IFOs, as well as third country waters.[229] Vessels must also transmit the required information to the FMCs of other Member States when they are in their waters.[230] Member States must also ensure that the Commission has, on request, remote access by online session to the data recorded by the FMCs.[231] In addition, twice a year they must report to the Commission on the functioning of the VMS system.[232]

2.1.2 Monitoring of catches

To properly implement the TAC system, it is essential to know when fishing opportunities have been exhausted and by whom. To this end, the Community has adopted several measures on the monitoring of catches, in their entirety, referred to as the Catch Registration System.[233] These measures require detailed documentation of the 'history' of catches by participants of the production process. Art. 6 of Regulation 2847/93 requires masters of all Community vessels over 10 metres long to keep a logbook.[234] This book shall indicate the quantities of each species caught and kept on board, the date and location of such catches, and the type of gear used. Any amount greater than 50 kg (in the Mediterranean 15 kg, except for highly migratory and small pelagic species) of live weight of any species retained on board must be recorded. Masters of vessels must also record fishing effort in the logbooks.[235] Data must be registered either on paper or in computer-readable form.

of the latter currently derogates from requirements laid down in Regulation 2244/03. Since Regulation 2244/03 is *posterior* based on the current *basic* regulation, it overrules Art. 3 of the *priori* Regulation 2847/93.

[227] Art. 3 and Art. 13 of Regulation 2244/03.

[228] Arts. 5 and 8 of Regulation 2244/03.

[229] Art. 10 of Regulation 2244/03.

[230] Art. 10 of Regulation 2244/03.

[231] Art. 14 of Regulation 2244/03.

[232] Art. 15 of Regulation 2244/03. Similar provisions apply to third country vessels fishing in Community waters, see Arts. 17 et seq. of Regulation 2244/03.

[233] Long and Curran, *Enforcing the Common Fisheries* Policy, p. 117.

[234] See also Commission Regulation (EEC) 2807/83 laying down detailed rules for recording on Member States catches of fish, OJ 1983 No. L276/1 as amended by Commission Regulation (EC) 1804/2005, OJ 2005 No. L290/10.

[235] Art. 19(e) of Regulation 2847/93.

To ensure that all landings are only sold after having been recorded on the registrations system, Art. 7 of Regulation 2847/93 requires vessels which land their catch outside their flag Member State to comply with the designated port schemes established by the Member State at the place of landing.[236] If a Member State does not operate such a port scheme, the vessel's master must inform the competent authorities at least four hours in advance of the landing location, the time of arrival and the quantities of each species landed. If the notification requirements are not met, the master may be subjected to appropriate sanctions by the competent authorities.

To gain accurate information on the quantities of fish landed (and to supplement logbook provisions),[237] Art. 8 of Regulation 2847/93 requires of Community fishing vessels to submit a so-called landing declaration to the competent authorities where the landing takes place. The declaration must be handed in within 48 hours of landing and include, as a minimum, the quantities of each species landed as well as the area where the fish were caught.

To ensure that all landings are recorded, Member States may require that the first marketing take place by sale at an auction centre.[238] Where landings are not placed on the market at auction, Member States shall ensure that the quantities involved are notified to the auction centres or other bodies authorised by the Member States.[239]

To prevent collusion between fishers and buyers of fisheries products, Art. 9 of Regulation 2847/93 also places reporting obligations on buyers. Art. 9(1) requires auction centres or other bodies or persons authorised to market fishery products to submit upon the first sale, a sales note to the competent authority of the Member State in whose territory the first marketing takes place.

Where the first marketing is not carried out as under Art. 9(1), Art. 9(2) demands that landed products are not collected unless either a *sales note*, or a *transport document*, or a *take over declaration* has been submitted to the competent authority of the Member State. Under Art. 9(2), the sales note must be submitted by the buyer, and the take over declaration by the holder of that document. According to Art. 9(3), sales notes shall include detailed information on the following:

- the relevant name of each species and its geographical area of origin;
- for all species, where appropriate, individual size or weight, grade, presentation and freshness;
- where appropriate, the relevant minimum fish size;
- the price and quantity for each species, and, where appropriate on a individual size or weight, grade, presentation and freshness basis;
- where appropriate, the destination of products withdrawn from the market;

[236] Long and Curran, *Enforcing the Common Fisheries* Policy, p. 122.

[237] Long and Curran, *Enforcing the Common Fisheries* Policy, p. 113.

[238] Art. 14 of Regulation 2837/93.

[239] See Art. 14(2) Regulation 2847/93, for exceptions, see Art. 14(3) and (4) of Regulation 2847/93.

- the name of both the seller and the buyer;
- the place and the date of sale;
- where possible, the reference number of the sales contract.

Sales notes must also include information on the external identification and name of the fishing vessel which has landed the products concerned, the name of the vessel's owner or master, the port and date of landing and, where applicable, reference to transport or take-over documents.[240] Member States must take measures that ensure sales notes correspond to the invoices. Art. 9(4)(b) provides that the take-over declaration must include the following information:

- the relevant name of each species and its geographical area of origin;
- the weight of each species, broken down by type of product presentation;
- where appropriate, the relevant minimum fish size;
- identification of the fishing vessels that has landed the products;
- identification of the master of the vessel;
- the port and date of landing;
- the place where the products are stored;
- where applicable, reference to the transport or take-over document.

Sales notes, take-over declarations and copies of the transport documents (explained below) must be submitted within 48 hours of the first marketing or the landing, to the competent authorities in accordance with the laws of the Member State in whose territory the operations have been carried out.[241]

Art. 11 of Regulation 2847/93 is concerned with transhipment: the transfer of the catch of one fishing vessel to another. Art. 11 tries to prevent Community vessels from circumventing the Community catch registration system by transferring catches to receiving vessels, which then transfer the catches to third countries.[242] Art. 11 is a framework provision which, in principle, allows transhipment. The Commission may adopt more detailed rules on this issue in the future under the Comitology management procedure.[243] The provision does not explicitly apply to pair trawling activities by Community vessels. Transhipping mainly occurs within waters of Ireland and the United Kingdom. In response, the United Kingdom, for example, adopted provisions on transhipping, *i.e.*, Section 3 of the Sea Fish (Conservation) Act 1992.[244] Section 3 entitles the Minister to subject transhipping to prior authorisation. Fishers can be required to obtain a specific license. Such licenses may then determine specific conditions under which transhipping may be carried out.

Art. 13 of Regulation 2847/93 pertains to the so-called transport document. New control requirements were introduced because fish were being caught in

---

[240] Art. 9(4)(b) of Regulation 2847/93.

[241] Art. 9(5) of Regulation 2847/93.

[242] See Long and Curran, *Enforcing the Common Fisheries Policy*, p. 131.

[243] Art. 11(2) of Regulation 2847/93.

[244] Amending the Sea Fish (Conservation) Act 1967.

the waters of one Member State 'Nr. 1', landed in the ports of a different Member State 'Nr. 2', transported to Member State 'Nr. 3' for processing and finally sent to Member State 'Nr. 4' for sale. According to Art. 13(1), all fisheries products landed in or imported into the Community, either unprocessed or after having been processed on board, and for which neither a sales note nor a take-over declaration was submitted, and which are transported to a place other than that of landing or import, shall be accompanied by a document drawn up by the transporter until the first sale has taken place. The document must include:

- the name of the vessel of provenance and its external identification (in case of import other than by vessel, the location where the consignment was imported);
- the place of destination;
- the identification of the transport vehicle;
- the quantities of fish transported;
- the consignee and the place and date of loading;
- the relevant name of each species, its geographical area of origin and, where appropriate, the relevant minimum fish size.[245]

The transporter must ensure that the documents include certain required information. Exemptions from this duty are listed in Arts. 13(4) and (5). Where fisheries products have been sold and transported to a location other than the place of landing, the transporter must show documental evidence that a sales transaction has taken place.[246] Member States are required to control documents, in concert with other Member States, by means of sampling on their territory.[247]

2.1.3 Monitoring effort limitations

The effort regime requires special control measures. According to the definition of fishing effort in the new effort framework, these measures require the Community to control fishing capacity and the activity of vessels. The control system also includes measures that permit the monitoring of the amount of fish caught and the use of gear.

Effort control requirements vary in regards to the areas and species of fish targeted. Some provisions, for example, apply solely to demersal species.[248] When Community vessels fish for demersal species in areas in which effort limitations apply (Western Waters, Irish Box),[249] the masters of these vessels

[245] Art. 13(2) of Regulation 2847/93.

[246] Art. 13(5)(a) of Regulation 2847/93.

[247] Art. 13(6) of Regulation 2847/93.

[248] See Art. 19a(1a) and (3) of Regulation 2847/93.

[249] Special effort control provisions for the Baltic can be found in Council Regulation (EC) 1098/07 establishing a multi-annual plan for the cod stocks in the Baltic Sea and the fisheries exploiting those stocks, amending Regulation (EEC) 2847/93 and repealing Regulation (EC) 779/97, OJ 2007 No. 248/1.

must compile an 'effort report'. This report must include the following information:

- the name, external identification mark, radio call sign and name of the masters vessel;
- the geographical location of the vessel to which the communication refers;
- the date and time of each entry into, and exit from, a part located inside the area, each entry into an area, and each exit from an area;
- the catch retained on board by species in kilograms live weight.[250]

Communication can be conducted by telex, VMS, fax, telephone or radio. Member States should ensure that the masters of vessels bearing their state flag comply with the reporting obligations and that the information is recorded in computer-readable form.[251] This information is sent simultaneously to the flag Member State and other Member State(s) where fishing is carried out in waters under their jurisdiction or under their sovereignty.[252] The information must be communicated immediately before each entry and exit into these waters.

With regards to all other fisheries subject to effort limitations, masters of vessels are required to record in their logbooks the time spent at sea as follows:[253]

For towed gear:

- the date and time of entry of the vessel into an area or of exit from a port located in that area;
- the date and time of entry of the vessel from that area or of entry into a port located in that area with regards to static gear;
- the date and time of entry of the vessel using the static gear into an area, or of exit from a port located in that area;
- the date and time of setting or re-setting of static gear in the area concerned;
- the date and time of the completion of fishing operations using the static gear;
- the date and time of exit of the vessel from that area or the entry into a port located in that area.[254]

Member States are required to collect and record data on fishing effort deployed by vessels flying its flag.[255] To enable the Commission to tell Member States

---

[250] Special rules apply to vessels conducting trans-zonal fisheries, see Art. 19b(2) of Regulation 2847/93.

[251] Art. 19b(3) and 19d of Regulation 2847/93.

[252] Art. 19c(1) of Regulation 2847/93; special requirements exist for vessels conducting trans-zonal fishing, see Art. 19c(2) of Regulation 2847/93.

[253] Art. 19e of Regulation 2847/93.

[254] Special requirements exist for masters of Community vessels conducting trans-zonal fisheries, or that fish for demersal species, as well as for vessels exempt from the obligation to keep a logbook; see Art. 19e(2) to (4) of Regulation 2847/93.

[255] Arts. 19f,g,h, of Regulation 2847/93.

when their quotas are exhausted, Member States provide this information to the Commission regularly and in a timely manner.[256]

2.1.4 Monitoring of Community vessels in third country-, high seas-, and Community waters

Art. 17(1) of Regulation 2847/93 states:

> [Member States] shall take the necessary measures to ensure monitoring of the catches of species made by their vessels operating in waters subject to the sovereignty or jurisdiction of third countries and on the high seas, and to ensure verification and recording of transhipment and landings of such catches.
>
> Art. 17(2) adds that the control and verification measures shall ensure that vessel owners and/or masters shall keep a logbook on board in which catches are recorded; submit landing declarations to the authorities of the Member State of landing during landings and that they inform flag Member States of details of each transhipment of fish on to third-country vessels and of landings carried out in third countries.[257]

This provision applies without prejudice to fisheries agreements and international treaties.[258] Art. 18 requires Member States to inform the Commission before the end of the first month of each calendar quarter of the quantities of fish caught and landed in the previous quarter.

Member States are also authorised to control vessels flying their flags in all Community waters.[259] They are also entitled to inspect fishing vessels flying the flag of another Member States in international waters.[260] Finally, Member States can be authorised by other Member States to carry out inspections in their waters.[261]

2.1.5 Monitoring of third country vessels in Community waters

Currently, the Community grants quotas to Norway, Iceland and the Faeroe Islands. In 2007, Norway obtained the right to take 204,982

---

[256] Art. 19i of Regulation 2847/93

[257] Art. 17(2) of Regulation 2847/93.

[258] Art. 17(3) of Regulation 2847/93.

[259] Art. 28(3), para.1 of Regulation 2371/02.

[260] Art. 28(3), para. 3 of Regulation 2371/02; Commission Regulation (EC) 1042/06 specifying detailed rules for the implementation of Art. 28(3) and (4) of Council Regulation 2371/02 on the conservation and sustainable exploitation of fisheries resources under the Common Fisheries Policy, OJ 2006 No. L187/14.

[261] Art. 28(3), para. 2 of Regulation 2371/02.

tonnes of different fish species from Community and Greenland waters.[262] In 2006, quotas for the Faeroes lay at about 64,500 tonnes and for Iceland lay at 45,930 tonnes in 2005.[263]

Art. 28a of Regulation 2847/93 defines 'third country vessels' as: (a) vessels used primarily or secondarily for fishing, (b) vessels that can take the fisheries products by transhipment from other vessels, and (c) vessels on which fish can be filleted, sliced, skinned, minced, frozen and/or processed. Vessels must fly the flag of a third country and be registered there.

Art. 28b provides that, in order to be granted access to Community resources, third country vessels must obtain a fishing licence and a special fishing permit from its flag state, which, in turn, receives permission from the Commission.[264] In addition, third country vessels can only conduct transhipment and processing in cases where they have obtained prior authorisation from the Member State in whose waters the operation would take place. Art. 28c requires that all third country vessels fishing in Community waters abide by the following obligations:

- to record, in a logbook, the information Community fishers have to record therein;[265]
- vessels exceeding 20 metres between perpendiculars or 24 metres overall length must be equipped with VMS position monitoring system for reporting movements;
- to comply with a system for reporting catches retained on board;
- to comply with the instructions of the authorities responsible for monitoring, particularly as regards the inspections prior to leaving the Community fishing zone;
- to comply with the rules on the marking and identification of fishing vessels and their gear.

The Commission, in cooperation with the Member States, is responsible for controlling the activities of vessels from third countries fishing in Community waters.[266] The Commission is responsible for declaring that quotas fished by foreign vessels are exhausted. Where the Commission deems the quota to

---

[262] Council Regulation (EC) 41/07 fixing for 2007 the fishing opportunities and associated conditions for certain fish stocks and groups of fish stocks, applicable in Community waters and, for Community vessels, in waters where catch limitations are required, OJ 2007 No. L15/1.

[263] Council Regulation (EC) 860/05 amending Regulation (EC) 27/05, as concerns fishing opportunities in Greenland, Faeroese and Icelandic waters and fishing for cod in the North sea, and amending Regulation (EC) 2270/04 as concerns fishing opportunities for deep-sea sharks and roundnose grenadier, OJ 2005 No. L144/1; Council Regulation (EC) 51/06, fixing for 2006 the fishing opportunities and associated conditions for certain fish stocks and groups of fish stocks applicable in Community waters and, for Community vessels, in waters where catch limitations are required, OJ 2007 No. L16/1.

[264] As provided under Arts. 1(b), 6(1) and 9 of Regulation 1627/94.

[265] Art. 6 of Regulation 2847/93.

[266] Art. 26(5) of Regulation 2371/02.

be exhausted, it informs the third country and the Member States concerned without delay. As from this date, fishing, the retention on board, transhipment, and the landing of fish taken after that date are provisionally prohibited.[267] Masters of third country vessels are required to provide specific information to national authorities whose ports they use, such as the time of arrival at the port of landing and certain particulars about the catch (particularly where it has been caught).[268] Within 48 hours after landing, they must provide a landing declaration to respective authorities. Member States will then forward this information to the Commission.[269] Where fish have been taken on the high seas, vessel masters must either prove that the catch has been caught outside the regulatory area of an IFO of which the Community is a member or that it has been caught in compliance with the management rules of those IFOs.[270]

2.1.6 Control of technical measures

Catches that are retained on board must comply with the species compositions set out in the technical regulations.[271] Nets on board must be stowed in accordance with the following conditions:

- nets, weights and similar gear shall be disconnected from their trawl boards and towing and hauling wires and ropes.
- nets which are on or above deck shall be securely lashed to some part of the superstructure.

Where vessels use different nets of varying mesh sizes during the same voyage, the species composition is calculated for each part of the catch taken under different conditions. Net changes and species compositions at the moment of that change must be entered into the logbooks and landing declarations.[272] The Council may decide to put in place new legislation requiring specific mesh sizes for different fisheries, and may impose a 'one net rule'.[273]

2.1.7 Control of structural measures

Considering all the different structural measures adopted under the European Fisheries Fund, provisions concerned with the control of structural policies under Regulation 2847/93 can be regarded as being some-

[267] Art. 28d of Regulation 2847/93.

[268] Art. 28e of Regulation 2847/93.

[269] Art. 28f of Regulation 2847/93.

[270] Art. 28g of Regulation 2847/93.

[271] See Art. 20(1) of Regulation 2847/93 still referring to Council Regulation (EEC) 3094/86. The latter has been replaced by Regulation (EC) 894/97 and provisions in Council Regulation (EC) 850/1998; Council Regulation (EC) 1967/06; Council Regulation (EC) 60/04.

[272] Art. 20(2) of Regulation 2847/93.

[273] Art. 20(3) of regulation 2847/93.

what limited in scope. Control measures primarily aim at providing a clear picture of the status and the development of the fleet and the aquaculture sector, particularly with a view to sector adjustments.

In order to ensure compliance with the Community fleet adjustment objectives, each Member State must organise regular checks of all persons concerned by the implementation of structural measures.[274] To this end, Art. 25 of Regulation 2847/93 requires that Member States carry out technical controls, particularly in the following areas:

- restructuring, renewal and modernisation of the fishing fleet;
- adaptation of fishing capacity by means of temporary or definite cessation;
- restriction of the activity of certain fishing vessels;
- restriction of the design and number of fishing gear and of the method by which it is used;
- development of the aquaculture industry and coastal areas.

If the Commission establishes that the Member States do not comply with these conditions, it may propose the adoption of 'general measures' to the Council.[275] It is not made clear in the legislation, however, what the term 'general measures' implies. According to *Long and Curran*, general measures could refer to the cancellation of Community financial aid, the imposition of penalties, or the adoption of inspection and control provisions.[276] The Commission may, in accordance with the management procedure, adopt detailed rules on the application of Art. 25 particularly in respect of the verification of engine power, tonnage, periods during which the vessels are immobilised, and gear specifications and the amount of gear allowed on board each vessel. Member States are required to inform the Commission and the appropriate authorities of the control methods used.[277] To facilitate monitoring, Art. 27 requires Member States to establish a system of validation. This system shall include the verification of information on fleet fishing capacity and other activities by crosschecking the logbooks, landing declarations, and a Community fishing vessel register. For this purpose, Member States must establish databases.[278]

In practice, control measures carried out by Member States may involve paper checks or actual physical control (*e.g.*, the assessment of tonnage, length and engine power measurements, etc.). Certain issues arise in this context.[279] First, the registration process often is too slow to properly reflect change within

[274] Art. 24 of Regulation 2847/93 still refers to the objectives set out under Article 11 of Regulation 3760/92 (the former basic regulation). Objectives were laid down in the Multi Annual Guidance Programmes III and IV. The latter two programmes have been replaced under Regulation 2371/02 by a simple entry/exit scheme.

[275] Art. 25(2) of Regulation 2847/93.

[276] Long and Curran, *Enforcing the Common Fisheries Policy*, pp. 179-180.

[277] Art. 26(2) of Regulation 2847/93.

[278] Art. 27(2) of regulation 2847/93

[279] Long and Curran, *Enforcing the Common Fisheries Policy*, pp. 176-180.

the fleets. Second, installed fishing capacity is difficult to monitor and easy to manipulate.[280]

2.1.8 Control of the market organisation

Art. 28 of Regulation 2847/93 provides:

> [I]n order to ensure compliance with the technical aspects of the rules regarding the organisation of the common market in fishery and aquaculture products, each Member State shall organise on its own territory regular checks of all persons involved in the application of the measures.[281]

Art. 28 aims at ensuring fair competition through the uniform application of marketing standards and price arrangements. To this end, checks concern the technical aspects of applying:

- marketing standards, and in particular minimum sizes;
- price arrangements, in particular withdrawals of products from the market for purposes other than human consumption and storage and/or processing of products withdrawn from the market.

Member States are to carry out comparisons between the documents relating to the first placing on the market of the quantities referred to in sales notes and landing declarations, particularly as regards their weight.[282] Where minimum sizes have been fixed for a certain species, operators responsible for the selling, stocking or transporting of batches of products of that species smaller than the minimum size must be able to prove their geographical area of origin or the provenance from aquaculture of the products at all time.[283] Member States are also obliged to notify the Commission of the control measures adopted, the competent authorities, the type of infringements discovered and actions taken.[284]

2.2 Community control over Member States

A uniform and coherent application of the Community fisheries regime is a prerequisite for effective fisheries management. Management rules must be applied and enforced in all Member States to ensure that no European fisher has an advantage over another. However, as stated above, the power

[280] *Ibid.*, pp. 176-180.

[281] See Art. 28(1) of Regulation 2847/93, still referring to Regulation 3759/92 which has been replaced by Regulation 104/00 on the common organisation of the markets in fishery and aquaculture products, OJ 2000 No. L17/2.

[282] Art. 28(2) of Regulation 2847/93.

[283] Art. 28(2a) of Regulation 2847/93.

[284] Art. 28(3) of Regulation 2847/93.

to implement, control, and enforce Community fisheries law lies primarily with the Member States.[285] Community law forms an integral part of national legal systems and must be applied effectively by their legislators, courts and administrations.[286] Throughout the history of the CFP, Member States have often shown themselves to be unable or unwilling to fulfil these obligations.[287] Against this background, the Community has developed a particular interest in monitoring Member States' control and enforcement practices. Thus, while it is the role of the Member States to apply, control and enforce fisheries rules, the Community monitors the effectiveness of the Member States in carrying out its duties.[288] This practice has been referred to as 'dual vigilance'.[289] The Community's competent organ in this respect is the Commission. The Commission's duty to monitor Member States is laid down in the Regulation 2371/02. Art. 26(1) states that the Commission 'shall evaluate and control the application of the rules of the Common Fisheries Policy by the Member States, and facilitate co-ordination and co-operation between them'. The Commission's control functions will be described in the following sections.

2.2.1 Evaluation and control of the application of management measures

The Commission has the power to monitor the Member States' application of management measures. Art. 27(1) of Regulation 2371/02 provides that 'for the purpose of evaluating and controlling the application of the rules of the Common Fisheries Policy by the Member States', the Commission may 'of its own accord and by its own means, initiate and carry out audits, inquiries, verifications and inspections concerning the application of the rules' of the CFP by the Member States. According to Art. 27(1), the Commission may 'in particular verify:

- the implementation and application of those rules by Member States and their competent authorities;
- the conformity of national administrative practices and inspection and surveillance activities with the rules;
- the existence of the required documents and their concordance with the

[285] Art. 23(1) of Regulation 2371/2002.

[286] Case 6/64, Costa/E.N.E.L Case 6/64 [1964] ECR 1141, para. 3; Case 14/83, von Colson und Kamann v. Land Nordrhein-Westfalen [1984] ECR 1891, para. 26.

[287] Report 1991 from the Commission to the Council and the European Parliament on the Common Fisheries Policy, SEC (91) 2288 of 18 December 1991; Report from the Commission on the monitoring of the Member States' implementation of the Common Fisheries Policy from 2002 to 2004, Com(2004) 849 final of 4 January 2005; Report from the Commission to the Council and the European Parliament on the monitoring of the Member States' implementation of the Common Fisheries Policy, Com(2007) 167 final of 10 April 2007; Court of Auditors, Special Report 7/2007.

[288] Art. 26(1) Regulation 2371/02.

[289] Long Curran, *Enforcing the Common Fisheries Policy*, p. 62.

applicable rules;
- the circumstances in which control and enforcement activities are carried out by Member States'.

Evaluation and control is mainly carried out by the Commission's inspectorate, which currently consists of 24 Commission inspectors.[290] Inspectors may examine fishing vessels, businesses and the premises of other bodies with activities relating to the CFP. They 'shall have access to all information and documents to exercise its responsibility'.[291] Since 2002, the Commission itself may initiate inspections without the assistance of the Member States concerned. However, Community inspectors have neither the policing and enforcement powers, nor any other powers beyond those of national inspectors. Furthermore, inspections conducted without the assistance of a national inspector may not be carried out where the inspected party objects. Inspections can take place 'only on fishing vessels and places of first landing or first sale, and be limited to areas and stocks which are subject to specific monitoring programmes decided under Art. 34(c) of Regulation 2847/1993'.[292] Member States are required to provide the Commission with such assistance as it needs to fulfil its tasks.

Art. 27(2) requires that inspection reports be made available to the Member States concerned, and that they comply with Community provisions on the protection of personal data. Member States must be given the opportunity to comment on the findings. If the Commission carries out inspections without being accompanied by a Member State's inspector, the latter must be informed within one day and receive a report on the findings within one month. That report, however, cannot force the Member State concerned to take action against violators.[293]

2.2.2 Monitoring of quota uptake and the closing of fisheries upon quota exhaustion

As described above, Masters of vessels are required to report their 'quota uptake' as recorded in logbooks, landing declarations and other means. The information gathered in these records is entered in national registers and used in monthly (or even more frequently) reports to the Commission. The Commission relies on this information for evaluating 'national quota uptake'.[294] Reporting must include the location of all catches, and a quota consumption forecast, which provides the anticipated date of exhaustion of 70% of the allotted quota. If the Commission finds that a Member State has not

---

[290] On the development of the Commission's inspectorate, see Astrid Berg, *Implementing and Enforcing European Fisheries Law*, pp. 98-101 and 286-287; see also Churchill, *EEC Fisheries Law*, pp. 142-143.

[291] Art. 27(1), paragraph 2 of Regulation 2371/02.

[292] Art. 27(1), paragraph 2 and 3 of Regulation 2371/02.

[293] Art. 27(2), paragraph 4 of Regulation 2371/02.

[294] See Art. 15 of Regulation 2847/93.

complied with the obligations to transmit catch data in a timely manner, it may set the date on which the catch is deemed to be exhausted at 70% of the quota or other quantitative restriction. The Commission's estimates are based on the records of the previous years.[295]

While, in principle, Member States determine the date on which vessels flying their flag have to stop fishing, the Commission ultimately decides when a particular quota is exhausted.[296] The reason for that is that a Commission decision gives Member States the jurisdiction to enforce the closure against the vessels of other Member States fishing in their waters.[297] It is the normal practice to close a fishery based on the information provided by Member States. However, this procedure requires complete information based on proper control and monitoring by the Member States. Given the difficulties associated with control and monitoring, the Commission also has the right to close a fishery on its own initiative.[298] In that case, the Commission will notify the Member State of the closure date of the fishery. The Commission sets a high threshold before it will take any action in order to avoid unnecessary legal action. Accordingly, the provision is limited in scope to apply only to those fisheries which are easy to assess, *i.e.*, fisheries that target stocks which require a minimal number of ships and a relatively small number of landings.[299] To date, the Commission has made use of this right to close a fishery on its own initiative only twice.[300]

According to the Commission 2005 Scoreboard, Member States were inconsistent in fulfilling their reporting obligations in 2004. Only ten Member States had fully complied with the obligation to send the monthly reports; and three Member States have not provided any reports at all.[301]

2.3 Cooperation requirements, mutual inspection rights, Community inspectors

Art. 28 of Regulation 2371/02 requires Member States to cooperate with each other and with third countries to ensure compliance with CFP rules. To this end, it establishes different obligations and rights for Member States. As a general requirement, Member States must provide other Member States and third countries with the assistance needed to ensure compliance with CFP rules.[302] With regards to the control and inspection of transboundary

[295] Long and Curran, *Enforcing the Common Fisheries Policy*, p. 146.

[296] Art. 21(3) of Regulation 2847/93; see also Art. 26(4) of Regulation 2371/02.

[297] Long and Curran, *Enforcing the Common Fisheries Policy*, p. 152.

[298] Art. 21(3) of Regulation 2847/94 and Art. 26(4) of Regulation 2371/02.

[299] Court of Auditors, Special Report 7/2007, para. 50.

[300] Commission Staff Working Document SEC(2008) 2760 accompanying the Proposal for a Council Regulation establishing a Community control system for ensuring compliance with the rules of the Common fisheries Policy, p. 11.

[301] Cyprus, Malta and Slovenia.

[302] Art. 28(1), paragraph 1of Regulation 2371/02.

fishing activities, Member States must coordinate their control and enforcement activities, even exchanging inspectors in certain cases.[303]

Art. 28(3) sets out Member States' mutual inspection rights. Without prejudice to the primary responsibility of the coastal Member State, other Member States are authorised to inspect vessels flying their flag in all Community waters.[304] Prior to taking such action, the Member State must notify the coastal Member State of its intention to inspect the vessel, providing the name and call sign of the vessel and of the time of entry into the waters under the jurisdiction of the coastal Member State.[305] Member States are also authorised to inspect vessels flying the flag of other Member States in all Community waters outside waters under their sovereign control, where they have been authorised by the flag Member State or where a specific monitoring programme has been adopted.[306] The request for authorisation must also include the name and call sign of the inspected vessel and of the time of entry into the waters under the jurisdiction of the coastal Member State. The coastal Member State has 24 hours to decide whether it will grant the request.[307] In addition, Art. 28(3) permits Member States to inspect vessels flying the flag of other Member States in international waters.[308] To guarantee the functioning of the inspection scheme, Member States are to designate authorities that will serve as contacts for notifications and requests and notify the other Member States and the Commission of their appointment.[309] After conducting an inspection in Community waters under the jurisdiction of another Member State, the inspecting Member States must submit a report to the coastal Member State concerned.[310] If an infringement has been detected, the inspecting Member State shall immediately submit a summarised inspection report to the coastal and flag Member State concerned. Furthermore, within seven days from the time of inspection, a full report shall be submitted.[311] Where vessels have been inspected in international waters, the flag state shall receive a report or, in case of an infringement, a detailed inspection report.[312] The report shall also be submitted to the Commission when it so requests.[313]

---

303 Art. 28(2), paragraph 1of Regulation 2371/02.

304 Art. 28(3), paragraph 1 of Regulation 2371/02.

305 Arts. 1(1) and (2) of Commission Regulation (EC) 1042/06 laying down detailed rules for the implementation of Article 28(3) and (4) of Council Regulation (EC) 2371/02 on the conservation and sustainable exploitation of fisheries resources under the Common Fisheries Policy, OJ 2006 No. L187/14.

306 Art. 28(3), paragraph 2 of Regulation 2371/02.

307 Art. 2(2) of Regulation 1042/06.

308 Art. 28(3), paragraph 3 of Regulation 2371/02.

309 Art. 3 of Regulation 1042/06.

310 Art. 4(1) of Regulation 1042/06.

311 Art. 4(2) of Regulation 1042/06.

312 Art. 4(3) of Regulation 1042/06.

313 Art. 4(5) of Regulation 1042/06.

The Commission, on the basis of appointments proposed by the Member States, establishes a list of 'Community inspectors, inspection vessels and inspection aircraft and other means of inspection authorised to carry out inspections [...].'[314] Inspectors are national inspectors designated by the Member States to participate in cooperative and co-ordinated inspection procedures under Art. 28 of Regulation 2371/02. Member States must ensure that they nominate persons who are fisheries inspectors under their national law, who have a thorough experience in fisheries control and inspection, and have an in-depth knowledge of Community fisheries legislation, etc.[315] The tasks of the Community inspectors is to carry out inspections pursuant to chapter V of the Regulation 2371/02. They may be assigned to any of the following duties:

- the implementation of specific control and inspection programmes under Art. 34c of Regulation 2847/93 on fisheries involving two or more Member States;
- international control and inspection programmes, where the Community is under an obligation to provide for inspection and controls;
- inspection programmes developed between Member States pursuant to Art. 34b(2) of Regulation 2847/93 regarding voluntary joint inspection programmes.

Community inspectors have the same powers as the national fisheries inspectors in which the inspection takes place.[316] They have no police or enforcement powers beyond the territory or outside the Community waters under the sovereignty or jurisdiction of their Member State of origin. Community inspectors shall produce written authority,[317] and are provided with an identification document by the Commission. Member States must assist them with the fulfilment of their tasks, for example, by granting them access to all necessary information and documents.[318] Community inspectors must submit a daily report on their activities, including the name and identification number of each vessel inspected and the type of inspection carried out, to the coastal Member States concerned.[319] If they detect an infringement, they shall immediately submit a summarised report to the coastal Member State concerned. Within seven days of the date of inspection, the flag state shall receive a full inspection report.[320] Where the inspection was carried out outside of Community waters, a full

[314] Art. 28(4) of Regulation 2371/02; see Commission Decision (EC) 266/07, adopting the list of Community inspectors and inspection means pursuant to Art. 28(4) of Council Regulation (EC) 2371/02 on the conservation and sustainable exploitation of fisheries resources under the common fisheries policy, OJ 2007 No. L76/22.

[315] Art. 5(2) of Regulation 1042/06.

[316] Art. 8(1) of Regulation 1042/06.

[317] Art. 8(2) of Regulation 1042/06.

[318] Art. 11(1) of Regulation 1042/06.

[319] Art. 9(1) of Regulation 1042/06.

[320] Art. 9(3) of Regulation 1042/06.

inspection report shall be submitted within seven days from the date of inspection to the flag state.[321]

The Commission has the right to request access to the reports.[322] Art. 10(1) of Regulation 1042/06 requires Member States to 'consider to act' on reports submitted by Community inspectors in the same way as they consider and act on reports from their own inspectors. The Member States of origin of the Community inspector must cooperate with the Member States acting on a report submitted by the Community inspector in order to facilitate judicial or administrative procedures.[323] The Community inspector shall, on request, assist and give evidence in infringement proceedings undertaken by any Member State.[324] Art. 28(5) of Regulation 2371/02 provides that inspection and surveillance reports drawn up by Community inspectors or inspectors of another Member State or Commission inspectors shall constitute admissible evidence in administrative and judicial proceedings of any Member State. For establishing facts they shall be treated equally to inspection and surveillance reports of the Member States.

2.4 Community Fisheries Control Agency

In April 2005, the Council established the 'Community Fisheries Control Agency' ('Agency') by Regulation 768/05 ('Agency Regulation'), which operates out of Spain.[325] The main objective of the Agency is to 'organise operational coordination of fisheries control and inspection activities by Member States and to assist them to cooperate so as to comply with the rules of the Common Fisheries Policy in order to ensure its effective uniform application'.[326] According to Art. 3 of the Agency Regulation, the mission of the Agency is as follows:

- to coordinate control and inspection by Member States relating to the control and inspection obligations of the Community;
- to coordinate the deployment of the national means of control and inspection pooled by the Member States concerned;
- to assist Member States in reporting information on fishing activities and control and inspection activities to the Commission and third parties;
- in the field of its competences, to assist Member States to fulfil their tasks and obligations under the rules of the CFP;
- to assist Member States and the Commission in harmonising the applica-

[321] Art. 9(2) of Regulation 1042/06.

[322] Art. 9(4) of Regulation 1042/06.

[323] Art. 10(2) of Regulation 1042/06.

[324] Art. 10(3) of Regulation 1042/06.

[325] Council Regulation (EC) No. 768/2005 establishing a Community Fisheries Control agency and amending Regulation (EEC) No. 2847/93 establishing a control system applicable to the common fisheries policy, OJ 2005 No. L128/1.

[326] Art. 1 of Regulation 768/05.

tion of the CFP throughout the Community;
- to contribute to the work of Member States and the Commission on research into and development of control and inspection techniques;
- to contribute to the coordination of inspector training and the exchange of experience between Member States;
- to coordinate the operations to combat illegal, unreported and unregulated fishing in conformity with Community rules.

The Agency has no control powers of its own in relation to fisheries to control. Thus, the regulation does also not change in any way the competences of the Commission to control Member States or the Member States ability to enforce CFP rules. Its main purpose is to increase cooperation and assist the Commission and Member States in fulfilling their obligations. The operational coordination of control and inspection activities by Member States is the Agency's central task. Operational coordination covers inspection and control of fishing activities, including importation, transport and landing of fishery products, up to the point that the first buyers receive the product.[327] The Agency shall, at the request of the Commission, coordinate control and inspection activities by Member States on the basis of international or specific control and inspection programmes by establishing joint deployment plans.[328] A joint deployment plan means 'a plan setting out operational arrangements for the deployment of available means of control and inspection'.[329] Deployment plans shall meet the following objectives:
- fulfil the requirements of the relevant control and inspection programme;
- give effect to the criteria, benchmarks, priorities and common inspection procedures determined by the Commission in control and inspection programmes;
- seek to match the existing national means of control and inspection with needs and organise their deployment;
- organise the use of human and material resources with regard to the periods and zones in which these have to be deployed, including the operation of teams of Community inspectors from more than one Member State;
- take account of the existing obligations of the Member States concerned of other joint deployment plans, as well as any specific regional or local constraints;
- define the conditions under which the means of control and inspection of a Member State may enter the waters under the sovereignty and jurisdiction of another Member State.

In principle, joint deployment plans must receive the approval of both the Agency and the Member States. If a Member State objects to the plans, the

---

[327] Art. 5(1) of Regulation 768/05.

[328] Arts. 8, 9 and 2(d) and (e) of Regulation 768/05.

[329] Art. 2(c) of Regulation 768/05.

Commission will adopt the plan in accordance with the Comitology management procedure.[330]

2.5 Community rules on enforcement

Where CFP rules have been infringed, Member States are required to take appropriate measures to effectively deprive the violators of the economic benefits gained from the infringement and discourage further offences.[331] Member States are free to choose their own means to accomplish this objective. They may, for example, choose to take administrative or penal actions, and determine how stringent sanctions should be.[332] To assist with this process, the Community has provided Member States with a list of sanctions, which can be taken in response to violations of CFP rules by their fishers. These include:

- fines;
- seizure of prohibited fishing gear and catches;
- sequestration of the vessel;
- temporary immobilisation of the vessel;
- license suspension;
- license withdrawal.

According to Art. 25(4), the Council shall establish a catalogue of measures to be applied by Member States for 'serious infringements' as defined in Regulation 1447/99.[333] Such infringements include failures to cooperate with monitoring authorities or observers and failures to observe gear, zone, time, recording or landing prescriptions.[334] To this day, however, the Council did not establish such a catalogue. This is particularly problematic since the lack of uniform enforcement of measures has been identified on many occasions as one of the main

---

330 Art. 15(4) and (4) of Regulation 768/05.

331 Art. 25(1) of Regulation 2371/02. Community law requires Member States to ensure that violations of Community law 'are penalized under conditions, both procedural and substantive, which are analogous to those applicable to infringements of national law of a similar nature and importance and which, in any event, make the penalty effective, proportionate and dissuasive. Moreover, the national authorities must proceed, with respect to infringements of Community law, with the same diligence as that which they bring to bear in implementing corresponding national law', see Case 68/88, Commission v. Greece [198] ECR 2965, paras. 24 and 25.

332 Art. 25(1) of Regulation 2371/02.

333 Council Regulation (EC) 1447/99 establishing a list of types of behaviour which seriously infringe the rules of the Common Fisheries Policy, OJ 1999 No. L167/5; see also Commission Regulation (EC) 2740/99 laying down detailed rules for the application of Council Regulation (EC) 1447/99 establishing a list of types of behaviour which seriously infringe the rules of the common fisheries policy, OJ 1999 No. L328/62.

334 See Regulation 1447/99, Annex I.

problems of the Community's control and enforcement system.[335] As it currently stands, the sanctions imposed by Member States are manifold and do not serve as an effective deterrent. Besides, in Member States, the chance of being caught at violating the fisheries rules is very low.[336] The Commission has recently stated that among Member States there 'remains a general lack of understanding of the relationship between compliance, risks, inspection activity and the effect of follow-up, including sanctions.'[337]

2.6 Community sanction powers

Arts. 16(1), 23(4), and 26(3) of Regulation 2371/02 allow the Commission to sanction certain types of bad conduct, laying down the conditions under which the Commission can deduce the future fishing opportunities of a Member State, take preventive action, and suspend financial assistance. In addition, the Commission can initiate formal infringement procedures before the ECJ. The following section will look at the Community's sanction competences. The quota deduction competence will be assessed in depth.

2.6.1 Quota deductions[338]

The content, objectives, effects and legal form of 23(4) of Regulation 2371/02 as well as a Commission implementation measure will be described in the following.

2.6.1.1 Art. 23(4) of Council Regulation 2371/02 and Commission Regulation 147/07

The Treaty authorises the Commission to respond to infringements of the CFP quota system. According to Art. 23(4), the Commission can deduct the future fishing opportunities of a Member State that has exceeded its fishing opportunities. The amount of quota that can be subtracted, however, is not specifically stated in the provision. Art. 23(4), para. 2 only provides that if, as a direct result of a Member State having exceeded its fishing opportunities, another Member State has not been able to exhaust its own fishing opportunities, fishing opportunities equivalent to those deducted under para. 1 may, in

---

335 Commission Communication Com(2001) 135 final Green Paper on the Common Fisheries Policy, vol. 1, p. 12 and vol. 2, p. 17.

336 Commission, 'A level playing field for better enforcement of CFP rules', p. 4.

337 Commission Report Com(2007) 167 final on the monitoring of the Member States implementation of the Common Fisheries Policy 2003-2005, p. 9.

338 Art. 23 (4), para 1 and (2) of Regulation 2371/02 had forerunners, see Art. 5 of Regulation 847/9 and in Art. 11 (4) of Regulation 2241/87. Art. 32 of Regulation 2847/93 provides a sanction power responding cases where Member States of landing or transhipment do not take penal or administrative actions where they have discovered infringements of conservatory measures.

total or in part, be reallocated to this Member State. When reallocating quota, the Commission must take into account the interest to conserve resources, as well as the 'interest in compensation of both Member States concerned'.

Any decision by the Commission to deduct quota must be made in accordance with the Art. 4 of Council Decision 468/1999,[339] referred to as the 'Comitology Decision'.[340] This article requires that the Commission's decision making power be exercised in conjunction with a management committee composed of representatives of the Member States. The committee is chaired by a representative of the Commission, whose job it is to submit to the committee a draft measure. The Committee then delivers its opinion on the draft within a time limit, which the chairman may lay down according to the urgency of the matter by a qualified majority as provided in Art. 205(2) of the Treaty. If the Committee does not deliver its opinion on time, the Commission is free to adopt the measure on its own.[341] The committee chair does not vote.[342]

On 15 February 2007, for the first time, the Commission made use of its quota deduction competence, adopting Commission Regulation (EC) 147/2007 adjusting certain fish quotas from 2007 to 2012.[343] The Commission's action in this regard provides some indications of how it interprets Art. 23(4) of Regulation 2371/02. Both the United Kingdom and Ireland had their quotas reduced for herring and mackerel, because they exceeded their quotas from 2001 to 2004.[344] When exercising its powers under Art. 23(4), the Commission tried to take into account the objectives laid down in Art. 2 of Regulation 2371/02, by ensuring that the Community's exploitation of living aquatic resources provides 'sustainable economic, environmental and social conditions'.[345] Thus, it deducted fishing opportunities equivalent to the amount overfished and reductions were adopted over a six-year period.[346] The Commission also applied a correction

339 OJ 1999, No. L184/23.

340 Arts. 23 (4), para. 3 and 30 (2) of Regulation 2371/02.

341 Josef Falke, 'Komitologie – Entwicklung, Rechtsgrundlagen und erste empirische Annäherung', in Joerges and Falke (ed.), *Das Ausschusswesen der Europäischen Union*, (Baden-Baden: 2000), p. 59.

342 The Commission has declared that it will follow the opinion of the Committee when adopting measures, see OJ 1999 Nr. C 203/1. However, if the Commission takes any action that is not in accordance with the committee's opinion, this will be communicated to the Council forthwith. In that event, the Commission may defer application of the measure for 20 working days, see Art. 30 (2), para.2 of Regulation 2371/02 and Art. 4(3) of Council Decision 468/1999. The Council then may take a different decision by a qualified majority.

343 Commission Regulation (EC) 147/2007 adapting certain fish quotas from 2007 to 2012 pursuant to Article 23(4) of Council Regulation (EC) 2371/2002 on the conservation and sustainable exploitation of fisheries resources under the Common Fisheries Policy, OJ 2007, No. L46/10. According to the Commission, the assisting Committee has not delivered its opinion within the time limit set by the Commission's chairman, see Reason (14) of Commission Regulation 147/2007.

344 Reason (2) of Commission Regulation 147/2007.

345 See Reason (6) of Commission Regulation 147/2007.

346 See Annex I and II of Regulation 147/07.

factor in order to take into account the decline of mackerel stock between 2001 and 2004.[347] Finally, it declared that, in principle, deductions would not exceed 15% of the regular annual quota.[348] Where the quantity to be deducted in any one year, nevertheless, exceeds 15% of the annual quota, the deduction period will be extended to reduce the quantity equal to or below 15%.[349]

2.6.1.2 Objectives and effects

Art. 23(4) primarily aims at improving the functioning of the Community' quota system.[350] However, given that Art. 23(4) does not explicitly set out the limits of the quota deduction power, it is unclear whether the Commission is entitled to penalise Member States by imposing a real legal punishment,[351] or if its duty is to ensure that the quota system is 'equalised' or 'compensated' by only reducing fishing opportunities equivalent to the amount overfished.

The answer to this question lies with Regulation (EC) 147/2007, which favours a compensatory (non-punitive) interpretation of Art. 23(4). Here, the Commission only deducted fishing opportunities equivalent to the amounts overfished, and did not penalise Member States by imposing reductions greater than what was taken in excess. It also applied a correction factor and spread out deductions over a six-year period in order to limit 'as much as possible the negative impact' on the relevant sectors.[352] This means that in the end the fishers do not loose any of their fishing opportunities. They are merely forced to reduce future quota by the amount that was taken illegally in advance.

Another effect of Art. 23(4), in addition to stabilising the quota system, is that stocks and ecosystems burdened by quota excesses will receive increased protection. Any reduction in fishing gives fish stocks a better chance to recover. In addition, if the Commission decides that quotas are not merely to be reduced but also reallocated to another Member State, 'conservation interests shall be taken into account'.[353]

---

347 See Reason (7) of Commission Regulation 147/2007.

348 See Reason (11) of Commission Regulation 147/2007.

349 See Reason (12) of Commission Regulation 147/2007.

350 See Case 9/89 Spain v Council [1990] I-1383, paras. 24-31 (30-31). This judgment refers to the quota reduction measure in Art. 32 of Regulation 2847/93.

351 Wolfgang Pühs, *Der Vollzug von Gemeinschaftsrecht*, (Berlin: 1997), p. 261; Stefanie Schreiber, *Verwaltungskompetenz der Europäischen Gemeinschaft*, (Baden-Baden: 1997), p. 172; Astrid Berg, refers to quota reduction measures as 'administrative penalties', see Astrid Berg, *Implementing and Enforcing European Fisheries Law*, p. 102-105; Siegfried Magiera, 'Durchsetzung des Europarechts', in Schulze and Zuleeg (ed.), *Europarecht* (Baden-Baden: 2006), paras. 82 to 84.

352 Which may, in practice, includes fishers who have fished within the law. See Reason (6, 11 and 12) of Regulation 147/2007; See also, Commission Press Corner of 15 February 2007.

353 Art. 23 (4), para. 2 of Regulation 2371/02.

2.6.1.3 Effects on fishers

According to Art. 24(3), quotas are deducted from future fishing opportunities 'of Member States'.[354] This is because Member States are responsible for the implementation of quotas and thus are also in charge of implementing quota deductions. The ECJ has recently confirmed that Member States are 'free to choose between a system of deductions equally distributed among all fishing undertakings or imposed solely on those fishing undertakings responsible for exceeding the quota' when making reductions to national quota.[355]

2.6.1.4 Legal form of the Commission's implementing measure – tertiary legislation

The Commission's deduction of Ireland's and UK's quotas under Art. 23(4) took the form of tertiary legislative act, *i.e.*, a Commission Regulation based on Council Regulation 2371/02. The Commission's regulation thus modifies future tertiary Council Regulations which lay down annual fishing quotas.[356]

2.6.1.5 Critique

In principle, quota deductions likely have the effect of a sanction. However, this disciplinary effect is somewhat reduced by the Commission's interpretation of Art. 23(4) of Regulation 2371/02, which falls short of imposing a real penalty on violating states. Deductions take place under the supervision of the Council through the Comitology procedure. The Commission has recently argued that it should have an increased mandate to reduce quotas 'as a normal act of administration'.[357] In addition, the Court of Auditors has observed that the Commission can only deduct quotas where it obtains sufficient evidence that the final catch figures submitted by the Member states are erroneous.[358] Given that obtaining such evidence is both time intensive and costly, the Commission has imposed a quota deduction only once as of January 2009.

---

[354] Art. 23(4), para. 1 of Regulation 2371/02.

[355] Case T-172/07, Atlantic Dawn Ltd. et. al v. Commission, OJ 2007 No. C170/29, para. 41; see also Art. 20(3) of Regulation 2371/02.

[356] On the legality of this modification, see Chapter 6.

[357] Commission Staff Working Paper SEC(2008) 2760 accompanying the Proposal for a Council Regulation establishing a Community control system for ensuring compliance with rules of the CFP, p. 11-12.

[358] Court of Auditors, *Special Report No. 7/2007*, para. 102.

2.6.2 Commission's right to take preventive measures

The 2002 Reform introduced a three-step Commission infringement procedure. Where the Commission establishes that there is 'evidence that rules on conservation, control, inspection or enforcement are not being complied with and that this may lead to a serious threat to the conservation of living aquatic resources or the effective operation of the Community control and enforcement system necessitating urgent action', it shall *inform in writing the Member State concerned* and set a deadline of no less than 15 working days to *demonstrate compliance, and to give its comments.* If the violating Member State fails to comply or provide comment, the *Commission may take preventive action,* which may include restricting fishing days.[359] Such measures may only be taken if there is evidence of a risk that fishing activities carried out in a given geographical area could lead to a serious threat to the conservation of living aquatic resources.[360] Adopted measures must be proportionate to the risk of the threat to the conservation of living aquatic resources, and cannot exceed three weeks in duration. If the Commission deems it necessary, the preventive measures may be prolonged up to six months by a decision in accordance with the Comitology management procedure.[361] In taking preventive measures, the Commission must take account of the Member States comments. The difficulty of collecting the necessary evidence and applying the preventive measure proportionately have to this day dissuaded the Commission from applying the preventive measures.[362]

2.6.3 Commission's right to suspend financial assistance

To improve the functioning of the CFP's fleet policy, Art. 16(1) of Regulation 2371/02 lays down the conditions for Community financial assistance and the reduction of fishing capacity. Member States will only receive structural assistance where they meet the following requirements:

- comply with the entry-exit schemes provided in Art. 11 to 13 of Regulation 2371/02;
- comply with the provisions on the fleet register in Art. 15 of Regulation 2371/02;
- provide the information required under the CFP provisions on structural aid.[363]

---

359 See Commission Press Release 11.11.2003.

360 Art. 26(3) of Regulation 2371/02.

361 Art. 26(3), paragraph 3 refers to Art. 30(2) which, in turn, refers to the procedures laid down in Arts. 4 and 7 of Council Decision (EC) 468/1999 laying down the procedures for the exercise of implementing powers conferred on the Commission, OJ 1999, No. L184/23.

362 Court of Auditors, Special Report 7/2007, para. 101.

363 Art. 16(1) of Regulation 2371/02 still refers to Regulation 2792/99, which was repealed by Regulation 1198/06 on the European Fisheries Fund, OJ 2006 No. L223/1.

After having afforded the Member States concerned with the possibility of being heard, the Commission may suspend financial assistance as far as proportionate to the degree of non-compliance. With regards to capacity, the following procedure applies: Where the Commission deems that the capacity of the fleet of a Member State exceeds the capacity limits set out in Arts. 11, 13 and 15, it shall inform the Member State concerned. The Member State must immediately reduce capacity, and forward to the Commission a reduction plan for verification. Verification of the Member State's reduction plan is conducted under the Comitology management procedure.

2.6.4 Infringement procedures

Where Commission's sanctions, *i.e.*, quota deductions, preventive measures or the suspension of financial assistance do not lead to the removal of the infringement of CFP rules by Member States, the Commission can still choose to formally initiate infringement procedures before the ECJ.[364] The infringement procedure is generally divided into three distinct stages.[365] If in the Commission's view the Member State fails to comply with CFP rules, it will formally notify this Member State of the specific infringement. The Commission will usually require a response within two months. If the Commission and the Member State do not agree after this period on the legal aspects of the case, the Commission will deliver a reasoned opinion to the Member State concerned. The opinion will include the Commission's arguments regarding the facts and legal grounds on which the infringement rests, and impose a time frame within which the Member State must comply. If the state decides to disregard the Commission's requirements within the stipulated period, the Commission may file a claim with the ECJ. The ECJ will then deliver a judgement, which is binding on the Commission and the Member State concerned. Where the offending Member State fails to follow conditions set out in the judgment, the Commission may, once again, bring the issue before the Court. If the Court finds that the Member State has not complied with its prior judgment, it can impose a lump sum or penalty payment.

According to the Commission's yearly evaluation of Member States' application of CFP fisheries laws ('Scoreboard 2005'), 49 out of 69 cases pending in

---

[364] Arts. 226 and 228 of the Treaty. The action under Art. 226 is 'objective in nature' as well as for the purpose of 'obtaining a declaration that a specific conduct of a Member States infringes Community law as well as to terminate that infringement', see Case 7/68, Commission v Italy [1968] ECR 423, at 428; Cases 15 and 16/76, France v Commission [1979] ECR 321, para. 27. Thus, the Commission's action before the ECJ is admissible where the obligation to fulfil Community law exists on the day of the expiry of the period of the reasoned opinion as provided in Art. 226 of the Treaty; see Case C-166/97 Commission v. France [1999] ECR I-1719, para. 18.

[365] Cremer, 'Art. 226 EGV' in Calliess and Ruffert (ed.) *EUV/EGV*, (München: 2007), p. 1981. Craig and De Bùrca include the pre-contentious informal phase as a first phase, see Craig and De Bùrca *EU Law*, p. 432

2005 were concerned with quota overfishing;[366] eight cases concerned Member States' failure to deliver required data on catch or fishing effort; others related to the unsatisfactory control of technical measures; six cases were pending at the stage of being referred to the ECJ. In 2005, the ECJ decided a groundbreaking case. The Court found in 1991 that France had failed to enforce Community technical measures for the conservation of fish resources.[367] Subsequently, the Commission claimed that France had not complied with this judgment and referred this matter, again, to the ECJ.[368] In response, France was ordered to pay the Commission a lump sum of €20,000,000, as well as an additional penalty payment of €57,761,250 for each six-month period from the delivery of the 2005 judgement on, at the end of which France has failed to comply. However, infringement procedures are inherently long and require extensive preparatory work on the side of the Commission. The above-mentioned judgement was the only case in the area of fisheries in which a Member State had to pay a financial penalty. It must also be noted that the decision was delivered in 2005 for non-compliance with a 1991 judgement.[369]

2.7 Implementation deficits

Payne assumes that the fact that Member States themselves implement CFP measures means that the regulations are frequently under enforced against their own fishermen.[370] In fact, despite the Community's continued efforts to overhaul and tighten its control and enforcement regime,[371] implementation deficits have continued to exist throughout the entire history of the CFP. Non-compliances result both from fishermen's infringements of quantitative and technical conservation measures or Member States ineffective application of control and enforcement measures.[372] Current evaluations of the implementation of CFP rules basically criticise the same implementation

---

[366] The Common Fishery Compliance Scoreboard 2005, pp. 50-55, available online at http://ec.europa.eu/comm/fisheries/scoreboard/index_en.htm.

[367] Case 64/88, Commission v. France [1991] REC I-2727.

[368] Case 304/02 Commission v. French Republic [2005] ECR I-6263.

[369] Court of Auditors, Special Report 7/2007.

[370] Payne, Policy-Making in Nested Institutions: Explaining the Conservation Failure of the EU's Common Fisheries Policy', p. 318.

[371] The first control regulation was adopted in 1982, consolidated in 1987 and repealed by the current Control Regulation in 1993, see Council Regulation 2057/82, OJ L220/1; Council Regulation (EC) 2241/87, OJ L207 and Regulation 2847/93.

[372] See also Case 290/87, Commission v. Netherlands [1989] ECR 3083; Case 64/88, Commission v. France [1991] ECR I-2727; 244/89, Commission v. France [1991] I-163; Case 258/89, Commission v. Spain [1991] ECR I-3977; Commission v France; Case 52/95 Commission v. France, para. 38.

deficits that were targeted twenty years ago.[373] The following two paragraphs will outline the substantial failures of the CFP's control system.

In 2007 and 2008, the Community's Court of Auditors and the Commission both published reports that exposed the substantial failure of the CFP's control system.[374] On 14 November 2008, the Commission forwarded a proposal to the Council that recommended substantial reforms to the Community control system.[375]

The Commission emphasised that the CFP's control system is 'inefficient, expensive, complex, and its continued failure will have significant consequences for the future of the fishing industry and the regions dependant on fishing'.

Although the basic provisions are laid down in Regulation 2847/93 and 2371/02, additional control provisions are included in a further 27 regulations.[376] The overall complexity and incoherence of these measures creates a certain degree of legal uncertainty.

In addition, Community measures on the collection, validation and compilation of catch data and their implementation by Member States generally fall short. For example, weighing the quantities of landings is not a general obligation. Discarding also creates uncertainties with regards to the scientific assessment of fish stocks.[377] However, a proper application of the TAC system requires

[373] Report by the Commission to the Council on the Enforcement of the CFP, Com(86) 301 final; Communication from the Commission to the Council and the European Parliament on the Common Fisheries Policy, SEC (90) 2244 of 6 December 1990; Report 1991 from the Commission to the Council and the European Parliament on the Common Fisheries Policy, SEC(91) 2288 of 18 December 1991; Report on Monitoring Implementation of the Common Fisheries Policy, SEC (92) 394 final of 6 March 1992. For an explanation of the last four reports, see Long/Curran, pp. 66-72. See also Report from the Commission on the monitoring of implementation of the Common Fisheries Policy – Synthesis of the implementation of the control system applicable to the Common fisheries Policy by Member States, Com(2001) 526 final of 28 September 2001; Report from the Commission on the monitoring of the Member States' implementation of the Common Fisheries Policy from 2002 to 2004, COM (2004) 849 final of 4 January 2005; Report from the Commission to the Council and the European Parliament on the monitoring of the Member States' implementation of the Common Fisheries Policy Com, Com(2007) 167 final of 10 April 2007. Art. 27(4) of Regulation 2371/02 requires the Commission to draw up every three years a report on the application of the CFP rules by Member States: Commission Report Com(2007) 167 final on the monitoring of the member states' implementation of the common fisheries policy 2003-2005. In addition, to improve compliance through transparency, the Commission publishes a yearly scoreboard on the implementation of CFP rules since 2003. Scoreboards are available at the web site of the DG Fisheries.

[374] Commission Staff Working Document SEC(2008) 2760 accompanying the Proposal for a Council Regulation establishing a Community control system for ensuring compliance with the rules of the Common fisheries Policy, p. 11; Court of Auditors, *Special Report 7/2007*.

[375] Commission Communication Com(2008) 721 final on a proposal for a Council Regulation establishing a Community system for ensuring compliance with the rules of the Common Fisheries Policy.

[376] Listed in Annex III of the Commission Staff Working Document SEC(2008) 2760.

[377] Court of Auditors, *Special Report 7/2007*, paras. 21 and 25.

reliable data. Currently, the Commission is not in a position where it can satisfactorily identify mismanagement and respond in a timely and effective manner.

Member States' application of Community control and enforcement measures is often of poor quality and the risk of detection of an infringement is low. In many cases, Member States' control efforts and services with regard to inspections are viewed as inadequate,[378] particularly at the stages of landing and marketing. Given the absence of a Community standard for 'inspections', Member States' control practices differ greatly. Consequently, the control systems put in place by the Member States do not guarantee that infringements will be detected effectively. This, in turn, makes Member States' judicial responses to violations as well as the Commission's evaluation of the effectiveness of national enforcement practices management difficult. In addition, underenforcement decreases the legitimacy of Community conservation measures, creating the perception that the rules have different levels of enforcement across the Community.

National sanction measures also deviate largely from each other and, in many cases, sanctions are not dissuasive. In 2006, fines paid by the industry amounted to only €9.2 million constituting only 0.04% of the number of active fishing vessels. Moreover, the average fine decreased from €4664 in 2003 to €1548 in 2006. Penalties like licence suspensions are hardly ever used in practice.

Given its limited competences and personnel capacity, it is difficult for the Commission to effectively control the implementation practices of Member States. The Commission only has 24 inspectors with very limited inspection powers. Even if inspectors find evidence of an infringement, a Member State is not obliged to initiate administrative or judicial proceedings against the violators. The Commission's power to close fisheries after quota has been exhausted is also limited. In principle, Member States are responsible to provide data on the state of the quota. Given that this data is often deficient, the Commission needs to collect the required data itself if it wants to close a quota on its own initiative. This process is time and cost intensive. The Commission's competence to reduce quota is also of limited use because in serious cases of overfishing the reduction process is influenced by political considerations. The Commission's power to take preventive as well as emergency measures, pursuant to Art. 26(3) and Art. 7 of Regulation 2371/02, is also limited by the requirement to gather sufficient evidence first. Finally, the most powerful instrument – the infringement procedure before the Court of Justice – is regarded as cumbersome and slow.

A further criticism is that overcapacities incite fishers to overfish, under declare their catches, and throw less profitable species overboard.[379] Commu-

---

[378] See also Commission Report Com(2007) 167 final on the monitoring of the member states' implementation of the common fisheries policy 2003-2005, p. 8.

[379] Court of Auditors, *Special report* 7/2007, paras. 107-110.

nity law regulating overall fishing capacities is ineffective at requiring Member States to reduce capacities and thus is unlikely to resolve these problems.

Finally, with the exception of a few positive developments, the co-ordination of Member States is weak overall. The competence of the Community's Fishery Control Agency does not mandate cooperation as regards following up and penalising violations of management measures.[380]

3 Small-scale coastal fisheries management

To be effective and efficient, different types of fisheries require different types of management regimes.[381] The literature makes a general distinction between *small-scale coastal* and *large-scale industrial fisheries.*[382] There are clear differences between fisheries carried out by single fishermen and small crews using small boats close to the shores, which produce primarily for local markets on the one hand, and fishing operations with large crews working on large efficient vessels, which are owned by companies with global reach. Since the inception of the CFP in the early 1970s, both the Council and the Commission have also noted the distinct character of small-scale coastal fisheries.[383] These needs and potentials are increasingly recognised by fisheries scientists, too.[384] The following sections will analyse the extent to which the current CFP management system really considers the particular needs and potential of small-scale coastal fisheries.

---

380 Court of Auditors, *Special report 7/2007*, para. 87.

381 See, for example, World Bank, *Saving Fish and Fishers – Toward sustainable and Equitable Governance of the Global Fishing Sector*, pp. 6-7; Michael King, *Fisheries Biology, Assessment and Management*, (Oxford: 2007), pp. 278 et seq.; OECD, *Towards Sustainable Fisheries*, (Paris: 1997), pp. 61 et seq.; Stephen Cunningham, 'Overview', in Cunningham and Bostock (ed.), *Successful Fisheries Management*, (Delft: 2005), pp. 9-20.

382 Distinctions are often blurred and categories, of course, vary; see, for example, Michael King, *Fisheries Biology, Assessment and Management*, p. 273; Kura *et al.*, *Fishing for Answers – Making Sense of the Global Fish Crisis*, pp. 8-9.

383 See Reason (3) of Regulation 2141/70, *i.e.*, the CFP's first structural regulation. See also 'Commission Report on the Situation in the Fisheries Sector of the EEC Member States and the Basic Principles for a Common Policy', (67/196/EEC), OJ 1967, p. 862; when the common management policy was first instituted, the Council adopted a declaration that recognised the distinct nature of small-scale coastal fisheries, see paragraph 2(b) of Council Declaration of 30 May 1980 on the common fisheries policy, OJ 1980, No. C158/1. See also Reason (10) of the first management Regulation 170/83 and Reasons (14) and (16) of the current management Regulation 2371/02.

384 IFREMER, *Small-Scale Coastal Fisheries in Europe*, (Brussels: 2007), p. 6; Pavel Salz, Erik Buisman, Jos Smit and Birgit de Vos, *Employment in the Fisheries Sector: current situation*, (Brussels: 2006), p. 15, available online at: http://ec.europa.eu/fisheries/publications.

3.1 Characteristics of Community small-scale coastal fisheries

Community fisheries take many different forms, varying with regards to the species they exploit, the marine or coastal environment in which they fish, the scale of production facilities, the types of vessel they use, and the socioeconomic background of the operators involved.

There are a large number of small-scale vessels in the Community's fishing fleet. It has been estimated that in 2007 about 81% of all Community vessels were less than 12 metres long and approximately 100,000 (of a total of 209,000) fishermen earned their livelihood by small-scale fishing.[385] Their vessels are relatively immobile, generally operating within the 12 nm zone off the coast.[386] Thus, they are particularly vulnerable to changes in catch abundance and competition from large-scale vessels and other interest groups claiming access to the narrow coastal strips. Furthermore, alternative opportunities for employment are often few in Community coastal regions.[387] Compared to other interested parties (including large-scale fisheries), small-scale fishers are often poorly organised and their interests politically marginalized.[388] However, given their use of less destructive fishing practices and the high quality of their products, small-scale fisheries have the potential to make a strong contribution to the Community fishing industry.[389]

3.2 Concepts and definitions of small-scale coastal fisheries under Community law

Concepts and definitions of small-scale fisheries can be found within different CFP measures. Structural measures, for example, have drawn a distinction between small scale-coastal fisheries and other fisheries since 1993. At that time, 'small local coastal fishing fleets' were defined as vessels 'under 220 kw, for which quotas have not been set at Community level'.[390] The follow up regulation stated that `small-scale coastal fishing means fishing carried on by

[385] Pavel Salz, Erik Buisman, Jos Smit and Birgit de Vos, *Employment in the Fisheries Sector: current situation*, (Brussels: 2006), p. 15.

[386] IFREMER, *Small-Scale Coastal Fisheries in Europe*, p. 21.

[387] Commission Communication Com(2001) 135 final Green Paper on the Future of the Common Fisheries Policy, pp. 24, 31-33.

[388] COFAD, *Die Küstenfischerei in Niedersachen – Stand und Perspektiven (Kurzfassung)*, (Tutzing: 2004), pp. 11-12.

[389] IFREMER, *Small-Scale Coastal Fisheries in Europe*, pp. 6-7; Commission Communication Com(2001) 135 final Green Paper on the Future of the Common Fisheries Policy, pp. 32-33.

[390] Art. 7(1) second indent of Council Regulation 3699/93 laying down the criteria and arrangements regarding Community structural assistance in the fisheries and aquaculture sector and the processing and marketing of its products, OJ 1993 No. L356/1.

vessels of an overall length of less than twelve metres.'[391] Art. 26(1) of the Regulation on the European Fisheries Fund from 2006 defines small-scale coastal fishing as 'fishing carried out by fishing vessels of an overall length of less than 12 metres and not using towed gear [...]'.[392]

Management measures only indirectly refer to small-scale coastal fisheries, recognising their economic interest in having preferential access to the 12 nm miles zone. The management regime allows Member States to restrict fishing to vessels that traditionally fish in waters up to 12 nm off their coasts from ports of the adjacent coasts.[393] The Regulation is justified on the basis that the 'rules in place restricting access to resources within the 12 nautical mile zones of the Member States have operated satisfactorily benefiting conservation by restricting fishing effort in the most sensitive part of Community waters and preserving traditional fishing activities on which the social and economic development of certain communities is highly dependant.'[394]

One may provisionally conclude that the Community views small-scale coastal fisheries as a fishing activity carried out by vessels of a small size or power, being somewhat bound to local coastal fishing grounds, and its operators (as with their locality) being socially and economically highly dependant on fishing.

3.3 Community measures

In principle, the CFP management measures cover all Community waters and relate to all fleet segments. TACs, effort limitations and technical measures do not explicitly distinguish between small-scale coastal and large-scale vessels. However, the CFP management regime leaves certain competences regarding the governance of 12 nm zones to the Member States. Since this zone is the primary geographical target of small-scale fisheries, Member States could use this competence to benefit small-scale fisheries. Community law provides a legal framework for the management of small-scale coastal fisheries, allowing Member States to (a) restrict fishing to fishing vessels that traditionally fish in waters up to 12 nm from their coasts from ports of the adjacent

[391] Art. 11 of Council Regulation (EC) 2792/99 laying down detailed rules and arrangements regarding Community structural assistance in the fisheries sector, OJ 1000 No. L337/1.

[392] Not using towed gear as listed in Table 3 of Annex I of Commission Regulation 26/2004 regarding the fishing vessels register of the Community, OJ 2004 No. L5/25.

[393] See Art. 17(2) of Regulation 2371/02.

[394] See Reason (14) of Regulation 2371/02; ten years later, Art. 6 of the first general management regulation, Regulation 170/83, was also included this exception. Reason 8 of that regulation stated that the exception was maintained 'to enable the inshore fishing industry to cope with the new fishing conditions resulting from the institution of 200-mile fishing zones'. The ECJ has recognised the Community's competence to adopt this exception rule on the basis of Art. 37 of the Treaty; see Case 103/91 Spain v. Council [2005] ECR I-2267, para. 30

coasts,[395] and (b) adopt, subject to certain conditions, management rules within their 12 nautical mile zone.[396] The following section will assess whether these provisions really serve the interests of small-scale fisheries.[397]

3.3.1 The exception to the principle of free access

In principle, Community vessels have free access to all Community waters.[398] However, as stated above, Member States are entitled to restrict fishing in waters up to 12 nm to 'fishing vessels that traditionally fish in those waters from ports on the adjacent coasts'.[399] The Member States can exclude fishers from other Member States, unless such fishers previously enjoyed traditional fishing rights in these areas.[400] According to the Council, this provision's aims are the support of local coastal economies and protection of the most sensitive areas of Community waters.[401]

In practice, Member States' application of the exception rule generally keeps away large, foreign vessels (with the exception of those which have traditional fishing rights).[402] However, it must be noted that several Member States do not specifically reserve these areas to their small-scale fisheries, but instead use this competence to exclude the fishers of other Member States from fishing within their 12 nm zone.[403] Germany, for example, has not adopted measures which specifically give preference to small-scale local fishers. Under the German Fisheries Code,[404] all German vessels have free access to the German territorial sea. Community fishing vessels may also apply for special fishing permits to fish in this area,[405] which are granted on the same basis as for German fishers.[406] All catches taken out of the territorial sea by the fishers of other Member

[395] Art. 17(2) of Regulation 2371/02.

[396] Art. 9 of Regulation 2371/02.

[397] It must be noted that the Community provides measures, which particularly aim at benefiting small-scale coastal fisheries, see Art. 26 of Regulation 1198/06. These measures will be described in Chapter 4.

[398] Arts. 17(1) and 3(a) of Regulation 2371/02.

[399] Art. 17(2) of Regulation 2371/02.

[400] These traditional rights are listed in Annex I of Regulation 2371/02. For example, traditional fishing rights in German territorial waters in the North Sea exist for Dutch and Danish fishers. UK fishers have right in the Zone around Helgoland. In the Baltic Sea, Danish fishers have traditional fishing rights in the German territorial sea.

[401] Reason (14) of Regulation 2371/02.

[402] Van Rijn, *'Fischereipolitik'*, para. 21.

[403] According to Churchill, Member States like Denmark, Ireland, Italy, and the UK exclude fishing from other Member States fishers from their 12 nm zones; see Churchill, *EEC Fisheries Law*, pp. 135-136; see also van Rijn, *'Fischereipolitik'*, para. 21.

[404] BGBl. I 1984, 876 as amended by BGBl. I 2004, 2198.

[405] § 5(2), paragraph 2 of the German Fisheries Code.

[406] § 3(1), (2), (3), paragraph 3 and (5) of the German Fisheries Code.

States holding a special fishing permit must land their catches in predetermined German ports.[407]

3.3.2 National conservation measures instituted within the 12 nm zones

Within 12 nm of the baseline, Member States have the right to take 'measures for the conservation and management of fisheries resources and to minimise the effect of fishing on the conservation of marine eco-systems'. However, this is permitted only to the extent that the 'Community has not adopted measures addressing conservation and management specifically for this area'.[408] In addition, national measures must be non-discriminatory, compatible with the objectives laid down in the basic management regulation, and 'no less stringent than existing Community legislation'.[409] The management of fisheries within the 12 nm mile zone in Germany's federal state of Lower Saxony will serve as an example here.

This section will describe Germany's and Lower Saxony's system of conservation measures adopted within the scope of Art. 9 of Regulation 2371/02.

3.3.2.1 Fisheries law in Germany

In Germany, the competence to regulate fisheries is divided between the federal government and the federal states (*Bund* and *Länder*, respectively). Fisheries are subject to the so-called 'competing competences' ('*Konkurrierende Gesetzgebung)*', so that as long as federal legislators have not chosen to regulate on certain matters, the federal states are free to legislate in these areas.[410] In Germany, federal legislators have adopted national rules for the German territorial sea and the EEZ. However, the law allows the federal states to adopt rules concerning fisheries in areas where federal legislation has not been adopted.[411] They also may adopt management rules that are more stringent than federal laws as long as these rules accord with Community law.[412] Federal

---

[407] § 4(1) of the German Fisheries Code in combination with Annex III (Anlage 3) of the Maritime Fisheries Regulation (Seefischereiverordnung BGBl. I 1989, 1485 as amended by BGBl. I 2005, 3499.

[408] *Ibid.*

[409] National measures, in principle, apply to all Community fishing vessels, see Reason (11) of Regulation 2371/2002. However, if the vessels of other Member States are affected by such measures, these Member States, the Commission and the Regional Advisory Councils must be consulted first, see Art. 9(2) of Regulation 23712/02.

[410] Arts. 72(1) and 74(17) of the German Constitution (*Grundgesetz*). The federal states have the sole competence to regulate inland fisheries; see A. Lorz, 'Naturschutz im Fischereirecht der Gegenwart' (1994) *Natur und Recht*, 2, p. 63.

[411] § 10 of the German Fisheries Code (Seefischereigesetz), BGBl. I 1984, 876 as amended by BGBl. I 2004, 2198.

[412] *Ibid.*

states, which have large coasts, are Niedersachsen, Mecklenburg-Vorpommern, and Schleswig-Holstein.[413] In practice, these states have limited the scope of their fisheries legislation to their respective share of the German territorial sea.

3.3.2.2 National conservation measures

The federal government has adopted only a single conservation measure which applies to German territorial waters.[414] The law provides that within German territorial waters in ICES areas III c and III d (*i.e.*, on the Baltic Sea coasts of Schleswig-Holstein and Mecklenburg-Western Pomerania) fishing for herring and sprat and using pelagic trawls is prohibited for vessels with more power than 588 kw. Vessels fishing for other species or using different gear must not exceed an engine power of 221 kw.

3.3.2.3 Conservation measures in Lower-Saxony

Lower Saxony has adopted fishing limitations and technical measures and established no take areas and periods for its part of the German territorial sea, both within its fisheries and environmental legislation.

The Fisheries Code of Lower Saxony (*Niedersächsisches Fischereigesetz*[415]) creates a legal framework for coastal fisheries management in German territorial and inland waters within federal state borders. Under the Code, the terms 'coastal waters' include the territorial zone as well as some closely connected waters landwards of the baseline.[416] The Code provides that, in principle, everyone is free to take fish and crustaceans from Lower Saxony's coastal waters.[417] Only fishing for mussels and mussel aquaculture requires prior authorisa-

---

[413] Fisheries Code of Niedersachsen (*Niedersächsisches Fischereigesetz*) Niedersächsisches Gesetz- und Verordnungsblatt, 1 February 1978, p. 81 as amended by Niedersächsisches Gesetz- und Verordnungsblatt 10 November 2005, p. 334; Fisheries Code of Schleswig-Holstein (*Fischereigesetz für das Land Schleswig-Holstein*) Gesetz- und Verordnungsblatt Schleswig-Holstein, 10 February 1996, p. 211; Mecklenburg-Western Pomerania Fisheries Code (*Fischereigesetz für das Land Mecklenburg-Vorpommern*), Gesetz- und Verordnungsblatt Mecklenburg-Vorpommern, 4 May 2005, Nr. 7, p. 153. Bremen and Hamburg have also adopted fisheries codes providing regulations for fishing in coastal waters within their small shares of German territorial waters: Gesetzblatt der Freien Hansestadt Bremen, 26 September 1991, Nr. 40; Hamburgisches Gesetz- und Verordnungsblatt, 28 May 1986, Teil I Nr. 22.

[414] § 2 (2), paragraph 2 of the Maritime Fisheries Regulation (*Seefischereiverordnung*), BGBl. I 1989, 1485 as amended by BGBl. I 2005, 3499. Other measures concern competences, implementation, licenses, control and sanctions.

[415] Niedersächsisches Gesetz- und Verordnungsblatt, 1 February 1978, p. 81 as amended by Niedersächsisches Gesetz- und Verordnungsblatt 10 November 2005, p. 334.

[416] § 16 (2) and (3) and Annex I of the L-S Fisheries Code.

[417] § 16 (1) of the L-S Fishery Code. See also W. Möller, *Umwelt, Wald, Planung, Naturschutz, Jagd u.a. Band IV*, (Hannover: 2004), p. 478.

tion.[418] According to § 57 (1), fishing operators are, nevertheless, required to fulfil specific registration requirements.[419] The Fisheries Code allows the competent executive ministry to adopt further fisheries management measures, such as no take periods, fishing limitations, minimum sizes, etc.[420] The Ministry has adopted the Coastal Fisheries Regulation of Lower-Saxony (*Niedersächsische Küstenfischereiordnung*),[421] which lays down specific technical measures and catch limitations for species for which no limitations have been adopted under the CFP.

§ 4 of the Fisheries Regulation determines the conditions for the use of fixed gears and weirs with an overall length of more than 100 metres close to the shores: Gears and weirs must be controlled and emptied every twenty-four hours,[422] as well as earmarked and their owner identified. § 5 explicitly prohibits beam trawling with beams longer than three metres or trawling with beams longer than four metres, as well as fishing with trawl nets and gillnets in coastal waters.[423] According to § 7, fishing with electronic current is forbidden, except where it is carried out for the purpose of stock or scientific surveys. Additionally, § 6(1) provides that certain fish species that have not reached a specific minimum size must not be caught in coastal waters. § 6(2) prescribes no-take periods for particular species. Furthermore, § 6(6) forbids fishing for North Sea crabs (crangon crangon) between 1 January and 30 June of each year. Fish caught in violation of these rules must be released immediately.[424] According to § 9, prior authorisation is required to introduce new species in coastal waters, and for mussel fishing. The competent authority can also impose further quantitative restrictions (TACs and effort limitations), no take periods and gear restrictions.[425] The Regulation also establishes no take periods and minimum sizes for specific mussel species.[426] The management rules of the Fishery Code and the Fishery Regulation are supported by a system of sanctions.[427]

---

418 § 17(1) and (2) of the L-S Fishery Code.

419 § 59 of the L-S Fisheries Code. Fishing vessels that want to carry out fishing for professional reasons in the coastal waters of Lower-Saxony (as well as fishing vessels used for fishing in waters beyond these waters) and which are registered in a ship register of Lower Saxony (Niedersachsen) or whose owners live in Lower Saxony must enrol at the competent fishery authority, § 2 of the Fisheries Regulation of Lower Saxony. They have to provide detailed information on their vessel's power, age, size, etc. The vessel then receives an identification badge, pursuant to § 2(3). Changes have to be reported according to § 3(1). If the vessel is withdrawn, the badge will be retracted by the competent authority in accordance with § 3(2).

420 § 53(1), paragraph 1 to 6 and 9 to 11 and (2) of the L-S Fisheries Code.

421 Niedersächsisches Gesetz- und Verordnungsblatt, 3 March 2006, p. 108.

422 § 4(4) of the Coastal Fisheries Regulation of Lower Saxony.

423 § 5(2) of the Coastal Fisheries Regulation of Lower Saxony.

424 § 6(4) of the Coastal Fisheries Regulation of Lower Saxony.

425 § 8(2), paragraphs 1 to 4 of the Coastal Fisheries Regulation of Lower Saxony.

426 §§ 8(3) to (6) of the Coastal Fisheries Regulation of Lower Saxony.

427 §§ 61 and 62 of the Fishery Code and § 12 of the Fishery Regulation of Lower Saxony.

German small-scale coastal fisheries are also affected by international, Community, federal and federal state laws regulating the protection of the coastal environment.[428] Such rules are primarily concerned with the protection and conservation of coastal landscapes, flora, fauna, ecosystems, etc., but not particularly with fisheries. Lower Saxony has established a large marine nature reserve on its North Sea coast (*Gesetz über den Nationalpark Niedersächsisches Wattenmeer*).[429] § 2 of that Code states that the main reason for the establishment of the nature reserve is to protect nature and biodiversity within the reserve and its natural development. The nature reserve is subdivided into three zones, each of them providing different levels of protection.[430] In principle, all activities that destroy, damage or alter the nature in the reserve are forbidden in Zones I and II. However, exceptions do exist, one of them being fishing. Capturing fish and crustaceans as well as blue mussels is allowed in certain parts of Zone I and II.[431] Rearing mussels is allowed only in cases where fishing and environmental authorities approved a plan that is in line with the objectives of the reserve. Recreational fishing as well as gathering worms (*arenicola marina*) is allowed only from particular spots in predefined areas of Zone I and both activities are permitted in Zone II.[432] In Zone III, only recreational angling is allowed.[433] Violations of these regulations can result in the imposition of severe fines.[434]

3.4 Small-scale fisheries interest representation

According to the CFP, the Community primarily bears the responsibility of regulating fisheries. Member States and local communities have relatively less regulatory power in comparison. The following section will review the ability of small-scale fisheries to represent their interests under Community law, using Lower Saxony's small-scale fishers as an example.

---

428 Wolff, *Fisheries and the Environment*, pp. 93-144; A. Lorz, 'Fischerei und Naturschutz – eine rechtliche Betrachtung' (1982) *Natur und Recht*, pp. 4-10.

429 Act establishing the National Park Wadden-Sea, Niedersächsisches Gesetz- und Verordnungsblatt, 11. July 2001, p. 443, as amended by Niedersächsisches Gesetz- und Verordnungsblatt, 23 June 2005, p. 210. See also the Act of Schleswig-Holstein on the establishment of nature reserve in Schleswig-Holstein's North-Sea territory, *i.e.*, the Wadden Sea (Gesetz zum Schutz des schleswig-holsteinischen Wattenmeeres Gesetz- und Verordnungsblatt Schleswig-Holstein, 17 December 1999, p. 518, as amended by Gesetz- und Verordnungsblatt Schleswig-Holstein, 16 September 2003, p. 503.

430 § 5 of the Act establishing the National Park Wadden Sea.

431 §§ 9(1), 13(1) and §§ 9(2) and 13(4) of the Act establishing the National Park Wadden Sea.

432 § 9(3), § 13(5) of the Act establishing the National Park Wadden Sea.

433 § 15(1) of the Act establishing the National Park Wadden Sea.

434 § 29 of the Act establishing the National Park Wadden Sea.

3.4.1 Situation of small scale fishers in Germany – Lower-Saxony

Coastal fishing in Lower Saxony is primarily carried out by small family businesses, which operate small beam trawlers with an engine power of less than 221 kw.[435] In 2002, about 140 beam trawlers and four vessels fishing for mussels were in operation. About 350 to 390 fishers were employed on board these ships. Some of these businesses have organised themselves in producer organisations and maintain membership in the German Cutter- and Coastal Fishers Association. Many, however, remain unorganised.

A comprehensive study on coastal fisheries in Lower Saxony from 2004[436] suggests that, in comparison to other interest groups, small-scale coastal fishers are not successful at representing their interest in gaining greater fishing opportunities.[437] The study finds that small-scale fishers suffer from the lack of exclusive fishing areas and long-term use rights. As a result, fishers have no ability to plan over the long-term and are often unsuccessful in defending or advancing their interests politically or judicially against other interest groups. Other groups, for example, those promoting nature conservation, such as wind farmers, mining firms, etc. have obtained more comprehensive legal rights in about 77% of the 12 nm zone in Lower-Saxony. Some of these activities severely restrict fishing. The remaining 23% of the waters are open to maritime traffic and thus cannot be assigned exclusively to fisheries.

3.4.2 Interest representation at the Community level

Due to varying and dissociated local interests and existing political structures, Community's fisheries industries and small-scale coastal fishers have not managed to jointly promote their interests at the Community level. Thus, interest promotion mainly takes place at the national level.[438]

However, recent CFP legislation gives special consideration to small-scale coastal fishers and their involvement in the political process. In order to enhance stakeholder involvement within the CFP, as a part of the 2002 Reforms, the Community introduced Regional Advisory Councils (RACs)[439] to include all the

[435] COFAD, Die Küstenfischerei in Niedersachsen – Stand und Perspektiven, pp. 11-12.

[436] *Ibid.*, pp. 1-23.

[437] *Ibid.*, pp. 4-5.

[438] Information on French, British and Spanish small-scale coastal fisheries organisations can be found in Lequesne, *Politics of fisheries in the EU*, pp. 43-50.

[439] Art. 31 of Regulation 2371/02; Council Decision (EC) 585/04 establishing Regional Advisory Councils under the Common Fisheries Policy, OJ 2004 No. L256/17; as amended by Council Decision (EC) 409/07, 2007 No. L155/68; see also Commission Decision (EC) 222/07, OJ 2007 No. L95/52; Commission Decision (EC) 206/07, OJ 2007 No. L91/52; Commission Decision (EC) 191/06, OJ 2006 No. L66/50; Commission Decision (EC) 668/05, OJ 2005 No. L249/18; Commission Decision (EC) 606/05, OJ 2005 No. L206/21; Commission Decision (EC) 774/04, OJ 2004 No. L342/28.

interests affected by the CFP.[440] The idea is that RACs are to be composed of representatives from the fisheries sector and other interest groups affected by the CFP, including small-scale coastal fishermen.[441] Two thirds of the seats in the general assembly and the executive committee must be allotted to representatives of the fisheries sector. At least one representative of the catching subsector from each Member State needs to be represented in the executive committee.[442] Nevertheless, the rules governing RACs do not guarantee sufficient representation of small-scale fishers for two main reasons. First, it does not define who small-scale fishermen actually are. Second, it leaves it up to the Member States to decide on the composition of the members of the RACs.

3.5 Analysis and interim summary

Community small-scale coastal fisheries face difficulties which are similar to those of small-scale coastal fisheries in other parts of the world. A major issue is that they are comparably immobile and thus economically dependent on fishing close to shores and their home ports. Furthermore, they have difficulties promoting their interests. They suffer from the lack of exclusive fishing areas or clearly defined use rights, which they can effectively defend by political or judicial means against other user interests.

The CFP rules do provide a definition of small-scale fisheries. The CFP management regime, however, does not create a system of conservation rules, which give special consideration to the needs of small-scale fisheries. The rules merely allow Member States (a) to restrict fishing to fishing vessels that traditionally fish in waters up to 12 nm from their coasts from ports of the adjacent coasts, and (b) to adopt, subject to certain conditions, management rules within their 12 nautical mile zone.

The right to restrict fishing to vessels that traditionally fish within the 12 nm zone from ports of the adjacent coasts is used by different Member States to exclude fishing vessels from other Member States. This, in fact, generally keeps away large, foreign vessels (with the exception of those which have traditional fishing rights). Above that, the rule does not require Member States to adopt fisheries management measures which take into account the specific needs and potential of small-scale coastal fisheries; and it does not appear that national legislators have taken full advantage of this competence.[443]

German and Lower Saxony's conservation and management measures adopted within the German 12 nautical mile zone do not specifically target the need and potential of small-scale coastal fisheries. Minimum sizes, no take periods and zones are primarily intended to protect fisheries resources and the marine environment rather than to support small-scale coastal fisheries

[440] Reason (4) of Decision 585/2004.

[441] Arts. 1(2) and 5(1) of Decision 585/2004.

[442] Art. 5(3) and (4).

[443] To gain a conclusive view, further investigations are required.

interests (even though these fishers will receive long-term benefits from these conservation measures). The lack of exclusive fishing areas and long-term use rights make it difficult for Lower Saxony's small-scale fisheries to defend their interests against other users. However, effort and gear size limitations for trawl and gillnets do help small-scale coastal fisheries operating in these areas by excluding large-scale fisheries.

One may conclude that under the current CFP management regime the needs of small-scale fisheries are not sufficiently taken into consideration. Community rules, which entitle Member States to adopt specific measures within their 12 nm zones, do not impose a clear duty on national legislators to protect and or promote the interests of small-scale fisheries. The same applies to the rules on the composition of the RACs. There is little evidence that Member States use their legislative competences to adopt access or conservation measures within the 12 nm zone specifically to the benefit of small-scale coastal fisheries. If the Community really wants to promote small-scale fisheries, unequivocal binding obligations would be required. Existing measures could be modified to achieve this. First, Art. 17(2) of Regulation 2371/02 could be amended; instead of restricting fishing to vessels that traditionally fish in the 12 nm zones from ports of adjacent coasts, Member States could be required to restrict fishing in these areas to small-scale vessels. Another way to do this would be to establish different fishing zones within the 12 nm zone for different categories of fishing vessels. Traditional fishing rights should not be prejudiced by this new rule. This change would not be a drastic change for current legislative practice in most Member States, but might do a better job of protecting small-scale vessels from large-scale vessels. Second, the Community should initiate a debate on the use of specific management instruments like exclusive fishing areas and long-term use rights in Community's small-scale fisheries. Finally, RAC rules should be amended to guarantee that Member States include small-scale fishers within the RAC assemblies and committees.

4 Finding the right combination of management instruments

The development of the CFP's management regime does not necessarily allow us to assume that there is an 'ideal type of management measure' or the *'right combination of management instruments'* which can equally apply to all fisheries. To this day, the Community has clearly not managed to find 'the perfect approach' to protect stocks from overexploitation. The performance of single or combined management measures and instruments depends on innumerable factors that vary from fishery to fishery.[444] Such factors include, i.a. the composition of the fleet, the equipment used, compliance traditions, socio-

[444] Kura *et al.*, *Fishing for Answers – Making Sense of the Global Fish Crisis* (Washington D.C.: 2004), p. 100; see also S.S. Gezelius, 'The Problem of Implementing Policies for Sustainable Fishing', p. 22.

economic pressures, control and enforcement practices, scientific monitoring of fish stocks, and of course the unpredictable nature of fish stocks and ecosystems. In the ongoing struggle to *improve* existing management systems, the Community's almost 30 years of experience should not be disregarded.

The above chapters have tried to explain and analyse the complex set of management rules and instruments that govern Community fisheries. Such measures regulate the utilisation of the fleet (e.g., space, time, power, gear restrictions), the outtake of fish resources (e.g., TACs), as well as the initial limitation of fishing power (e.g., structural measures, which will be explained in-depth in the following chapter). Management instruments take different forms and fulfil different purposes, and each approach has specific advantages and disadvantages.

It is clear that in many cases management measures are only effective where they are supplemented by other instruments. For example, where TACs are not complemented by measures protecting nursery grounds or requirements on minimum-mesh-size to protect young fish, the reproduction of stocks would seriously be endangered. Moreover, where the initial number of vessels is too high, TACs, TAEs and technical measures are difficult to enforce. It is suggested here that the use of a single management instrument cannot be evaluated in isolation from other existing management measures. Success or failure of management systems in most cases depends on the right combination and co-ordination of different management measures.[445]

What can be learned from the CFP with regard to choosing the right combination of measures? As stated above, the CFP refers to TACs as its main management instrument. Member States chose TACs in the 1970s because they are easy to negotiate, difficult to manipulate (in terms of definition), and it is simple for industry members to predict TAC-quotas.[446] The choice of TACs, however, also created major problems. Setting TACs at proper levels has proven to be both difficult and cost intensive. In addition, TAC-quotas are unsuitable for the management of the Community's multi-species fisheries. They also motivate fishers to overfish and upgrade their catches. Besides, given that the Community's TACs system refers to amounts which can be *landed* (not caught), fishers are also inclined to discard the so-called incidental catches.[447] Finally, the choice to make Member States responsible for the *implementation* of TACs sets the ground for future failure. In response to these problems, the Community has continuously tried to improve and supplement its TAC-system by introducing input restrictions (TAEs and fleet reductions),[448] reinforcing technical measures, and

[445] G. Winter, *Towards a legal clinic for fisheries management*, pp. 45-50; House of Lords – European Union Committee, *The Progress of the Common Fisheries Policies* (London: 2008), p. 11.

[446] S.S. Gezelius, 'The Problem of Implementing Policies for Sustainable Fishing', pp. 9-10.

[447] See definition in Art. 3(m) of Regulation 2371/02.

[448] See Chapter 4 of this book as well as Hegland and Raakjær, 'Recovery Plans and the Balancing of Fishing Capacity and Fishing Possibilities', in Gezelius and Raakjær (eds.), *Making Fiseries management Work* (Berlin: 2008), chapter 5.

fortifying its control and enforcement obligations. The Community has faced different challenges with its adoption of supplemental measures.

The Community has also encountered difficulties as a consequence of introducing TAEs. As stated earlier, it is particularly difficult to determine precisely the actual fishing power of a fleet and then translate this power into fishing mortality. In addition, where TAEs are adopted in combination with TACs, scientific advice must provide the basis for the integration of these two systems. Given these difficulties, Community law requires that where TAE-quotas are not sufficient to fish up the assigned TACs, Member States are entitled to file a request to the Commission to modify the effort levels (to enable fishers to fish up given TACs).[449] Second, it is difficult to assess vessel engine power and the overall fishing potential of a vessel and these measurements are furthermore susceptible to fraud. These challenges have prevented the adoption of an effort regime before 2002.[450] *Hegland* and *Raakjær* have pointed out that during the negotiation on the 2002 reform, Member States like Spain, France, Greece, Portugal, Italy, Ireland, and Finland also doubted the usefulness of effort limitations, particularly when used in conjunction with TACs.[451] Only recently, has the Commission acknowledged that up to now the effort limitations that have been adopted have failed to compensate for over-capacity in the fleet.[452]

Technical measures mainly aim at limiting the impact of fishing on marine ecosystem as well as on juvenile and non-target species. Technical measures thus supplement TACs and TAEs, increasing the 'sustainability factor'. One example of a technical measure is the requirement that fishing gear be selective with regards to non-target species, juvenile fish, and seabirds. No take periods and no take areas aim at protecting the water body and the seafloor (particularly nursery areas and fragile ecosystems). Minimum catch and landing sizes are adopted to protect young fish to the day they have spawned. The Community's example makes clear that technical measures are inevitably necessary to supplement quantitative instruments. Adopting TACs or TAEs without imposing additional technical requirements would ultimately result in the indiscriminate exploitation of all marine resources, without recognising the specific character of the surrounding ecosystems, the conservation requirements regarding non-target species, and the reproduction patterns of fish stocks. Due to the specific features of the different marine areas of exploitation, the Community has adopted several tailor-made technical regulations covering these different areas (e.g., technical measures governing the Baltic Sea and the Mediterranean).

---

449 See Art. 12 of Regulation 1954/03.

450 Communication Com (2002) 135 final Green Paper on the Future of the Common Fisheries Policy Vol. 2, p. 6; For other reasons, see Holden, *The Common Fisheries Policy*, pp. 196-198.

451 Hegland and Raakjær, 'Recovery Plans and the Balancing of Fishing Capacity and Fishing Possibilities', p. 152.

452 Commission Report Com (2007) 167 final on the monitoring of the Member States implementation of the Common Fisheries Policy 2003-2005, pp. 9-10.

Where initial fishing power is limited through reductions in fishing allowances, fleet size, or financial interventions, it important that fishing power actually corresponds to the available resources. Such structural policies, however, are a very special case in fisheries management, *i.e.*, politically, practically and theoretically. The CFP is a particularly interesting example in this regard and will be examined in-depth in the next chapter.

Further important lessons that can be learned from the CFP are that management measures need to be cohesive with each other and *all* measures must aim at sustainable exploitation. If one management measure allows for unsustainable fishing practices, the other measures are likely to be ineffective. The CFP provides at least three important examples in this regard. First, where the overall quantity of fish that can be taken from the waters is unsustainable, supplemental measures are unlikely to improve conservation. The Council's practice of setting TACs at unsustainable levels for twenty-five years is a key reason for the current state of fisheries resources in the European Community. Second, since the late 1970s, the Community's structural policies allowed too many vessels to fish in Community and third country waters. The CFP's persisting overcapacities have maintained fishing pressure at unsustainable levels and control and enforcement measures remain largely ineffective. In addition, huge investments in the sector have created political pressure to uphold high exploitation rates and subsidies programmes, which, again, have maintained overcapacities. Third, the Community's technical measures have been criticised for allowing fishing under conditions that harm the marine environment or the reproduction fish stocks, as well as for being too complex to be implemented, and too difficult to monitor. Consequently, the overall impact of technical measures and TACs and TAEs is substantially reduced.

Too many factors determine the performance of single or combined management instruments, making it impossible to draw up a universally applicable set of management measures. It is proposed here that the search for the right management system is thus a learning process based on trial and error. Since many errors have been committed over the life of the CFP, the study of this policy can contribute to this learning process. CFP experiences are therefore useful for the development of a methodological approach to fisheries management referred to as the 'legal clinic', which can be employed to diagnose problems in a fisheries management system and, from this basis, propose the right mixture of measures.[453]

[453] G. Winter, *Towards a legal clinic for fisheries management*, pp. 45-50.

CHAPTER 4

Community Instruments Promoting Fisheries

Since its inception in 1970, the CFP has pursued conflicting objectives. On one hand, it has tried to manage fisheries, maintaining exploitation at sustainable levels by establishing, implementing, and enforcing input and output limitations. On the other hand, it has strived at securing food supplies, increasing employment, the sector's competitiveness and furthering economic development in coastal regions. As a result, support was granted under the structural policy to build up the fisheries fleet, processing industry and aquaculture sector. In addition, the Community has intervened in the market to stabilise the price of fisheries products.

Politicians and scientists have become increasingly aware that such promotional activities can negatively affect management efforts and ultimately the overall state of resources. Promotion reduces production costs, encouraging people to enter, remain and invest in the industry. This, in turn, increases fishing and production capacities. High capacity (*i.e.*, a large number of vessels, employees and huge capital investments) necessitates high catch rates, putting pressure on marine capture resources and ecosystems. Besides, real world management regimes are never perfect in terms of design, implementation and enforcement. Thus, in many instances, promotional activities fuel the 'race to fish', exerting increased pressure on marine capture resources. This undermines the effectiveness of management regimes and the promotional measures aimed at supporting such management efforts.

It is only recently that the CFP has really begun to adjust its support practices to meet the situational and legal management requirements by stopping aid to the sector for the most problematic purposes. Nevertheless, harmful subsidisation continues even under the new EFF rules. Conflicts between promotional and management measures give rise to legal considerations. The Community's legal system, in general, and the CFP regime, in particular, aim at consistency and coherence.[1] This basically implies that secondary CFP measures should not be inconsistent with each other and those of other policy areas.[2] Moreover, promotional measures, as they currently exist, may be incompatible with Community primary law and the Community's international commitments and obligations, *e.g.*, commitments to sustainable fleet management.[3]

This chapter has two main purposes. First, the effects of the most common CFP subsidies will be explained in light of existing theoretical and empirical knowledge. Second, the promotional regime will be explained by reviewing

---

1 See Arts. 1, paragraph 3 and, Art. 3, paragraph 1 TEU and Arts. 1(2) and 2(2)(d) of Regulation 2371/02, Reasons (9) and (10) and Art. 6 of Regulation 1198/06. C. Calliess Art. 1 EU-Vertrag, p. 23; S. Bertea 'Looking for Coherence within the European Community Law' (2005) 11 *European Law Journal*, 2, pp. 154-172. pp. 154-172.

2 Nemitz, 'Fischereipolitik – Nach Art. 34 EGV', p. 575.

3 Such as the Art. 5(h) of the Fish Stocks Agreement, Arts. 7.1.8. and 7.2.2. of the FAO Code of Conduct on Responsible Fisheries, paras. 25 and 26 of the International Plan of Action for the Management of Fishing Capacity, Art. 30 (f) of the Johannesburg Plan of Implementation or Paras. 39-43 of UN GA Resolution 59/25 of 17 November 2004.

historical measures and providing a detailed description of the current CFP structural and market rules. Other sources of funding will also be described briefly.

This study will focus on aid allotted under the Community's structural policy and the market organisation. These constitute the main sources of Community aid and, due to the practice of 'co-financing'[4], determine a major share of Member State support. Structural and market measures are also the most problematic subsidies with regards to overfishing.

1 Sources of promotion to the Community's fisheries sectors

Community promotion of marine fisheries has developed significantly over the last 38 years. The complexity of promotional measures has increased and the law has undergone substantial change. Financial aid to the Community's fisheries sector flows from two main sources: the Community and the Member States. Community support is granted primarily under the CFP's structural policy. Major contributions are also provided through fisheries access agreements. Comparatively small amounts have been allocated under the CFP's market organisation, the European Regional Fund, the European Agriculture Guidance and Guarantee Fund and the European Social Fund. The Member States also provide large amounts of subsidies, mainly by co-financing Community measures, providing expensive general services (which they mostly do not recover from the fisheries sector), and through the grant of additional state aid.

2 Forms and impact of promotion on fisheries resources – theories and facts

Current political and academic discussions on public promotional activities in marine fisheries focus on terms like public 'assistance, aid, support, transfers or financial grants'. Most of these terms are also referred to as subsidies. Many different definitions of subsidies exist. A subsidy may, for example, be defined as 'a grant paid by a public entity to a private person, business or enterprise that benefits the public and does not require reciprocation'[5] or as the 'monetary value of government interventions associated with fisheries policies'.[6] However, forming a general definition of 'fisheries subsidies' in scientific, politi-

---

[4] As explained below under 3.

[5] Jörn Axel Kämmerer, ‚§ 124 Subventionen', in Isensee and Kirchhof (ed.), Hand Buch des Staats Rechts V, (Heidelberg: 2007), p. 1399; Hartmut Maurer, *'Allgemeines Verwaltungsrecht'* (München: 2006), § 17, para. 5.

[6] OECD, *Transition to Responsible Fisheries – Government Financial Transfers and Resource Sustainability: Case Studies,* (Paris: 2000), p. 129.

cal and legal contexts has proven difficult.[7] The reasons for this include (a) the large number of different forms and functions of public support to the fisheries sector, (b) the difficulties in determining with certainty the long-term effects of public support,[8] and (c) the fact that defining fisheries subsidies is not a policy-neutral undertaking.[9]

Despite these challenges, in this author's view, economics and existing empirical studies provide a sufficient analytical framework for the assessment of the promotional system of the CFP.[10] Subsidies have different effects on marine fisheries, including:

- increasing competitiveness and production capacities of fisheries sectors (affecting competing sectors and policies);
- increasing exploitation capacities which, in turn, reduce the effectiveness of management regimes and pressure fish stocks as well as the marine environment;[11]
- the distortion of trade in fisheries products;[12]

---

7 Gareth Porter, *Analysing the Resource Impact of Fisheries Subsidies: A Matrix Approach (UNEP)*, (Nairobi: 2004), p. 5; W.E. Schrank, 'Introducing fisheries subsidies' (Rome: 2003), pp. 1-6; the same author, 'Subsidies for Fisheries: A Review of Concepts' in FAO (ed.), *Papers presented at the Expert Consultation on Economic Incentives and Responsible Fisheries*, (Rome: 2000), pp. 1-58; L. Westlund, *Guide for identifying, assessing and reporting on subsidies in the fisheries sector – FAO Fisheries Technical Paper 438*, (Rome: 2004), pp. 7-12; Kämmerer, '§ 124 Subventionen', p. 1398.

8 *Ibid.*, OECD, *Transition to Responsible Fisheries – Government Financial Transfers and Resource Sustainability: Case Studies*, pp. 149 et seq.

9 John Virdin and David Schorr, *Hard Facts, Hidden Problems – A Review of Current Data on Fishing Subsidies – A WWF Technical Paper* (Washington D.C.: 2001), p. 12.

10 Overviews over existing studies are available in G. Porter, *Analysing the Resource Impact of Fisheries Subsidies*, p. 1 et seq.; A.S. Khan, U.R. Sumaila, R. Watson, G. Munro and D. Pauly, 'The nature and magnitude of global non-fuel subsidies', in Sumaila and Pauly (ed.), *Catching more bait: a bottom-up re-estimation of global fisheries subsidies*, (Vancouver: 2007); OECD, *Transition to Responsible Fisheries – Government Financial Transfers and Resource Sustainability: Case Studies*. OECD '*Financial Support to Fisheries – Implications for Sustainable Development*' (Paris: 2006) and OECD '*Liberalising Fisheries Markets: Scope and Effects*' (Paris: 2003); Holden, *The Common Fisheries Policy*, pp. 16-38; Erik Lindebo, 'Role of Subsidies in EU Fleet Capacity Management' (2005) 20 *Marine Resource Economics*, pp. 445-466; A. Hatcher and C. Robinson, *Overcapacity, overcapitalisation and subsidies in European fisheries. Proceedings of the first Concerted Action workshop on economics and the Common Fisheries Policy*, (Portsmouth: 1999); Hans Frost and Peder Andersen, 'The Common Fisheries Policy of the European Union and fisheries economics' (2006) 30 *Marine Policy*, pp. 737-746; A. von Moltke *et al.*, 'Fisheries Subsidies', in Najam, Halle and Meléndez-Ortiz (ed.), *Trade and Environment – A Resource Book*, (Geneva: 2007), pp. 101-114.

11 G. Porter, *Analysing the Resource Impact of Fisheries Subsidies*, p. iX; Court of Auditors, Special Report 7/2007, paras. 107-136.

12 PricewaterhouseCoopers, *Study into the nature and extend of subsidies in the fisheries sector of APEC members economies*, (Singapore: 2000), pp. 41-55; F.D. Schorr, Healthy Fisheries, Sustainable Trade: Crafting New Rules in the World Trade Organisation, (Washington D.C.: 2004).

- stabilisation of markets and incomes;
- lowering retail prices and thus increasing consumer demand.[13]

In this study, the effects of subsidies on the exploitation of fisheries resources will be of primary relevance. When analysing the effects of subsidies, one must consider that they largely depend on variables like the state of the existing management system, as well as the type of fishery, the functioning of control and enforcement measures, the biological status of stocks as well as the socio-economic, political and commercial context in which they are applied.[14]

Most empirical or theoretical studies assume that subsidies, which make fishing more profitable, create an economic incentive for fishers to enter or remain in the industry, even where fishing costs exceed the total potential income earned; this is referred to as 'bionomic equilibrium' under which catch rates exceed the maximum economic (and the maximum sustainable) yield.[15] Such studies conclude that unless an *effective management* regime reduces or eliminates all socio-economic incentives to overfish, for example, through the establishment of property or exclusive territorial use rights, most subsidies will continue to create technical and political inducement for fishers to increase their fishing potential. Based on this assumption and on empirical evidence, *Porter* suggests a matrix approach be used.[16] The matrix includes eight different types of subsidies. The effects of subsidies are investigated in respect to two important factors: the *state of fleet capacity* and the *type of management regime*. With regards to the state of fleet, Porter suggests three ideal type categories:

- overcapacity;
- full capacity;
- less than full capacity.

Overcapitalisation can be assumed where 'the level of fishing capacity in the fishery exceeds what is required for the desired level of catch'.[17] The desired level of catch in different management systems may lie somewhere between MEY, MSY, the bionomic equilibrium or, realistically, even beyond bionomic equilibrium itself.

---

[13] S. Ludicello, M. Weber and R. Wieland, Fish, *Markets and Fishermen – The Economics of Overfishing*, (London: 1999), p. 60; G. Porter, *Analysing the Resource Impact of Fisheries Subsidies*, pp. 19-41; A.S. Khan *et al.*, 'The nature and magnitude of global non-fuel subsidies', p. 13-16. OECD '*Financial Support to Fisheries – Implications for Sustainable Development*' (Paris: 2006) and OECD '*Liberalising Fisheries Markets: Scope and Effects*' (Paris: 2003).

[14] G. Porter, *Analysing the Resource Impact of Fisheries Subsidies*, p. ix; Khan *et al.*, 'The nature and magnitude of global non-fuel subsidies' p. 13.

[15] S. Ludicello *et al.*, *Fish, Markets and Fishermen – The Economics of Overfishing*, pp. 41-57; A.S. Khan *et al.*, 'The nature and magnitude of global non-fuel subsidies', pp. 13-16; Holden, *The Common Fisheries Policy*, pp. 181 and 230.

[16] G. Porter, *Analysing the Resource Impact of Fisheries Subsidies*, pp. 13 et seq., particularly 18 and 48.

[17] *Ibid.*, p. 15.

Management regimes are also divided into three ideal type categories:

- the '*open access regime*' does not provide catch, effort and technical fishing restrictions, a control system, or a rights-based management system, which eliminates the socio-economic incentives to overfish;
- the '*catch control regime*' does provide catch, effort, and technical restrictions as well as a control system. However, implementation deficits exist and the regime lacks a rights-based management system which eliminates the socio-economic incentives to overfish;
- the '*effective management regime*' provides catch, effort, and technical restrictions as well as a control system. In addition, it provides some kind of rights-based management system which eliminates the socio-economic incentives to overfish.[18]

At present, like most other OECD countries,[19] the Community can be categorised as a '*catch control regime*', which has to overcome the overcapitalisation of its fleets.[20] Thus, the following matrix can be applied to the Community policy:[21]

Table 4.1: Effects of CFP subsidies on fisheries resources (catch control & overcapacity' situation)

Subsidy category	Effects
Fisheries infrastructure • Harbour facilities • Fishing ports infrastructure • Support for producer organisation (except price support)	*Harmful to fisheries resources* Reducing the costs of infrastructure makes fishing more profitable. It provides an incentive for fishers to enter or remain in the business, even where total costs of fishing would normally exceed the total income.
Management services • Stock enhancement and re-stocking • Control and enforcement • Assessment and identification and development of new fisheries • R&D to develop new fisheries technologies • Artificial reefs	*Not harmful to fisheries resources* In principle, these 'implicit' subsidies are not considered harmful. The main argument is that government funded management services are likely to be more effective than services financed by the industry. Nevertheless, some states have chosen to recover costs from the industry (Australia, Canada, New Zealand).

18 *Ibid.*, p. 17.

19 *Ibid.*, p. 17.

20 Commission Com(2001) 135 final, Green Paper on the Future of the Common Fisheries Policy; European Environment Agency, *Europe's environment – the fourth assessment*, (Copenhagen: 2007), pp. 221-228; see also the first, second and third edition of the Common Fisheries Compliance Scoreboard, available online at http://ec.europa.eu/fisheries/index_en.htm.

21 The description of the effects of subsidies as provided in the chart below is a summary of Porter's analyses as understood by the author.

Subsidy category	Effects
Access to foreign waters • Concluding access agreements and supporting access financially	*Harmful to fisheries resources* Where access to foreign waters is subsidised, costs of fishing are reduced for the industry. This creates incentives to increase fishing in the waters of third countries. Depending on the third country's management system and the state of the stocks, subsidies may be harmful. Since fishing in the waters of a third country is neither effectively controlled by the Community nor by third countries, subsidies must realistically be regarded as harmful in most cases.[1] The financing of access agreements also gives fishers an advantage over competitors from third countries.[2]
Subsidy category	Effects
Decommissioning • Payments for the permanent withdrawal of vessels, fishing permit or licenses	*Potentially harmful to resources* Decommissioning may, in principle, contribute to capacity reductions. However, in catch control systems, decommissioning subsidies may increase capacities. First, additional capital is introduced in the sector, which can be used for capital investments. Second, operators are attracted to the fisheries sector, because investment risks seem low. Third, operators may stay longer in the business than they would without subsidies. Only where re-investments are severely restricted and capital-stuffing by remaining operators is prevented, will the effects be potentially positive.
Capital costs • Any grant, loan, loan guarantee or tax-relief which supports the renewal and modernisation of vessels and gear	*Harmful to fisheries resources* Subsidies for the construction of vessels further the race to fish by attracting operators. Subsidies for the modernisation of vessels and gears increase fishing power, unless an effective decommissioning scheme is provided.
Variable costs • Support for operating costs • Fuel tax exemptions • Income tax deferral for fishers compensation • Compensation for damaged vessels and gears • Transport subsidies • Support for energy saving devices on board	*Harmful for fisheries resources* Respective subsidies provide incentives for fishers to use more powerful and fuel consuming vessels, extend fishing journeys and using more refrigeration. All these effects lead to an increase in catch and thus pressure fish stocks.
Subsidies to income • Payments beyond the norm of other sectors • Payments for unemployed fisheries workers going beyond social insurance in other sectors • Payments fishermen who are idled by restrictions on fishing • Payments to vessel owner for temporary cessation of fishing	*Potentially harmful (except temporary cessation = harmful)* The effects of income support on fish resources are rather unclear. However, aid for the temporary cessation of fishing provided in response to collapsed stocks is supposed to have negative effects. Fishers are likely to pressure managers to re-open fisheries. Moreover, due to reduced losses, fishers are motivated to remain in an already overfished fishery.

Subsidy category	Effects
Price support subsidies • Ensuring prizes by withdrawals or support payments	*Harmful to fisheries resources* Prize support provides incentives for operators to increase production, i.e., pressure resources. Besides, increased income and, in parts, a guaranteed income, reduces economic risk, which makes the sector more attractive to enter or to remain in.

1 G. Porter, *Analysing the Resource Impact of Fisheries Subsidies*, p. 23; A.S. Khan *et al.*, 'The nature and magnitude of global non-fuel subsidies', p. 24.

2 European Commission, *Study on the European External Fleet* (Brussels: 2008), p. 9.

Porter (and most other studies) leave out most of the so-called 'indirect' subsidies which arise from government inaction; besides support for the processing and marketing sector and support for small-scale fisheries and rural development. Nevertheless, *Khan et al.* explain that, theoretically, public support for these activities results in reduced production costs and increased revenue, creating incentives for fishers and processors to enter or remain in the business and placing increased pressure on fish stocks.[22]

3 Promotional measures within the Community's structural policies

The Community's fisheries structural policy concerns the production side. This includes all equipment required to catch and process fisheries goods.[23] In the past, promotion in this sector has aimed at building up or modernising the catching and the processing industries. Today, management, environmental and social objectives increasingly play an important role. Under III. to VIII. below, the development of the CFP promotional policy from 1970 until today will be described and explained.

3.1 Administrative aspects of the CFP's structural policy

To understand how the CFP's structural policy works, it is important to understand two important features of the system: co-financing and the administration of the Community aid.

22 A.S. Khan *et al.*, 'The nature and magnitude of global non-fuel subsidies', pp. 15, 25

23 Churchill refers to the object of the structural policies simply as the 'catching side of the industry'; see R.R. Churchill, *EEC Fisheries Law*, p. 203.

3.1.1 The Co-financing System

In principle, the CFP's structural policy aims at providing a uniform and centralised allocation of financial aid to the fisheries sector. Structural measures set down the conditions under which aid can be requested by the fisheries sector for specific purposes. This does not imply that the Community provides all the funding. Since the early days of the CFP's structural policy, the availability of Community subsidies has been dependant on contributions of Member States and beneficiaries ('principle of joint funding/co-financing')[24]. The following example will help to illustrate this point:

Under the former structural measure, the 'Financial Instrument for Fisheries Guidance (FIFG)', fishers were eligible for public aid for the construction of vessels.[25] If a fisher intended to build a vessel in the fleet segment '0 < 10 GT', support was granted under the following conditions: First, an overall investment sum was determined. For example, for values between '0 < 10 GT', the investment sum was €22,000 per GT + €4,000.[26] This means that if a fisher wanted to build a vessel of 10 GT, the overall investment sum was €260,000. The FIFG regulation required that the Community contribute 15% of this amount if at least 5% was provided the Member State, and at least 60% by the beneficiary.[27]

This approach has two important consequences. First, any public aid granted by the Community is in most cases increased substantially by the contributions of the Member States. Second, participation rates determine significantly the attractiveness of private investments. While large contributions from the Community and Member States make investments attractive, high beneficiary participation rates have the opposite effect. Altering the participation rates is an important instrument for steering investments to specific parts of the sector and the fleet.

3.1.2 The administration of CFP structural aid

Throughout the history of the CFP, the competence to decide on the amount and the available forms of assistance has increasingly shifted towards the Community. In 1993, however, it was made clear that Member States are primarily responsible for the implementation of Community structural measures and the determination of how aid is actually to be used.[28] From

---

[24] Court of Auditors, Special Report No. 3/93 concerning the implementation of the measures for the restructuring, modernisation and adaptation of the capacities of fishing fleets in the Community together with Commission's replies ('Court of Auditors Report – 1993'), OJ 1994 No. C2/1, para. 1.29.

[25] See Art. 9 of Regulation 2792/99.

[26] See Art. 9(4) in combination with Annex IV of Regulation 2792/99; note that Art. 9(4)(a) says that aid for the construction of vessels may not exceed *twice the scales in table 1 of Annex IV.*

[27] See Table 3 of Annex IV of Regulation 2792/99.

[28] Reason 6 of Regulation 2082/93. Prior to 1993, in many cases, the Commission itself decided on individual aid applications, see, for example, Case 514/93, Cobrecaf and other v. Commission, [1995] ECR, II- 621.

this point onward, Community law has referred to the principle of 'shared responsibility'.

Under the current system, Member States must submit to the Commission a *national strategic* plan on their fisheries industry. This requirement is set out under the European Fisheries Fund (EFF-regulation), which entered into force on 1 January 2007.[29] These plans have served as the basis for dialogue between the Commission and Member States in regards to structural support schemes over 2007 to 2013.[30] Member States are to consult with relevant stakeholders when drawing up their plans. The plans should set out national priorities, objectives, estimates financial resources required, and timelines for putting measures into effect.[31] Member States have also been required to draw up an *operational programme* for implementing the policies that will be co-financed by the EFF. The programme has to be in line with the national strategic plan objectives and include, *inter alia*, tables which detail the financial contributions of the EFF and the Member States and where they will be allocated.

The Commission then evaluates, *e.g.*, whether the programmes contribute to EFF objectives and the guiding principles for the operational programmes laid down in Art. 19 of Regulation 1198/06, and whether they take into account the national strategic plans.[32] If, in its opinion, the programme is incongruent with CFP and EFF rules, it will ask the Member State to amend its programme accordingly. If the proposal is approved, the Commission will adopt an approving decision.[33]

Thus, the EFF leaves Member States a substantial margin of initiative and discretion as to how funds are actually allocated. For example, Germany's current (approved) operational programme states that about €12 million of EFF and national contributions will be set aside for the adaptation of its fleet.[34] This money is designated for modernising approximately 300 vessels.[35] When granting aid, Germany must follow the strict conditions (e.g., participation rates and the capacity ceilings) laid down in the EFF regulation.

In the aid distribution process, Member States designate a managing and a certifying authority.[36] Aid applicants turn to the managing authority, which decides whether the aid is granted or not and monitors the use of the funds. The certifying authority then draws up and submits applications for payment to

---

29 Council Regulation (EC) 1198/06 on the European Fisheries Fund ('Regulation 1198/06'), OJ 2006 No. L223/1.

30 Art. 15(1) of Regulation 1198/06.

31 Art. 15(2) of Regulation 1198/06.

32 Art. 4 of Regulation 1198/06.

33 Art. 17(6) of Regulation 1198/06.

34 See German Operational Programme, available online at the homepage of the DG Fisheries, pp. 72, 80 and 120.

35 *Ibid.*, p. 81.

36 Arts. 59 and 60 of Regulation 1198/06. Member States also install an audit authority reviewing the management and control of the operational programme, see Art. 61 of Regulation 1198/06.

the Commission. When the Commission accepts the application, the certifying authority receives the EFF funds from the Commission and hands them over to the applicant.[37]

3.2 Objectives and principles of the structural policies

The core objectives governing the Community's structural policy from 1970 onwards have basically remained the same until today. The first structural regulation stated that:

> Structural policy shall seek to promote harmonious and balanced development of this industry within the general economy and to encourage rational use of biological resources of the sea and of inland waters.[38]

The objectives are divergent, allowing for developing fisheries exploitation on one hand, and requiring rational use on the other. Subsequent aims, inserted into the legislation at various times, have supported either one or the other or both of these objectives. Art. 4 of the current structural regulation, Regulation 1198/2006 on the European Fisheries Fund, requires that assistance meets the following aims:

- support the common fisheries policy so as to ensure exploitation of aquatic living resources and support aquaculture in order to provide sustainability in economic, environmental and social terms;
- promote a sustainable balance between resources and the fishing capacity of the Community fishing fleet and promote a sustainable development of inland fishing;
- strengthen the competitiveness of the operating structures and the development of economically viable enterprises in the fishing sector;
- foster the protection and the enhancement of the environment and natural resources where related to the fisheries sector;
- encourage sustainable development and the improvement of the quality of life in areas with activities in the fisheries sector;
- promote equality between men and women in the development of the fisheries sector and fisheries areas.

Objectives are accompanied by so-called 'principles of assistance', which are laid down in Arts. 6 to 11 as follows:

- the Commission and the Member States shall ensure that assistance from the EFF is consistent with the policies and activities of the Community and complementary to other financial instruments of the Community [...].
- operations financed by the EFF shall comply with the provisions of the Treaty and the acts adopted thereunder.

---

37 A different procedure applies under the market organisation; see below.

38 Art. 1(1) of Regulation 2141/70.

- operations financed by the EFF shall not increase fishing effort.
- Member States and the Commission shall ensure that equality between men and women and the integration of the gender perspective are promoted during the various stages of the implementation of the EFF.[39]

These objectives are, again, complemented by procedural principles. According to Art. 8, the objectives of the EFF are to be pursued within a framework of close co-operation between the Commission, Member States and competent regional and local public and private partners. Arts. 9 and 10 require that the distribution of competences regarding the implementation of the operational programmes and the EFF budget are governed by the principles of proportionality and shared management.

It could be argued that some of these objectives and principles are incompatible. However, it will be explained in this work that all of these objectives can be pursued and promoted as long as the political players in the Community accept that promotion must not erode management efforts.

3.3 Promotion from 1970 to 1982 – the build-up

The adoption of 'Regulation 2141/70 laying down a common structural policy for the fishing industry' ('Regulation 2141/70') marked the beginning of the common structural policy.[40] This Regulation aimed at co-ordinating the existing national structural policies, and also empowered the Community to support the Fisheries sector. It was repealed in 1976 by 'Council Regulation (EEC) 101/76 laying down a common structural policy for the fishing industry'.[41] However, this new Regulation did not bring about any substantial changes. Community assistance at that time was also granted on the basis of various other regulations, described below.

3.3.1 Structure, scope and objectives of the central regulation

The scope of the first Regulation was somewhat limited. Member States were simply required to co-ordinate their structural policies by providing and exchanging information.[42]

The role of the Member States was to provide financial assistance,[43] and the Community merely established a regulatory framework setting out the conditions under which support could be granted.[44] Aid had to meet the objectives laid down in Art. 10, which were as follows:

---

39 See Art. 6(2), (3) and (5) and Art. 11 of Regulation 1198/2006.

40 OJ 1970 No. L2361/1.

41 OJ 1976 No. L20/19.

42 Arts. 1 and 6 Regulation 2141/70.

43 Churchill, *'EEC Fisheries Law'*, p. 222.

44 Art. 9(1) Regulation 2141/70.

- increasing productivity of fleets through restructuring fishing fleets and other means of production, in keeping with technical progress, and intensification of the search for new fishing grounds and new methods of fishing;
- adaptation of production and marketing conditions to market requirements;
- improvement, in step with technical progress, of the standard and condition of living of the population which depends on fishing as their livelihood.

Additional grants by the Member States (*i.e.*, other than provided under this regulation) were subject to Arts. 87-89 of the Treaty, which set out the Community's rules on state aid. These provisions aimed at maintaining fair competition between the national industries of Member States.[45]

Though structural aid was mainly financed and carried out by Member States, the Community also provided some support to the fisheries sector. Art. 10(2) allowed the Community to adopt measures on public assistance, financed by the European Agricultural Guidance and Guarantee Fund ('EAGGF'). Such measures served the same purpose as Art. 33(a) of the Treaty (ex Art. 39(a)), which was to increase agricultural productivity.[46]

3.3.2 Fleet renewals, modernisation of vessels and improvements to fishing ports

Requirements for EAGGF-funding of fleet renewals, modernisation of vessels and improvements to fishing ports were set out in the Guidance Section of 'Council Regulation (EEC) 17/64 on the conditions for granting aid from the EAGGF'.[47] Primarily designed to serve the agricultural sector, the wording of the provision was somewhat vague with respect to the funding of the fisheries sector.

According to Arts. 1(3) and 11(1), aid was available for projects designed to increase productivity, adapt and improve production conditions, adapt production to market outlets, improve the product quality, *etc.* Art. 14(1)(C) required that supported projects offered an adequate guarantee of lasting economic benefits.

---

[45] C. Nonnenmacher, *Die Gemeinsame Fischereipolitik der Europäischen Wirtschaftsgemeinschaft*, (Tübingen: 1990), p. 247; see also Art. 7 of the current central structural regulation 1198/2006 on the European Fisheries Fund, OJ 2006 No. L223/1.

[46] Art. 10(2) refers to Art. 6(1) of Council Regulation (EEC) No. 729/70 on the financing of the common agricultural policy, OJ 1970 No. L94/13.

[47] OJ 1964 No. 34/586.

From 1971 to 1977, Community financial aid totalled almost 65.2 million ecus for the modernisation and renewal of vessels; and an additional 2.6 million ecus had been spent on improving port infrastructure.[48]

In 1978, 'Council Regulation (EEC) 1852/78 on an interim common measure for restructuring the inshore fishing industry' entered into force. It established a funding programme that lasted until 1982.[49] The programme's aim was to promote the 'rational use of resources as well as the best use of production factors'.[50] In actual fact, however, it was designed to expand existing fisheries capacity.[51] Aid was granted for the purchase or construction of fishing vessels between 6 (12 in 1980) and 24 metres in length as well as for fishing vessels between 25 and 130 GRT.[52] From 1980, financial transfers were granted for the modernisation or conversion of existing vessels between 12 and 24 metres long for the purpose of rationalising fishing operations, better preserving catches or saving energy.[53] All projects had to demonstrate adequate proof of profitability, and funds were reportedly only available where fishing potentials in the relevant region allowed for this.[54] Underdeveloped regions enjoyed priority treatment.[55]

The total amount spent by the Community amounted to approximately 72 million ecus. According to the Commission, approximately 6% of the fleet segment has been renewed under the programme.[56]

3.3.3 Promoting the cod-fishing sector

Between 1973 and 1977, a special programme was set up to promote the global competitiveness of the European cod-fishing sector.[57] This

---

48 Churchill, *EEC Fisheries Law*, p. 210.

49 See OJ 1978, No. 211/30. The programme was supposed to last for one year, but was extended by Regulations 592/79, 1713/80, 2992/82 and 31/83 amending Regulation (EEC) No. 1852/78 on an interim common measure for restructuring the inshore fishing industry, OJ 1979 No. L78/5, OJ 1980 No. L167/50, OJ 1981 No. L299/24, OJ 1983 No. L5/1. The regulations introduced minor amendments, such as requiring vessels to be between twelve (and not six) and twenty-four metres in length.

50 See Reason (2) of Regulation 1852/78.

51 Churchill, *EEC Fisheries Law*, pp. 210-211.

52 Arts. 2 and 3 of Regulation (EEC) 1852/78.

53 See Art. 1 of Regulation 1713/80.

54 Arts. 1 and 4(1) of Regulation (EEC) 1852/78.

55 Art. 4(2) of Regulation (EEC) 1852/78.

56 The total amount of financial aid provided in this programme was 95 million ecus. 23 million ecus were allocated to aquaculture projects, Churchill, *EEC Fisheries Law*, pp. 211 and 217.

57 Regulation (EEC) 2722/72 on the financing by the EAGGF, Guidance Section, of the conversion projects in the salt cod-fishing industry, OJ 1972 L291/30 and Commission Regulation (EEC) 1462/73 concerning requests for aid to the EAGGF, Guidance Section, for conversion of the cod fishing fleet, OJ 1973 No. L145/11; Commission Regulation (EEC) 645/74 on procedures for implementing decision by the Guidance Section of the EAGGF to grant aid to the salt cod fishing industry, OJ 1978 No.L78/19.

was necessary because tariffs on foreign cod imports were abolished.[58] Aid was granted for projects to restructure production, processing and marketing activities in the sector.[59] Funding was also made available for retraining fishers affected by the changes.[60] In fact, 9.8 million ecus were spent on projects in France and Germany, with most of this money going towards vessel construction and modernisation.[61]

3.3.4 Aquaculture

The Community also promoted aquaculture to foster economic growth and increase export rates for the high value fish products of poor coastal regions. From 1971 to 1978, the EAGFF provided Community funding under Regulation 17/64 worth 5.53 million ecus. In 1978, the Community based its subsidies to the aquaculture industry on Regulation 1852/78, granting aid for the construction, equipment or modernisation of aquaculture establishments for the breeding or rearing of fish, crustaceans or molluscs in salt or brackish water.[62] Project approval was dependent upon adequate proof of profitability.[63] Regions lagging behind in their development enjoyed priority treatment.[64] The total amount spent by the Community amounted to 23 million ecus.[65]

3.3.5 Processing and marketing sector

Financial transfers from the Community were also allotted for processing and marketing of fish. Prior to 1978, only a small amount of aid, 12.6 million ecus, was allocated in accordance with Regulation 17/64. In 1977, 'Council Regulation (EEC) 355/77 on common measures to improve the conditions under which agricultural products are processed and marketed' was adopted.[66] The Regulation provided aid for projects relating to the treatment, processing or marketing of fisheries products.[67] According to Art. 6, such projects could include investments relating wholly or in part to buildings and equipment for improving the processing of fish products, the improvement of marketing channels and market analysis. None of them, however, applied to retail sales and

---

[58] See Reason (2) of Regulation 2722/72.

[59] Art. 1(3) of Regulation (EEC) 2722/72.

[60] Art. 8 of Regulation (EEC) 2722/72.

[61] Churchill, '*EEC Fisheries Law*', p. 210.

[62] Arts. 2 and 3 of Council Regulation (EEC) 1852/78 on an interim measure for restructuring the inshore fishing industry, OJ 1978 No. L 211/30.

[63] Art. 4(1) of Regulation 1852/78.

[64] Art. 4(2) of Regulation 1852/78.

[65] Churchill, '*EEC Fisheries Law*', pp. 217.

[66] OJ 1977 No. L51/1.

[67] Art. 1(1) of Regulation 355/77.

adequate guarantees of profitability were still required.[68] Overall, more than 80 million ecus of aid were distributed under Regulation 355/77 between 1978 and 1986.[69]

Table 4.2: Shares of Community, Member States and beneficiary contributions to the total amount of co-financed projects in the fisheries sector from 1970 to 1982[70]

Area	Legal base	Community share	National contribution	Beneficiary contribution	Total million ecus
Fleet renewal & modernisation from 1970 to 1977	Council Regulation 17/64	max. 25%	(-)	min. 30%	65.2
Fleet renewal & modernisation from 1978 to 1982	Council Regulation 1852/78	max. 25% max. 50% l.d.r.	(-)	min. 50% min. 25% l.d.r.	72
Port infrastructures	Council Regulation 17/64	max. 25%	(-)	min. 30%	2.6
Restructuring the cod-fishing sector	Council Regulation 2722/72	max. 25%	(-)	min. 50%	9.8
Aquaculture from 1970 to 1977	17/64	max. 25%	(-)	min. 30%	5.53
Aquaculture from 1978 to 1983	Council Regulation 1852/78	max. 25% max. 50% l.d.r.	(-)	min 30% min. 25% l.d.r	23
Processing and marketing from 1970 to 1977	Council Regulation 17/64	max. 25%	(-)	min. 30%	12.6
Processing and marketing from 1977 to 1982	Council Regulation 355/77	25%	min. 5%	min. 50%	about 40

l.d.r. = less developed regions

3.4 Promotion from 1983 to 1986: Build-up, territorial expansion and restructuring

In 1983, two Regulations and a Directive were adopted under the main structural Regulation 101/76.[71] These were: Council Regulation (EEC) No. 2908/83 on a common measure for restructuring, modernising, and developing the fishing industry and for developing aquaculture ('Regulation 2908/83');[72] Council Regulation (EEC) 2909/83 on measures to encourage

68 Art. 10(B) of Regulation (EEC) 355/77.

69 Churchill, *'EEC Fisheries Law'*, p. 218.

70 Sources: Regulations 17/64, 2722/72, 1852/78, 355/77, 1713/80 and Churchill, *EEC Fisheries Law*, 203-226.

71 Regulation 101/76 was repealed by Art. 34 of Regulation 2371/02.

72 OJ 1983 No. L290/1.

exploratory fishing and co-operation through joint ventures in the fishing sector ('Regulation 2909');[73] and Council Directive (EEC) 515/83 concerning certain measures to adjust capacity in the fishing sector ('Directive 515/83').[74] All Community aid was based on these Regulations and financed by the EAGGF.

3.4.1 Changes and Innovations

During the second phase, new types of assistance were adopted. Assistance was granted for the protection of coastal areas, exploratory expeditions and the temporary and permanent withdrawal of vessels. In addition, a programme to adjust fleet capacities to correspond to current resources was introduced, the so-called First Multi Annual Guidance Programme ('MAGP I').

3.4.2 Adjusting capacity – Multiannual Guidance Programme I (1983-86)

Arts. 3 to 7 and 14 of Regulation 2908/83 introduced a programme that would, for the two succeeding decades, serve as the main instrument for adjusting the size of the fisheries sector to correspond to the available resources. MAGP I and its successors established framework conditions for the promotion of the fleet. To be eligible for public funding, investments for the purchase, construction or modernisation of vessels as well as aquaculture projects had to be 'consistent with guidelines provided for in the MAGPs'.[75] MAGPs had already been mentioned in Regulations 2141/70 and 101/76, but had not been utilised up to this point. They were defined under Art. 2 of Regulation 2908/83 as follows:

> Multiannual Guidance Programme means a set of objectives, together with a statement of the means for attaining them, designed to restructure, modernise and develop certain fishing fleets and to develop aquaculture, in a Member State.

According to Art. 3 of Regulation 2908/83, Member States had to draw up MAGPs for the period of 1983 to 1986, which specified the methods, measures and facilities that would be used to attain a 'satisfactory balance between fishing capacity and the resources that were expected to be available'.[76] According to Art. 4, the MAGPs needed to include the following information:

- the initial situation and discernible trends as regards to the various categories of vessels of the fleet;

---

[73] OJ 1983 No. L290/9.

[74] OJ 1983 No. L290/15.

[75] Art. 8(1)(A) Regulation 2908/83. See First MAGPs of Member States at OJ 1985 No. L157/1; the Belgian, Spanish and Portuguese MAGPs are published in OJ 1985, No. L44/44; OJ 1986, No. L279/46 and OJ 1986, No. L205/46.

[76] Art. 3(a) of Regulation 2908/83.

- an overall estimate of the fishing capacity;
- an estimate of the future capacity of the fleet, including estimates of the number of vessels to be withdrawn, laid-up and commissioned, with an indication of their fishing capacity.

After drafting the MAGPs, Member States were to forward them to the Commission for approval. The Commission would then determine whether the particular MAGP constituted a 'suitable framework for projects which may qualify for financial support from the Community', on the basis of the following factors[77]:

- production potential;
- the measures for the conservation and management of fish stocks;
- demand for products concerned;
- Common Fisheries Policy guidelines.

The Commission's decision was to be made in consultation with the Standing Committee for the Fishing Industry, which was composed of Member States representatives. The Commission's decision applied immediately.[78] The result of this procedure for most Member States was that investments and removals had to be balanced. Practically, this meant that Member States were required to maintain capacities at 1983 levels.[79] However, despite the efforts taken under MAGP I, fishing power increased over the ensuing years. The reasons for this included: the inconsistency in methods for the measurement of capacity; the absence of annual targets, the lack of experience in the structural policy, and the absence of objectives for permanent withdrawals of existing vessels to compensate for new constructions and the low mandatory force of MAGP I.[80] It was for these reasons that the MAGP I is widely regarded as unsuccessful.[81]

3.4.3 Construction, modernisation and conversion of the fleet – Regulation 2908/83

The objective of Regulation 2908/83 was to restructure those segments of the Community fleet that were negatively affected by the establishment of the EEZ at international law. The Community tried to achieve this 'by renewal and, where necessary, economically appropriate expansion in line with actual catch potential'.[82] Financial incentives were also provided to encourage

---

77 Art. 5(2) of Regulation 2908/83.

78 Art. 21(3) of Regulation 2908/83.

79 A. Hatcher, 'Subsidies for European fishing fleets: the European Community's structural policy for fisheries 1971-1999' (2000) 24 *Marine Policy*, p. 130.

80 Y-H. Song 'The Common Fisheries Policy of the European Union, p. 556; Commission SEC(91) 2288 final, pp. 25-29.

81 Commission's SEC(91) 2288 final, p. 25.

82 See Reason (3) of Regulation 2908/83.

the growth of the aquaculture sector, based on the assumption that this industry would increasingly contribute to fish supplies.[83]

To be eligible for funding, construction, modernisation and conversion projects had to meet certain requirements. The applicant's fishing vessel had to be between 9 and 33 metres in length.[84] Aid was granted for modernisation and conversion only if the work to the vessel was substantial, undertaken to rationalise fishing operations, or to improve conditions of storage of catches. The costs of the projects had to be at least 20,000 ecus.

Aquaculture projects had to have enough capacity to ensure sustained commercial operation.[85]

Under to Art. 11, the following renewal, modernisation and aquaculture projects were given priority consideration for receiving aid:

(A) In respect of the purchase or the construction of new fishing vessels, the commissioning of vessels, money was given for those projects that were:
- intended to replace vessels more than 12 years old;
- intended to replace vessels lost as a result of accidents or wrecks, irreparably damaged, broken up or permanently withdrawn from the use of fishing vessels, or
- based on coastal areas where fishing is traditionally an important economic activity in particular in the region referred to in Annex VII to the Council Resolution 3 November 1976.

(B) In respect of the modernisation of fishing vessels, money was more likely to be granted for projects designed to encourage a more rational use of fuel or fuel-saving types of fishing and projects which are co-ordinated in their economic and technical aspects and projects aimed at improving the processing of catches.

(C) In respect of aquaculture, funds were to be granted for the launching of pilot projects designed to guide and develop production in the sector and, where appropriate, to facilitate retraining of fishermen.

When granting aid, the Commission was to consider factors like the diversification of the fishing sector, living, working and safety conditions on board, the membership of beneficiaries in producer organisations, environmental protection and consumer interests.[86]

The Regulation remained in force for three years, from the 1 January 1983 to 1 January 1986. The estimated total costs to the Community were 156 million ecus:[87] 118 million ecus were allocated to the construction and modernisation of fishing vessels, 34 million ecus for aquaculture and 4 million ecus for artificial

[83] See Reason (4) of Regulation 2908/83.

[84] Art. 9(1) of Regulation 2908/83.

[85] Art. 9(2) of Regulation 2908/83.

[86] Art. 11(2) of Regulation 2908/83.

[87] Art. 14(2) of Regulation 2908/83.

reefs. When Spain and Portugal gained Community membership, this total was increased to 230 million ecus.[88]

3.4.4 Protect and making fuller use of coastal areas – Regulation 2908/83

Regulation 2908/83 also introduced financial aid for protecting Mediterranean coastal populations who were dependant on fisheries.[89] The regulation provided for the installation of artificial structures for the purpose of restocking exploited species. Projects were to be set up within three nautical miles of the baseline.[90] To be eligible for aid, Member States were required to deliver a descriptive summary and outline on how the project would benefit the coastal population.[91] According to Art. 9(3), the projects had to target 'the installation of fixed or moveable components intended to constitute the boundaries of protected areas and to create the substrata on which or around which stocks of fish, crustaceans or molluscs may develop'. Additionally, all fishing activity within these protected areas was banned for the first three years of the project. Investments for aquaculture and artificial reefs amounted to approximately 44 million ecus.[92]

3.4.5 Promoting exploratory expeditions and joint-ventures – Regulation 2909/83

Regulation 2909/83 aimed at adjusting capacities to new or under-utilised fishing opportunities.[93] The Community encouraged the fishing industry to engage in two new areas by granting financial incentives. The first concerned exploratory fishing expeditions. The Community hoped to discover under-exploited stocks both within and outside of Community waters. Development in the second area aimed at establishing joint ventures between Community and national commercial fishing operators, particularly those working in third countries on the Mediterranean or at the West African coast.[94]

The regulation on exploratory fishing expeditions only applied to vessels over 24 metres in length, whose fishing activities in Community waters had been severely curtailed.[95]

---

88 See Art. 1(2) of Council Regulation (EEC) 3733/85 amending Regulation (EEC) No. 2908/83 on a common measure for restructuring, modernizing and developing the fishing industry and developing aquaculture, OJ 1985 No. L 361/78.

89 See Reason (5) of Regulation 2908/83.

90 Art. 2(c) of Regulation 2908/83.

91 Arts. 7 and 8(2) of Regulation 2908/83.

92 Commission CFP Report 1991, Annex XI, p. 34.

93 See Reasons (2) and (3) of Regulation 2909/83.

94 Arts. 1 and 2 of Regulation 2909/83.

95 See Reason (4) and Arts. 5 and 6(A) of Regulation 2909/83.

The measure on joint ventures allowed Member States to grant a 'co-operation premium' to their nationals, in instances where they participated in joint ventures with fishing operators from a third country on the Mediterranean or the West African coast. This premium was limited to the following:

- 25 ecus per gross registered ton per period of three consecutive month for temporary vessel transfers;
- 400 ecus per gross registered ton for the first 300 Tons, and 200 ecus per gross registered ton for each additional ton, for permanent vessel transfers.

The total level of subsidies allocated for exploratory fishing operations was estimated at 11 million ecus, while the total costs of the measure on joint ventures was estimated at 7 million ecus.[96] Both figures were later increased when Spain and Portugal gained Community membership to 13 million ecus for expeditions and 10 million for joint ventures.[97]

3.4.6 Subsidising temporary or permanent withdrawals – Directive 515/83

To adjust the overcapitalised Community fleet, Directive 515/83 encouraged Member States, on a voluntary basis, to establish a system of financial aid for measures relating to the temporary or permanent reduction of fishing capacity.[98]

Assuming that stocks would eventually recover, the Directive encouraged the temporary laying-up of vessels. Thus, fleet capacity was supposed to be retained until optimal production was again possible.[99] Arts. 3 and 4 regulated the financial aid scheme for the temporary reduction of fleet capacity. Art. 3 stated that a temporary reduction was to be attained by temporarily laying-up vessels, which were longer than 18 metres and which had engaged in at least 120 days of fishing over the preceding year. The aid provided by Member States was based on a daily laying-up premium, which was determined according to a formula set out in Art. 3(2). The total costs to the Community were estimated at 44 million ecus for these three years.[100] When Portugal and Spain gained Community membership, this amount was increased to 60 million ecus.[101]

Concerning permanent capacity reductions, Member States could grant their nationals financial aid for the scrapping of vessels, the permanent transfer of

96 Art. 11 of Regulation 2909/83.

97 Council Regulation (EEC) 3727/85 amending Regulation (EEC) No. 2909/83 in measures to encourage fishing and cooperation through joint ventures in the fishing sector, OJ 1985 No. L361/56.

98 Arts. 1 and 3 to 5 of Directive 515/83; see also Reasons (1) to (3) of Directive 515/83

99 See Reasons (4) to (6) of Directive 515/83.

100 Art. 9(4) of Directive 515/83.

101 Art. 1 of Council Directive (EEC) 590/85 amending Directive 515/83 concerning certain measures to adjust capacity in the fisheries sector, OJ 1985 No. 372/49.

vessels to third countries, or the assignment of vessels for purposes other than fishing. A final cessation premium was paid for vessels that were 12 metres or more in length, and which had engaged in fishing for at least 100 days during the preceding calendar year.[102] Expenditures were limited 650 ecus per gross registered ton. Estimated costs for the Community were 32 million ecus.[103] With the accession of Portugal and Spain, this amount was increased to 46 million ecus.[104]

3.4.7 Support for sardine canning plants

Upon accession, Portugal and Spain introduced their efficient sardine canning industries to the common market. As a result, sardine canning industries of the old Member States were put under severe pressure. In response, a special three-year programme was adopted,[105] which was aimed at promoting competitiveness of the old Member States sardine canning industries.[106] The total programme budget was set at 10 million ecus.

3.4.8 Integrated Programmes for the Mediterranean

In 1985, the Community adopted the Integrated Mediterranean Programme. This measure remained in force for seven years and aimed at supporting the socio-economic structures of southern regions in France, Italy and particularly Greece.[107] Financial aid was made available to all sectors of the economy, in particular, agriculture and the fisheries.[108] Art. 2(2) and Annex II (b) provided a detailed list of projects in the fisheries sector that were eligible for aid, and included funds for:

- restructuring, converting and modernizing parts of the fleet;
- improving infrastructure and port installations, including the biological protection of marine areas and the creation of marine parks;
- expanding aquaculture, including lagoon construction work;
- improving storage and processing facilities;
- encouraging the sale of fishery products in particular by advertising campaigns and
- intensifying research and vocational training and providing technical assistance.

---

102 Art. 5(1) to (3) of Directive 515/83.

103 Art. 9(4) of Directive 515/83.

104 Art. 1 of Directive 590/85.

105 Council Regulation (EEC) 3722/85 on a system of structural aid for the conversion of sardine canning plants, OJ 1985 No. L361/38.

106 Art. 4 of Regulation 3722/85.

107 Art. 1 and Annex I of Council Regulation (EEC) 2088/85 concerning the integrated Mediterranean programmes, OJ 1985 No. L197/1.

108 Art. 2 and Reason (5) of Regulation 2088/85.

To receive Community funds, Member States had to inform the Commission on the so-called Integrated Mediterranean Programmes.[109] The programmes were required to include multi-annual measures providing investments in the production sector, the creation of infrastructure, and the better use of human resources in the fisheries sector.[110] After examining the programmes, the Commission decided whether they would be approved. In determining the amount of assistance, the Commission would consider the actual needs of the region and of the economic and social development conditions. Less developed regions enjoyed preferential treatment.[111] Other criteria were, *i.a.*, the usefulness of the projects, the size of the Member States' contribution to the programmes, and the internal consistency of the plan. The Commission was supported by the Advisory Committee on Integrated Mediterranean Programmes, composed of Member State representatives and a representative of the European Investment Bank.[112]

The overall programme budget was set at 4,100 million ecus of which 466 million were allocated to fisheries projects.[113]

Table 4.3: Shares of Community, Member States and beneficiary contributions to the total amount of co-financed projects in the fisheries sector from 1983 to 1986.[114]

Area	Legal base	Community share	National contribution	Beneficiary contribution	Total million ecus
Fleet renewal & modernisation	Council Regulation 2908/83	max. 25% max. 50% (ldr)	(-)	min. 50% min 25% (ldr)	230
•Aquaculture •Artificial reefs	Council Reg. 2908/83	•max. 25% max. 50%(ldr) •max. 50%	(-) •min. 50%	•min. 50% min 25% (ldr) •min. 5%	44
Exploratory expeditions	Council Reg. 2909/83	50%	(-)	50%	13
Joint ventures	Council Reg. 2909/83	50%	(-)	50%	10
Adjustment of capacity	Council Directive 515/83	50%	Remain	remain	106 (permanent: 60, temporary: 46)

---

109 Art. 5(1) of Regulation 2088/85.

110 Arts. 1(1) and 2(1), (2) and (3) of Regulation 2088/85.

111 Art. 6(2) of Regulation 2088/85.

112 Art. 7 of Regulation 2088/85.

113 Art. 10 of Regulation 2088/85; Churchill, *EEC Fisheries Law*, p. 214.

114 Regulations 355/77, 2908/83, 2909/83, 2088/85, 3722/85, 3727/85, 3733/85 and Directives 515/83, 590/85; Commission of the European Communities, Report 1991 on the Common Fisheries Policy, SEC(91) 2288 final, Brussels, 18 December 1991; Churchill, *EEC Fisheries Law*, pp. 203-226.

Supporting sardine canning plants	Council Reg. 3722/85	50%	min. 5%	min. 25 or 35%	10
Integrated Mediterranean Programmes	Council Reg. 2088/85	max. 70%	Remain	remain	466
Processing and marketing	Council Regulation 355/77	25%	min. 5%	min. 50%	80.2

3.5 Promotion from 1987 to 1993 – Build-up, territorial expansion and restructuring

On 1 January 1987, Regulations 2908/83, 2909/83 and Directive 515/83 were replaced by a single Regulation: 'Council Regulation (EC) 4028/1986 on community measures to improve and adapt structures in the fisheries and aquaculture sector'.[115] The intention was for the Regulation to remain in force for ten years.[116] However, it underwent important modifications in 1990, and was finally replaced in 1993 with the general reform of the Community's structural funds.[117]

3.5.1 Changes and innovations

The Community adopted a second Multi Annual Guidance Programme. Under this initiative, Member States were not merely required to maintain the existing level of fleet capacity, but to reduce it by 2% in tonnage and 3% in engine power. Member States also had to consider the socio-economic consequences of their policies when setting up such programmes. Minor changes had been introduced to the promotion schemes on the construction, modernisation and conversion of the fleet. Assistance for the construction of trawlers was only available for vessels 12 to 30 metres long (before 9 to 30 metres), and preferential treatment for replacement support was only granted for vessels older than 15 years (compared to 12 years under Regulation 2908/83). The Community increased its subsidies for joint ventures, providing funding for projects on the Mediterranean and West African coasts, primarily serving Spanish and Portuguese fishers. The Community also granted money for finding and developing new markets. Aid for processing and marketing had to comply with sectoral plans, which included a wide range of views and comprehensive information on the fisheries and aquaculture sectors. In 1990, these aid schemes were amended. Assistance for the construction and modernisation of vessels was extended to other fleet segments and participation rates were lowered. Aid for temporary fleet reductions and fishing in non-Community waters was also

[115] OJ 1986 No. L376/7.

[116] See Art. 40 of Regulation 4028/86.

[117] Council Regulation (EEC) No. 3944/90 amending Regulation (EEC) No. 4028/86 on Community measures to improve and adapt structures in the fisheries and aquaculture sector, OJ 1990 No. L380/1.

increased. The Community also began supporting campaigns promoting the consumption of aquaculture products. Projects to protect and make fuller use of coastal areas were now supported on all Community coasts (not only the Mediterranean).

3.5.2 Adjusting capacity – Multi-annual Guidance Programme II (1987-91)[118]

The second Multi-annual Guidance Programme was a bit more ambitious with regard to adjusting the fleet's fishing capacity. Reductions of 3% in tonnage and 2% for engine power were instituted.[119] The programme design needed to take into account the following objectives:

- the establishment of a viable fishing fleet in line with the economic and social needs of the regions concerned and the foreseeable catch potential in the medium term;
- the adjustment of fishing activities to changes in consumer demand and to provide regular supplies for the market;
- the socio-economic consequences and the regional impact of developments foreseen in the sector concerned;
- the development of technically viable and profitable facilities for farming fish, crustaceans and molluscs.

Member States were also required to provide more detailed information to the Commission than was required under the preceding Regulation 2908/83 (see Annex I), including estimates on the importance of the fishing industry to the national economy, the current state of their fisheries resources and future estimates, the impact of the present situation on the fishing industry, and market trends and developments. To improve monitoring of the implementation of the MAGPs, Art. 5(1) required Member States to report annually. This information served as the basis for programme evaluations and amendments.[120] Enhanced transparency was achieved with the establishment of the Community's fishing vessel register in 1989.[121]

---

[118] Art. 3(2) of Regulation 4028/86 provided that the programme should cover the period from 1 January 1987 to 31 December 1991.

[119] Commission's CFP Report 1991, p. 26.

[120] This has been done quite often; see A. Hatcher, 'Subsidies for European fishing fleets: the European Community's structural policy for fisheries 1971-1999', p. 131, fn. 8.

[121] Commission Regulation (EEC) 163/89 concerning the fishing vessel register of the Community, OJ 1989 No. L20/5, replaced by Regulation (EC) 109/94, OJ 1994 No. L19/5; amended by Regulation (EC) 493/96, OJ 1996 No. L72/12; replaced by Regulation (EC) 2090/98 concerning the vessel register of the Community, OJ 1998 L 266/27 and by Regulation 2091/98 concerning the segmentation of the Community and fishing effort in relation to the multi-annual guidance programmes (MAGP), OJ 1998 No. L266/36; amended by Commission Regulation 839/2002 amending Regulation 2090/98 (...), OJ

Member States like Denmark, Germany, Spain, Ireland, Italy and Portugal were successful in achieving MAGP II reduction objectives, and capacity was generally stabilized. However, the Commission regarded MAGP II as a failure[122] for the following reasons:

- there was an insufficient classification of the fleet into categories related to the species caught, fishery zones and methods of fishing;
- the monitoring of the fleet was based on a limited number of physical capacity parameters only, without any consideration of the remaining parameters and fleet activity (fishing effort);
- there was a lack of short- and long- term objectives relating to the actual health of particular stocks;
- there were deficiencies in statistical data and inadequate measures to control fishing capacity and fishing effort;
- the programmes were non-obligatory, *i.e.*, to 'enforce' the MAGPs, the Commission could only refuse to provide Community aid;
- the programmes suffered from the fact that individual decisions were not subject to an overall strategy;
- the execution and compliance of measures supporting fleet renewal were insufficient;
- provisions regarding fleet modernisation did not consider the increase in fishing power due to technological progress;
- subsidies for the adjustment of capacities were too small.[123]

3.5.3 Promoting construction, modernisation and conversion of the fleet

Arts. 6 to 9 of Regulation (EC) 4028/86 resembled those in Regulation 2908/83, which provided funds for the purchase and the construction of new vessels. Projects were still supported if they formed a part of an approved MAGP and as long as they offered a satisfactory guarantee of yielding profits. All ships 9 metres or longer were eligible for aid. However, if the vessel was capable of trawling, the limit was raised to 12 metres.[124] Another minor change was introduced within the priority schemes: the replacement of vessels older than 15 years (compared to 12 under Regulation 2908/83) enjoyed preferential treatment. Such treatment was abolished for vessels based in areas where fishing was a traditionally important economic activity.

Aid was still available for modernisation projects that related to vessels over nine metres in length. Vessels capable of trawling were eligible for aid if they

---

2002 No. L135/5; replaced by Commission Regulation (EC) 26/2004 on the Community fishing fleet register, OJ 2004 No. L5/25.

[122] Commission's CFP Report 1991, p. 26.

[123] *Ibid.*, pp. 28-29.

[124] Art. 6 of Regulation 4028/86.

were over 12 metres long.[125] As under the preceding regulation, the purpose of projects had to relate to the rationalisation of fishing operations, better storage of catches, energy savings or the improvement of working conditions and crew safety.[126] Moreover, Art.9(3)(c)-(g) required Member States to ensure that projects 'did not exceed 50% of the value of a new vessel of the same type as the vessel concerned, relate to vessels having the necessary equipment for fishing operations and crew safety, and are implemented by qualified persons'.

Between 1986 and 1993, the Community spent about 232 million for the construction of new fishing vessels. Expenditures for the modernisation of vessels were approximately 143 million during that period.[127]

Measures to promote aquaculture were also financed under the new legal framework.[128] These projects were still required to form part of the MAGPs and some assurance of profitability was still required.[129] Community expenditures for aquaculture projects and artificial reefs from 1987 to 1990 amounted to approximately 149.4 million ecus.[130]

3.5.4 Protecting and making greater use of coastal areas

The territorial scope of measures for structural works in coastal areas was expanded under the new regime in comparison to the earlier Regulation 2908/83. Financial aid was no longer limited to coastal projects in the Mediterranean, but was also made available for projects on all Community shores. Under Art. 11(1)(b), the Commission was entitled to grant subsidies to public, semi-public or private projects relating 'to measures to protect and make fuller use of coastal marine areas by the installation, not deeper than 50 metre isbobath, of fixed or moveable structures for the delimitation of the protected areas and for the protection or development of fishery resources'. To qualify for aid, the projects needed to provide adequate scientific monitoring and institute a three-year fishing moratorium within the protected area. Moreover, projects had to be implemented by recognised producer organisations or similar bodies.[131]

3.5.5 Promoting exploratory expeditions and joint-ventures

Art. 13 of the new main Regulation changed the definition of 'exploratory fishing voyage', shifting the emphasis from 'prospecting new or

---

[125] Art. 9(3)(a) of Regulation 4028/86.

[126] Art. 9(3)(a) of Regulation 4028/86.

[127] Hatcher, 'Subsidies for European fishing fleets', p. 135.

[128] Art. 10 of Regulation 4028/86.

[129] Art. 11 of Regulation 4028/86.

[130] M. Holden, *The Common Fisheries Policy*, (Oxford: 1996), p. 30.

[131] Art. 11(5) of Regulation 4028/86.

underutilised fishing grounds' to 'assessing the profitability of regular, long-term exploitation of the fishery resources in a given area'.[132]

Under Art. 14, aid was eligible for exploratory fishing voyages on the high seas, in Community and in third country waters, where fishing was permitted under fisheries agreements. Under the previous Regulation, aid was available for projects relating to fishing vessels longer than 24 metres and which lasted at least 30 days.[133] The new measure merely required that the vessels be a minimum of 18 metres long, however, journeys now had to last at least 60 days.[134] In addition, expeditions had to target zones which promised profitability, and at least one scientific observer had to be on board the vessel. Priority was given to projects that were organised by ship owners or owners of processing or marketing companies.[135]

Joint ventures continued to be financed; however, under Art. 18, they were no longer limited to third countries located on the Mediterranean or West African coast, and were extended to all third countries that maintained relations with the Community on fishing matters. According to Art. 19(1), eligible projects had to relate to 'the catching and, where appropriate, processing and/or marketing of the species concerned, together with the supply of know-how or the transfer of technology where the latter were relevant to the fishing operations in question'. Regulation 4028/86 raised the amount of available subsidies for joint ventures, setting a co-operation premium at 40 ecus per gross registered ton for a period of three consecutive months (compared to 25 ecus provided under Regulation 2909/83).[136]

3.5.6 Subsidising temporary and permanent withdrawals

Regulation 4028/86 basically referred to the same reduction schemes that had been adopted under Directive 515/83. Member States were encouraged to provide premiums for temporary or permanent withdrawals of fishing vessels under almost identical conditions; only minor changes were introduced. This notwithstanding, lay-up periods and premiums were set out more clearly in Annex IV and Community reimbursement of Member States for final cessation or laying-up premiums was reorganised. The Commission had to decide by 1 April of each year, after receiving Member States' estimated expenditures, the extent to which it would provide reimbursements.[137] In total,

---

[132] Art. 2 of Regulation 2909/83 and Art. 13 of Regulation 4028/86.

[133] Art. 6 of Regulation 2909/83.

[134] Art. 14(2) of Regulation 4028/86.

[135] Art. 14(4) of Regulation 4028/86.

[136] Provisions regarding the definitive transfer of vessels to third countries had been placed under the provision on the permanent withdrawal of vessels, see Art. 24(1)(b).

[137] Arts. 25 and 26 Regulation 4028/86.

the Community paid about 496 million ecus for permanent withdrawals and 67 million ecus for temporary withdrawals.[138]

3.5.7 Financial aid for material investment in facilities at fishing ports

Financial aid for the material investment for facilities at fishing ports was available according to the conditions laid down in Art. 27(1) and (2) of Regulation 4028/86. Project investments needed to be 'designed to bring about lasting improvement in conditions for the production and initial sale of fish products'.[139] They formed a part of national programmes for improving the treatment, processing or marketing of fisheries products. The programmes were proposed by producer organisations or other competent bodies authorised by Member States, and then needed Commission approval.[140]

3.5.8 Financial aid for the search of new market outlets

To promote the consumption of fish products derived from surplus and under-fished species, Art. 29 allowed the Commission to grant financial aid for specific projects. This provision required that these projects were a) proposed by public, semi-public or private bodies representing the fisheries sector in one or more Member States and implemented under the direct supervision of such bodies, and (b) related to collective measures which were not oriented towards any commercial brands and did not make reference to a particular country or region.

3.5.9 Specific measures

The Commission was encouraged to implement 'specific measures' relating to structures of the fishing industry. Such measures aimed at removing or alleviating handicaps relating to fishing activities throughout the Community. The broad wording of this provision allowed the Commission to craft a flexible response to potential structural problems.[141] As a safeguard, Art. 47 required the Standing 'Committee for the Fishing Industry' to be consulted before such measures could be instituted.[142]

---

[138] Hatcher, 'Subsidies for European fishing fleets', p. 135.

[139] Art. 27(2)(c) of Regulation 4028/86.

[140] Art. 27(2)(a) refers to Art. 2 of Regulation 335/77.

[141] See Reason (13) of Regulation 4028/86.

[142] The Regulation did not specifically state how much funds were available for such measures, and it was not possible to find out whether the Commission had made use of its competence.

3.5.10 Subsidising the processing and marketing segment

The promotion of processing and marketing structures was governed by Regulation 355/77 until 1989. The Regulation was then replaced by 'Regulation (EEC) 4042/89 on the improvement of the conditions under which fishery and aquaculture products are processed and marketed.'[143] The Community's central objective in adopting this new measure was to promote the development of a Community processing industry for fishery and aquaculture products, which particularly aimed at producing high-value processed fish products. This strategy was adopted in response to a huge Community trade deficit for fish.[144]

Aid was granted for projects which formed an integral part of the so-called Community support framework.[145] This framework was based on the above-mentioned sectoral plans. According to Arts. 2 and 3, the plans had to include perspectives on the development of the fisheries and aquaculture sectors, in particular, looking at ways to:

- create a viable industry;
- develop and adapt facilities for the processing and marketing of fishery and aquaculture products and improve their quality and hygiene conditions in order to increase productivity and added value; and
- account for the socio-economic needs of the fishing and aquaculture industry.

Based on the sectoral plans, the Commission adopted, in agreement with the Member States, a support framework for fishery and aquaculture products.[146] Projects eligible for aid were listed in Art. 9. Investments could be provided, for example, for the improvement of facilities for cutting, salting, drying, smoking, packing, storing, etc., auction markets, new processing techniques, and facilities for improving hygiene, etc. Costs for the construction and acquisition of immoveable property (other than purchase of land), the purchase of machinery and equipment, fees for architects, engineers and consultants, etc. could also be reimbursed.[147] Certain projects (listed in Art. 11) enjoyed preferential treatment. All projects had to offer adequate assurance of technical and economic viability as well as guarantee that the bulk of raw materials originated in the Community.[148] The total amount of aid provided for processing and marketing activities between 1986 and 1990 was about 184 million ecus.[149]

---

[143] OJ 1989 No. L388/1.

[144] See Reason (12) of Regulation 4042/89.

[145] Art. 10(2) of Regulation 4042/89.

[146] Art. 5(2) of Regulation 4042/89.

[147] Art. 10(4) of Regulation 4042/89.

[148] Art. 10(2) of Regulation 4042/89.

[149] Commission's CFP Report 1991, Annex XI, p. 34.

3.5.11 1990 Reform – Changes and innovations

Existing measures underwent significant amendments as a result of the adoption of 'Council Regulation (EEC) 3944/90 amending Regulation (EEC) No. 4028/86 on Community measures to improve and adapt structures in the fisheries and aquaculture sector'.[150] Changes were introduced in four areas. First, aid schemes for the construction and modernisation of vessels were extended to other fleet segments, again lowering Community participation rates. Distribution of the aid was based on a more differentiated allocation key. Second, Community participation rates concerning permanent and temporary fleet reductions were raised. Third, fishing in non-Community waters was increasingly promoted. Fourth, promotion campaigns aiming at raising the consumption of aquaculture products were introduced.[151]

According to amended Art. 6(2)(b) of Regulation 4028/86, fishing vessels over 5 metres long were eligible for construction and modernisation aid (in the past, vessels needed to be 9 metres long). However, Community and Member States' participation rates for rebuilding, renewing and modernisation of the fleet were lowered (see Table 3b).

The conditions under which aid was granted for fleet reductions were altered as well. Under amended Art. 23(a), premiums for temporary lay-up were provided for Community vessels less than 12 metres (formerly, 18 metres). The total lay-up period was extended from a maximum of 300 days to 400 days per vessel at the end of 1993.[152] Regarding the permanent withdrawals of fishing vessels, funding was extended to vessels less than 9 metres in length.[153] The Community participation rate was increased to 70% for scrapping premiums.[154]

Art. 17a introduced 'redeployment operations', which were defined as 'any fishing operation carried out for commercial purposes in a given area with a view to the exploitation of fishery resources, primary consideration being given to the supply of the Community market'. According to Art. 17b, Community financial aid would be granted for redeployment operations carried out in third country waters or the high seas. Operations had to last between 60 and 220 days and needed to be carried out by vessels which were longer than 12 metres in length and in operation for more than 5 years. The new element here was that financing was provided for fishing in the waters of third countries with which the Community did not yet have fisheries agreements[155] and for fishing in

[150] OJ 1989, No. L380/1.

[151] See Reasons (4), (12) and (15).

[152] Art. 23(d) of Regulation 4028/86 as amended by Regulation 3944/90.

[153] Art. 24(1) of Regulation 4028/86 as amended by Regulation 3944/90.

[154] Art. 26(5) of Regulation 4028/86 as amended by Regulation 3944/90.

[155] Art. 17b(1)(a) of Regulation 4028/86 as amended by Regulation 3944/90 provides: '[...] with which the Community had not concluded a fisheries agreement but maintained relations'.

third countries or high seas waters beyond the existing agreements and fishing limitations.[156]

Joint-enterprises were also introduced.[157] Under this article, 'joint enterprise' was defined as a 'company incorporated under private law comprising one or more Community shipowners and one or more partners from a third country with which the Community maintains an access agreement'. Joint enterprises were eligible for aid only if they related to vessels over 12 metres long, which were transferred permanently to a third country and which had been in operation for more than five years.[158]

Art. 29 also introduced funds promoting the disposal of excess supplies of aquaculture products, which saw rapid growth in production in the prior years. The Commission was to adopt detailed rules assisted by the Standing Committee for the Fishing Industry involving representatives of the Member States.

Table 4.4.a: Shares of Community, Member States and beneficiary contributions to the total amount of co-financed projects in the fisheries sector from 1987 to 1990 (to 1993)[159]

Area	Legal base	Community share	National contribution	Beneficiary contribution	Million ecus
Fleet renewal & modernisation	Council Reg. 4028/86	• 20% vessels ≤ 33m • 35% vessels ≤ 33m (ldr) • 10% vessels ≥ 33m • 25% vessels ≥ 33m (ldr)	• 10–30% • 10–30% • 10–30% • 10–30%	remainder	375 (to 1993), (construction: 232; modernisation: 143)
• Aquaculture • Artificial reefs	Council Reg. 4028/86	• 25% 40% (ldr) • max. 50%	• 10–25% 10–30% •10–35%	remainder	149.4 (to 1990)
Exploratory expeditions	Council Reg. 4028/86	20%	10–20%	remainder	?
Joint ventures	Council Reg. 4028/86	20%	10–20%	remainder	?

156 Art. 17b (1)(b) and of Regulation 4028/86 as amended by Regulation 3944/90 provides: '[...] with which the Community has concluded fisheries agreements in so far as the project does not qualify for a Community aid with the same purpose under the common fisheries policy', 'waters which do not fall within the sovereignty or jurisdiction of any State, in so far as such operations do not involve the catching of species subject to a quota assigned to the Community'.

157 Art. 21a of Regulation 4028/86 as amended by Regulation 3944/90.

158 Art. 21b of Regulation 4028/86 as amended by Regulation 3944/90.

159 Sources Regulations 355/77, 4028/86 and 4042/89; Commission of the European Communities, Report 1991 on the Common Fisheries Policy, SEC(91) 2288 final, Brussels, 18 December 1991; Hatcher, 'Subsidies for European fishing fleets', p. 135.

Adjustment of capacity	Council Reg. 4028/86	50%	Remain	remainder	563 (to 1993), (permanent: 497: temporary: 66.7)
Material investment in facilities in fishing ports	Council Reg. 4028/86	max. 25% max. 35 or 50% (ldr)	5–25% 5–30% (ldr)	remainder	?
Search of new market outlets	Council Reg. 4028/86	max. 50% (double Member States contribution)	max. 25%	remainder	?
Processing and marketing	Council Regs. 355/77 and 4042/89	max. 30% max. 50% (ldr)	min. 5%	min. 45% min. 25% (ldr)	184 (to 1990)

l.d.r. = less developed regions

Table 4.4.b: Modifications of Shares of Community, Member States and beneficiary contributions for the period from 1990 – 1993 in Regulation 4028/86 by Regulation 3944/90[160]

Area	Legal base	Community share	National contribution	Beneficiary contribution
Fleet renewal & Modernisation	Regulations 4028/86, 3944/90	•20% vessels ≤ 9m •35% vessels ≤ 9m (ldr) •15% vessels 9–33m •30% vessels 9–30m (ldr) • 5% vessels ≥ 33m •20% vessel ≥ 33m (ldr)	•5–25% vessels ≤ 9m •5–25% vessels ≤ 9m (ldr) •5–25% vessels 9–33m •5–25% vessel 9–33m (ldr) •5–25% vessels ≥ 33m •5–25% vessels ≥ 33m (ldr)	(-) (-) (-) (-) (-) (-)
Adjustment of capacity	Regulations 4028/86, 3944/90	•50% •70% for scrapping	(-)	remain

l.d.r. = less developed regions

3.6 Promotion from 1994 to 1999 – Build-up, territorial expansion and restructuring

Regulation 4028/86 was replaced on 1 January 1994 with the adoption of the Single European Act ('SEA'). This required co-ordination of all Community structural funds. From 1 January 1994 onwards, all structural measures were granted under the so-called Financial Instrument for Fisheries Guidance ('FIFG').[161] Pertinent FIFG measures were laid down in four new

160 Regulations 4028/86 and 3944/90.

161 See Fischer, 'Fischereipolitik', p. 26; See also wordings of Art. 161 (ex Art. 130 d) of the Treaty as well as Reason (3) of 'Council Regulation (EEC) No. 2052/88, OJ 1988 No. L185/9.

regulations. The general framework was set out in 'Council Regulation (EEC) No. 2052/88 on the tasks of the structural funds and their effectiveness and on the coordination of their activities between themselves and with the operations of the European Investment Bank and other existing financial instruments'[162], and 'Council Regulation (EEC) 4253/88 laying down provisions for implementing Regulation 2052/88 [...]'.[163] Under these Regulations, 'Council Regulation (EEC) No 2080/93 laying down provisions for implementing Regulation (EEC) No. 2052/88 as regards the financial instruments of fisheries guidance' was adopted. This regulation provided more details on the provision of Community structural aid in the fisheries sector.[164] Concrete terms and conditions for the application of structural measures were provided by 'Council Regulation (EC) No. 3699/93 laying down the criteria and arrangements regarding Community structural assistance in the fisheries and aquaculture sector and the processing and marketing of products'.[165]

3.6.1 Changes and innovations

Important changes were introduced under the new regime. In 1994, all assistance to the fisheries sector became subject to the newly introduced 'Community Programmes', which had to be consistent with the tasks and objectives of the FIFG, the CFP and the MAGPs. Two new MAGPs were also adopted, covering the periods 1992 to 1996 and 1997 to 2002. Both required Member States to follow more stringent monitoring obligations and more ambitious fleet reduction objectives. The promotion of the construction, modernisation and conversion of the fleet also underwent important changes. Payments for the construction of vessels was made subject to a differentiated distribution pattern. The fleet was segmented and different categories of vessels received different amounts of financial aid. Financial support for the modernisation of vessels shifted towards rationalisation, more selective fishing gear, and safety and control measures; and aid for the exploration of fishing opportunities ended. Money was also provided for the reduction of fleet capacities, and grants became available for scrapping vessels, the permanent transfer of vessels to third countries and the permanent re-assignment of ships for uses other than fishing. Payments for the reduction of the fleet became subject to a more differentiated distribution pattern. Aid for market studies relating to the prospects for marketing Community products in third countries was made available. The Community also introduced two new categories of subsidies: first, it promoted operations carried out by trade members (requiring that operations serve the collective interest); and, second, in 1994 and again in 1998, social measures

162 OJ 1988 No. L185/9.

163 OJ 1988 No. L374/1.

164 OJ 1993 No. L193/1.

165 OJ 1993 No. L346/1.

were introduced to cushion the effects of the restructuring of the Community's fishery sector.

3.6.2 Sectoral Plans, aid applications and Community Programmes

The introduction of the FIFG changed the procedures for establishing structural aid schemes under the CFP. Under Art. 2 of Regulation 3699/93, the distribution of subsidies was now subject to the so-called Community Programmes. These programmes were negotiated between the Commission and the Member States in two steps, resulting in a binding Commission decision.[166] First, Member States forwarded proposals on their future national structural policies to the Commission, the so-called 'sectoral plans' and 'aid applications'. They covered a period of six years, starting from 1 January 1994.

Sectoral plans were required to provide information on the status of the sector, a summary of existing funds and their impacts, and had to include a strategy for the adjustment of the fishery structures (including an anticipated impact assessment) and a description of the means to attain this adjustment.[167] The plans could relate to measures providing legal and financial Community assistance for the adjustment of fishing effort, the renewal and modernisation of the fleet, aquaculture, for enclosed seawater areas, for fishing port facilities, product processing or marketing and product promotion etc.[168]

Aid applications described all the concrete measures that were planned. They also specified the form of assistance required.[169]

After receiving the plans and the aid applications, the Commission would evaluate them in terms of their consistency with FIFG objectives (*i.e.*, achieving a balance between resources and their exploitation, strengthening competitiveness of structures, viable enterprises, improving market supply and adding value to fisheries and aquaculture products).[170] The plans also had to comply with the general objectives of the CFP as well as with the provisions of the MAGPs.[171] Finally, the Commission had to give consideration to the effectiveness of the administrative and financial mechanisms to ensure proper implementation and determine the precise arrangements for providing assistance.[172]

The Commission's decision had to be based on negotiations with the Standing Management Committee for Fisheries Structures, and involve Member States and with the competent national, regional and local authorities. This

[166] Art. 4(2) of Regulation 3699/93.

[167] Art. 3(2) and Annex I of Regulation 3699/93.

[168] Art. 3(2) and Annex I(3) of Regulation 3699/93

[169] Art. 3(2), second paragraph of Regulation 3699/93 in combination with Art. 5 of Council Regulation (EEC) 2052/88 and 14(1) and (2) of Council Regulation (EEC) 4253/88.

[170] Art. 4(1) of Regulation 3699/93 and Art. 1(2) of Regulation 2080/93.

[171] Art. 4(3) of Regulation 3699/93.

[172] Art. 4(1) of Regulation 3699/93 and Art. 14(3) of Regulation 4253/88.

approach was referred to as the 'partnership procedure'.[173] The Commission's decisions regarding the proposals could be appealed to the Council, so that Member States had the final say.[174]

3.6.3 The third and fourth Multiannual Guidance Programmes (1993-02)

There were several critical reviews of fleet policy that had been published prior to the adoption of MAGP III. The first was written by an independent expert group which recommended that in order to achieve a 'competitive Community fishing fleet which is in balance with available fish resources at the latest by 2002' capacity reductions of not less than 40% were required.[175] Second, in 1991 and 1993, the Commission and the Court of Auditors released the results of their inquiries into fleet policy under MAGP I and II. They were particularly critical of the ineffective and inadequate targets for fleet reductions.[176] In this regard, MAGP III and IV were considered to be more ambitious. They were adopted under Regulations 3699/93. MAGP III entered into force on the 1 January 1993 and expired on 31 December 1996, and MAGP IV was adopted on 1 January 1997 and was in force until 31 December 2002.[177] MAGP III and IV both tried to reduce fleet size in accordance with the specific conservation needs of the fisheries. To this end, fleet segments were assessed in the light of their impact on fishing mortality.[178]

---

[173] Art. 4(1) of Regulation 2052/88. The Standing Committee for Fisheries Structures replaced the Standing Committee for the Fishing Industry under Art. 11 of Regulation 101/76; see Art. 7(2) of Regulation 2080/93.

[174] Art. 8 of Regulation 2080/93.

[175] The expert group released the so called 'Gulland Report' in 1990, see H.Y. Song 'The Common Fisheries Policy of the European Union', pp. 559.

[176] See Com SEC(91) 2288 final, Report 1991 from the Commission to the Council and the European Parliament on the Common Fisheries Policy ('Commission Report – 1991'); Court of Auditors, Special Report No. 3/93 concerning the implementation of the measures for the restructuring, modernisation and adaptation of the capacities of fishing fleets in the Community together with Commission's replies ('Court of Auditors Report – 1993'), OJ 1994 No. C2/1, pp. 18. et seq.

[177] See Art. 5(3) of Regulation 3699/93 and Council Decision (EC) 15/94 relating to the objectives and detailed rules for restructuring the Community fisheries sector over the period 1 January 1994 to 31 December 1996 with a view to achieving a lasting balance between the resources and their exploitation, OJ 1994 No. L10/20. See Art. 9 of Council Decision (EC) 413/97 concerning the objectives and detailed rules for restructuring the Community fisheries sector for the period from January 1997 to 31 December 2001 with a view to achieving a balance on a sustainable basis between resources and their exploitation, OJ 1997 No. L.175/27; Council Decision (EC) 70/2002 amending Decision 97/413 concerning the objectives and detailed rules for restructuring the Community fisheries sector for the period from 1 January 1997 to 31 December 2001 with a view to achieving a balance on a sustainable basis between resources and their exploitation, OJ 2002, No. L031/77.

[178] See Reasons (1-3) of Council Decision 15/94 and Reasons (1-2) of Council Decision 413/97.

The Council decided at the end of 1993 that, under MAGP III, each Member State's fishing effort in terms of fleet size had to decrease by the end of 1996, as follows:

- 20% for trawlers bottom trawling for demersal stocks;
- 15% for dredgers and beam trawlers targeting benthic stocks;
- no increases for the other segments of the fleet.[179]

Art. 1(2) of the Decision states that at least 55% of the required effort reductions had to be achieved through a reduction in capacity. The remaining 45% could be attained through activity reductions. However, not a single Member State made use of the latter option.[180]

In the end, fleet reductions under MAGP III amounted to approximately 15% in terms of tonnage and about 10% for engine power. At the end of 1996, only the United Kingdom and the Netherlands had failed to meet their fleet reduction obligations.[181] On this basis, the programme was regarded as relatively efficient and successful.[182] Nevertheless, preparing MAGP IV, a follow-up report by an independent expert and as second Commission report both stated that some aspects of the MAGP III were still inadequate, and that overcapacity was still one of the main reasons for overfishing in the Community.[183] Accordingly, the Commission proposed fleet cutbacks of up to 40% under MAGP IV.[184]

The Council laid down fleet reduction targets for MAGP IV in 1997.[185] To increase effectiveness, MAGP IV provided for more elaborate and specific provisions and requirements regarding different fleet segments, different stocks, multi-species fisheries, gear types (*i.e.*, active and passive) and specific guidelines for effort reduction schemes.[186] Art. 2(2) stipulated that by December 2001, at the latest, the fishing effort of each Member State needed to be reduced to the following pilot reduction rates:

---

179 See Council Decision 15/94.

180 H.Y. Song 'The Common Fisheries Policy of the European Union', pp. 562.

181 See Annual Report to the Council and to the Parliament on the Progress of the Multi-annual Guidance Programmes for the Fishing Fleet at the End of 1996, COM(97) 352 final, p. 55-56.

182 Commission Communication Com(2001) 135 final Green Paper on the Future of the Common Fisheries Policy, vol. I, p. 11; see also H.Y. Song The Common Fisheries Policy of the European Union', pp. 560-561; Annual Report to the Council and to the Parliament on the Progress of the Multiannual Guidance Programmes for the Fishing Fleet at the End of 1996, COM(97) 352 final, p. 57.

183 The expert group released the 'Lassen Report' in 1996; see H.Y. Song The Common Fisheries Policy of the European Union', pp. 562; Commission Communication Com(97) 352 final.

184 H.Y. Song, The Common Fisheries Policy of the European Union', pp. 563.

185 Adopted under Art. 11 of Regulation 3760/93 and Art. 5(2) of Regulation 3699/93 in mid-1997; see Council Decision (EC) 413/97 concerning the objectives and detailed rules for restructuring the Community fisheries sector for the period from 1 January 1997 to 31 December 2001 with a view to achieving a balance on a sustainable basis between resources and their exploitation, OJ 1997 No. L175/27.

186 See Reasons 2, 6, 7, 9, 13 and Art. 1 and Art. 2(5) of Council Decision (EC) 413/97.

- 30% for the stocks defined as depletion risk in Annex I;
- 20% for the stocks defined as overfished in Annex I;
- no increase in fishing effort for stocks defined as fully exploited in Annex I;
- No increase in fishing effort in respect of stocks for which the situation is unknown (except in cases where Member States can identify additional fishing opportunities of these stocks).[187]

The starting point for reductions in targets were the objectives fixed in previous programmes.[188] However, Member States could exempt fishing vessels over 12 metres long (with the exception of trawlers) from the reduction scheme.[189] They were also allowed to choose their preferred means of reducing fishing effort, cutting either capacity or fishing activity.[190]

Art. 7(1) of Regulation 3699/93 provides that where the reductions in capacity financed by official aid lead to overachievement of the MAGP-objectives, the new situation brought about solely as a result of that aid may not be invoked to bring into service new capacity. This provision did not apply to small local coastal fishing fleets made up of vessels under 220 kw, for which fisheries quotas have not been set at Community level. With regards to such vessels, Member States were permitted to grant some state aid according to the schemes provided in Annex IV. Another important measure was adopted in Art. 7(2): it required that Member States ensure that, for each fleet segment, aid allocated for modernisation and construction would not result in an increase in fishing effort.

During the period of MAGP IV, the entire Community fleet was reduced to about 19% in terms of tonnage and 13% in terms of engine power.[191] However, MAGP IV was also not regarded as a success. The Commission stated that MAGP IV had not been ambitious enough in terms of reduction objectives, and that MAGPs, in general, are difficult and costly to administer.[192] Reductions had not translated into significant decreases of fishing pressure. Others have attributed the lack of success to the adverse effects of subsidies (which increase fishing pressure), the continuous change in fisheries dynamics, complex fleet segmentation, the technological progress and management failures.[193]

---

[187] These provisions can be seen as an application of the 'precautionary approach'; see also Reason 8 of Council Decision 413/97.

[188] Art. 7(1) of Council Decision 413/97.

[189] Art. 3 of Council Decision 413/97.

[190] Art. 5 and Reason 14 of Council Decision 413/97.

[191] Report from the Commission to the Council and the European Parliament on the intermediate results of the multi-annual guidance programmes fort the fishing at 30 June 2002, Com(2002) 483 final, p. 23.

[192] Commission Communication Com(2001) 135 final Green Paper on the Future of the Common Fisheries Policy, vol. I, p. 11.

[193] Erik Lindebo, 'Role of Subsidies in EU Fleet Capacity Management' (2005) 20 *Marine Resource Economics*, p. 454.

3.6.4 Promotion of construction, modernisation and conversion of the fleet

The new set of regulations continued to provide the opportunity for Member States to adopt measures that supported the renewal and modernisation of fishing vessels.[194] However, conditions had changed significantly.

When granting aid for the construction of vessels, Member States had to consider the requirements of regulations and directives on hygiene and safety. Priority was given to vessels that used the most selective fishing gear and methods.[195] Assistance was provided according to the allocation key laid down in Annex IV of Regulation 3699/93.

Table 4.5.:

Fleet segment (GRT)			Premium for a fifteen year old vessel (ecus)			
0	≤	25	6,215	/GRT		
25	≤	50	5,085	/GRT	+	28,250
50	≤	100	4,520	/GRT	+	56,500
100	≤	400	2,260	/GRT	+	282,500
400 and over			1,130	/GRT	+	734,500

Community aid did not exceed an additional 37.5% of these sums. For vessels with a steel or glass fibre hull, the amount could be increased to 92.5%.

Investments for modernising vessels were required to relate to the rationalisation of fishing operations, in particular, the use of more selective gear and methods, the improvement of preserving techniques on board, the improvement of working conditions, and improved safety and equipment to monitor the vessel's fishing activities. In addition, investments would only be financed, if they covered vessels less than 30 years old.[196]

Overall, construction and modernisation projects were required to comply with the MAGP's reductions schemes.[197] Exceptions existed only for modernisation projects that were eligible for aid under the prior Regulation 4028/86 and which were applied for before 1 January 1994.[198]

Financial transfers were also available to encourage capital investments in aquaculture projects, basically under the same conditions as before.[199] More detailed conditions were prescribed in Annex III. Community participation rates were increased (see Table 4).

[194] Art. 10(1) and (2) of Regulation 3699/93.

[195] Annex III, 1.3 of Regulation 3699/93.

[196] Annex III, 1.4 of Regulation 3699/93.

[197] Art. 10(1) and (2) of Regulation 3699/93.

[198] Art. 10(2) of Regulation 3699/93 and Art. 9 (1) of Regulation 2080/93.

[199] Art. 11(1) of Regulation 3699/93.

3.6.5 Protecting and making fuller use of coastal areas

Member States were permitted to take measures to encourage capital investment for the 'protection and development of marine resources in coastal waters, in particular by installation of fixed or moveable facilities to enclose protected underwater areas'.[200] Investments needed to include scientific monitoring for at least five years, and had to be carried out by public or publicly recognised institutions or bodies. Furthermore, the investment needed to create lasting economic benefits and offer an adequate guarantee of technical and economic viability.

3.6.6 Promoting exploratory expeditions and joint-ventures

Regulation 3699/93 discontinued aid schemes for the exploration of new fishing opportunities. Financial transfers for temporary joint ventures and permanent joint enterprises were continued.[201] According to Annex III, such operations had to meet the following requirements:

- they had to involve vessels of more than 25 GRT, registered in a Community port and operating for more than five years under the flag of a Member State;
- the relevant vessel had to fly the Member States' flag throughout the joint venture;
- the joint venture needed to last between six months and one year;
- for joint enterprises, a permanent transfer of the vessel was necessary.

3.6.7 Supporting the scrapping, permanent transfer or re-assignment

Regarding the reduction of fishing effort, Art. 8 required Member States to achieve the objectives laid down in the MAGPs. To this end, they were required to take measures to permanently stop certain fishing activities or restrict them in some way.

Respective measures were scrapping, permanent transfer to third countries, or permanent re-assignments of vessels to uses other than fishing. Member States were also permitted to achieve the targets by restricting fishing days.[202] The latter restrictions were not eligible for public funding.

Vessels less than 25 GRT were eligible for aid for scrapping only. Permanent withdrawals only concerned vessels which carried out a fishing activity for at least 75 days at sea, 'in each of the two periods of 12 month preceding the date of their requests for permanent withdrawal or, as the case may be, a fishing activ-

---

[200] According to Art. 11(1) ss.2 and Annex II, 2.2. of Regulation 3699/93.

[201] Art. 9 of Regulation 3699/93.

[202] Art. 8(3) of Regulation 3699/93.

ity for at least 80% of the number of days at sea permitted by current national regulations. Vessels also needed to be older than 10 years'.[203]

The premium for scrapping of a 15 year old vessel or the transfer of such a vessel in the course of a joint enterprise was provided according to the following chart.[204]

Table 4.6

Fleet segment (GRT)			Premium for fifteen year old vessel
0	<	25	6 215/BRT
25	<	50	5 085/BRT + 28 250
50	<	100	4 520/BRT + 56 500
100	<	400	2 260/BRT + 282 500
400 and over			1 130/BRT + 734 500

Vessels less than 15 years old received the amount described in the above chart increased by 1.5% per year less than 15; vessels older than 15 years received the same amount increased by 1.5% per year over 15.

The premium for the permanent transfer to a third country or for the permanent re-assignment for other purposes than fishing could not exceed 50% of the relevant scrapping premium.

The premiums for laying-up vessels or temporarily engaging them in joint ventures, were set as follows:[205]

Table 4.7

Fleet segment (GRT)			Premium per vessel (ecus/day)
0	<	25	4.52/GRT + 20
25	<	50	4.30/GRT + 25
50	<	70	3.50/GRT + 65
70	<	100	3.12/GRT + 88
100	<	200	2.74/GRT + 120
200	<	300	2.36/GRT + 177
300	<	500	2.05/GRT + 254
500	<	1000	1.70/GRT + 372
1000	<	1500	1.50/GRT + 565
1500	<	2000	1.34/GRT + 764
2000	<	2500	1.23/GRT + 956
2500 and over			1.15/GRT + 1137

[203] See Annex IV in combination with Annex III 1.1. and 1.2.

[204] Annex IV of Regulation 3699/93.

[205] Annex IV of Regulation 3699/93.

3.6.8 Financial aid for the material investment in fishing ports

Art. 11 provided funds for the promotion of capital investments in fishing port facilities. Annex III set out the conditions under which aid could be granted. Eligible investments would be provided for installations and equipment improving landing and storing conditions in ports, supporting fishing vessels, and improving jetties to increase safety during landing and loading operations. Priority was given to investments which were of interest to all fishermen and that contributed to the general development of the ports. There were no limits placed on the amount of investment that was permitted under Regulation 3699/93.

3.6.9 Financial aid for the search of new market outlets

Measures which promoted new market outlets under Art. 29 of Regulation 4028/86 were further elaborated upon in Art. 12 of Regulation 3699/93. Member States have taken a broad range of initiatives in this area, including promoting consumer surveys, market analyses, promotion campaigns and providing sales advice to wholesalers and retailers. Market studies also dealt with the prospects for marketing Community products in third countries. Aid covered expenditures arising from advertising campaigns such as payments to advertising agencies and other promotional campaign service providers. Priority was given to campaigns encouraging the sale of surplus and underexploited species. There were no explicit limits stipulating the total amount of aid available.

3.6.10 Promotion of operations by trade members

Member States could also promote operations carried out by trade members themselves. Such initiatives needed to be 'short-term operations of collective interest' which serve the overall objectives of the CFP.[206] Aid was provided to support 'producer organisations', defined under Art. 4 and 7(1) and (2) of Regulation 3759/92, to encourage their formation and facilitate their operation during the first years of their existence.[207]

3.6.11 Subsidising the processing and marketing segment

Under the new Regulation, the processing and marketing of fishery and aquaculture products were financed pursuant to the requirements set out in Art. 11 and Annex III of Regulation 3699/93.[208] Investments must relate to the construction and acquisition of buildings and installations, equip-

---

[206] Art. 13 of Regulation 3699/93.

[207] Council Regulation (EEC) 3759/92 on the common organisation of the market in fishery and aquaculture products, OJ 1992 No. L388/1.

[208] Regulation 4042/89 was repealed by Art. 9 of Regulation 2080/93.

ment required between the time of landing and the end product-stage, and the application of new technologies intended to increase competitiveness and added value. No funding was granted for fishery and aquaculture products to be used for purposes other than human consumption (except for processing and marketing product wastes), including retail trade.

3.6.12 Social measures

Under Regulation 3699/93, Member States could grant financial aid to buffer the social effects arising from the temporary cessation of fishing activities. However, FIFG funding was only available for losses of income 'suffered as a result of a temporary cessation of fishing activities caused by unforeseen and non-repetitive events resulting from biological phenomena in particular'.[209]

3.6.13 1998 Amendments – Extension of social measures

In 1998, Regulation 3699/93 was replaced by 'Council Regulation (EC) 2468/98 laying down the criteria and arrangements regarding community structural assistance in the fisheries and aquaculture sector and the processing and marketing of its products'.[210] The conditions, scale, and the participation rates for aid largely remained the same. Nevertheless, two important modifications were introduced. First, due to the ongoing decline of the fishing sector,[211] Art. 15 of the new Regulation extended socioeconomic aid. The FIFG supported Member States, providing the following assistance:

- It provided partial financing for national *early-retirement* schemes for fishermen, on the condition that the beneficiaries were not more than 10 years from the national retirement age or at least 55 years. Beneficiaries also had to prove that they worked for a minimum of 10 years as fishermen. The FIFG would not make any contribution to the normal national retirement schemes.[212] The number of beneficiaries did not exceed the number of on-board-jobs eliminated as a result of those vessels, which had permanently stopped fishing, nor could they transfer vessels within the following six years;[213]
- It granted individual *compensatory payments* to fishermen who *permanently* withdrew from fishing. Aid was limited to 7,000 ecus per beneficiary, provided the vessel on which the beneficiary had been employed was permanently withdrawn or transferred, pursuant to Art. 8(2) or 9(3).

[209] Art. 14 ss.2 of Regulation 3699/93.

[210] Art. 20 Regulation 2468/98, OJ 1998 No. L312/19.

[211] See Reason (9) of Regulation 2468/98.

[212] Art. 15(3)(a) Regulation 2468/98.

[213] Arts. 15 (3)(a), 8(2), 9(3) and 3 Regulation 2468/98.

Member States were to ensure that beneficiaries received only one type of benefit and that they would permanently give up work as fishermen, otherwise they would have to return the funds.

The second important change instituted under the new regulation was increased structural aid for the temporary and permanent withdrawal of fishing vessels over 24 metres long. Under Annex III, the amount of assistance for the permanent withdrawal and joint enterprises for such vessels was set as follows.

Table 4.8

Fleet segment (GT)			Premium for 15 year old vessel (ecus)			
0.2	<	10	8,130/	GT	+	1,200
10	<	25	4,100/	GT	+	41,500
25	<	100	3,520/	GT	+	56,000
100	<	300	2,348/	GT	+	173,200
300	<	500 and over	1,912/	GT	+	304,000
500 and over			1,045/	GT	+	737,500

The amount of aid for temporary cessation of fishing activities and temporary joint ventures for vessels over 24 metres long was set as follows.

Table 4.9

Fleet segment (GT)			Premium per vessel (ecus/day)			
0	<	10	5,2/	GT	+	20
10	<	25	4,3/	GT	+	30
25	<	50	3,2/	GT	+	55
50	<	100	2,5/	GT	+	90
100	<	250	2,0/	GT	+	140
250	<	500	1,5/	GT	+	265
500	<	1,500	1,1/	GT	+	465
1,500	<	2,500	0,9/	GT	+	765
2,500 and over			0,67/	GT	+	1,340

3.6.14 The Community Initiative PESCA

The European Community Initiative PESCA supplemented the existing structural intervention mechanisms, and was in place from 1994 to 1999.[214] The Commission adopted this programme pursuant to its competence under Art. 11 of Regulation 4253/88.

PESCA was a response to the severe crisis faced by the fisheries sector, which was mainly due to the chronicle fleet overcapacity, economic difficul-

[214] OJ 1994 No. C180/1.

ties faced by businesses arising from the imposition of new management and technical regulations, and the completion of the Common European Market. There was a strong need for financial support to transform and diversify the fishing industry, make workforce adjustments, and cope with the difficult socio-economic consequences arising from industry decline, particularly in coastal regions. The beneficiaries could be collective, public or private bodies, as well as individual operators. The activities eligible for aid included the diversification of economic initiatives (tourism, start-up, etc.), business support services, new job creation (vocational training, training for teachers, assistance with geographical mobility), financial engineering, support of the aquaculture sector, increasing productivity, etc.

Member States had to draw up proposals on their national fisheries sector. The Commission would consider these proposals, taking into account the impact of the aid, monitoring procedures, participation of private and public entities as well as the suitability of the proposed funding. Community funding from 1994 to 1999 was set at 250 million ecus, of which 125 million ecus were allocated to less developed regions.

Table 4.10: Shares of Community, Member States and beneficiary contributions to the total amount of co-financed projects in the fisheries sector from 1994 to 1999[215]

Area	Legal base	Community share	National contribution	Beneficiary contribution	Total million ecus
Fleet construction & modernisation	Council Regs. 2052/88, 4253/88, 2080/93, 3699/93	max. 30% max. 50% (ldr)	min. 5% min. 5% (ldr)	min. 60% min. 40% (ldr)	640.71
Aquaculture	Reg. 3699/93	max. 30% max. 50% (ldr)	min. 5% min. 5% (ldr)	min. 60% min. 40% (ldr)	285.73
Protect marine coastal areas	Reg. 3699/93	30% 50% (ldr)	min. 5% min. 5% (ldr)	min. 50% min . 25% (ldr)	32.24
Adjustment of fishing effort/ capacity	Reg. 3699/93	25–50% 50–75% (ldr)	min. 50% min. 25% (ldr)	(-) (-)	707.36
Facilities at fishing ports	Reg. 3699/93	25–50% 50–75% (ldr)	min. 50% min. 25% (ldr)	(-) (-)	190.72
Search of new market outlets	Reg. 3699/93	30% 50% (ldr)	min. 5% min. 5% (ldr)	min. 50% min . 25% (ldr)	87.55
Processing and marketing	Reg. 3699/93	30% 50% (ldr)	min. 5% min. 5% (ldr)	min. 50% min . 25% (ldr)	601.61

215 Regulations 2080/93 and 3699/93; Commission of the European Communities, Fisheries information – Fact sheet IV-1 ('The Financial Instrument for Fisheries Guidance' (FIFG), June 1996); printed in Y.H. Song, 'The Common Fisheries Policy of the European Union: Restructuring of the Fishing Fleet and the Financial Instrument for Fisheries Guidance' (1998) 13 *The International Journal of Marine and Coastal Law*, 4, p. 575.

PESCA	Commission Decision 15/7/94	25–50% 50–70% (ldr)	Remain	Remain	250
Other measures	Reg. 3699/93	Depending	Depending	Depending	129.72

3.7 Promotion from 2000-2006 – A turnaround?

On 1 January 2000, the central Regulation 2052/88 on Structural Funds, as well as Regulation 2080/93 establishing the objectives of FIFG and Regulation 2468/98 setting out detailed rules on structural assistance, were repealed. This regime was replaced with Council Regulation (EC) 1260/1999 laying down general provisions on the Structural Funds,[216] Council Regulation (EC) 1263/99 on the Financial Instrument of Fisheries Guidance[217] and Council Regulation (EC) 2792/99 on Community structural assistance in the fisheries sector.[218] The latter two regulations set down the specific duties and conditions for the application of structural measures in the fisheries sector. These measures remained in force until 31 December 2006. In 2002, Regulation 2792/99 was amended by two new regulations: Council Regulation (EC) 2369/02 amending the central Regulation 2792/99, and Council Regulation (EC) 2370/02 establishing an emergency measure for scrapping fishing vessels.

3.7.1 Changes and innovations

Regulations 1263/99 and 2792/99 modified existing requirements and introduced new measures. For example, the Member States' sectoral plans now had to demonstrate in greater detail that public aid was necessary to attain MAGP objectives. Between the years, 2000 and 2002, more stringent requirements concerning capacity management were imposed on Member States. Capacity increases were only allowed where they did not increase exploitation rates. Member States were only allowed to support new capacity if MAGP objectives had been met and as long as the same amount of capacity was withdrawn without public assistance. Joint ventures were abolished, now leaving joint enterprises as the sole form of subsidised private co-operation with third country private entities. Since 1999, permanent transfers of old vessels to third countries needed to be consistent with conservation obligations at international law. Inland fisheries and small-scale coastal fisheries were increasingly supported. The scope of socioeconomic measures was broadened; aid was now provided for the diversification of commercial activities of fishermen outside the fisheries sector and to encourage young fishermen to enter the industry, thereby lowering the average age of fishermen in the fleet. They also adopted new environmental requirements for aquaculture projects and new initiatives regarding the search for alternate market outlets. Financial compensation for reductions in

216 OJ 1999 No. L161/1.

217 OJ 1999 No. L161/54.

218 OJ 1999 No. L337/10.

supply caused by the imposition of emergency measures, and the suspension of fisheries agreements was made available to the processing sector.

In the course of the 2002 reform, the Community ended its efforts to directly oblige Member States to reduce their capacities. Fishing activities were supposed to be regulated essentially through the reduction of fishing efforts. The overly complicated and ineffectual MAGPs were replaced by a simple entry-exit rule that employed 2002 MAGP capacity objectives as maximum value.[219] Aid for the construction of vessels was to be phased out by the end of 2004. Support for the modernisation of fishing vessels was only granted for improvements in safety, working conditions, hygiene and product quality, and only on the condition that such aid did not increase catch capacity. Additional support was granted to vessel owners who were affected by recovery plans. Funds for the permanent transfer of vessels were also phased out, ending in 2004.

3.7.2 The advancement of Community Programmes and sectoral plans

Compared to Regulation 3699/93, the programming procedure basically remained the same. However, sectoral plans drawn up by Member States had to provide a more elaborate fleet restructuring strategic plan.[220] This entailed demonstrating the following in greater detail:

The necessity of public aid for the attainment of fleet objectives

- the steps required to prevent counterproductive effects, particularly regarding the creation of surplus capacity;[221]
- fleet development indicators relating to MAGP objectives (required until the end of 2002);
- descriptions of promoted fishing gear and techniques, the naming of management authorities;
- the steps taken to ensure efficient and effective implementation, particularly with regards to monitoring, evaluation, checks, penalties and public measures.[222]

3.7.3 Promoting construction, modernisation and conversion

Regulation 2792/99 introduced new rules on fleet renewals and the modernisation of vessels (some of which only remained in force until the end of 2002), aimed at meeting objectives of MAGP IV.

The requirements set down in Art. 6 applied to both fleet renewal and modernisation measures. Member States were obliged to draw up so-called

---

219 See Arts. 6-11 of Commission Regulation (EC) 1438/03 laying down implementing rules on the Community Fleet Policy as defined in Chapter III of Council Regulation (EC) 2371/02, OJ 2003 No.L204/21.

220 Arts. 3(1),(2) and Annex I of Regulation 2792/99, Arts. 9(a) and (b) of Regulation 1260/99.

221 Art. 3(3) and Annex I (2)(c) of Regulation 2792/99.

222 Art. 3(3) and Annex I (4) Regulation 2792/99.

'permanent arrangements for monitoring fleet renewal and modernisation'. Within this framework, Member States had to demonstrate that fleet entries and exits would be managed in such a way that capacity would be reduced in reference to the levels set out in MAGP IV. They particularly had to ensure that capacity, excluding fishing vessels of less than 12 metres in length, withdrawn with public aid, was not replaced by newly introduced capacity. Capacity increases would be permitted only where they resulted in improvements in safety, navigation, hygiene, product quality and improvements to on-board working conditions, and as long as they did not increase exploitation rates.

Regarding renewals, Art. 9 states that where MAGP objectives for fleet segments are met, Member States must ensure that 'the entry of new capacity with public aid is compensated by the withdrawals of a capacity without public aid which is at least equal to the new capacity introduced in the segment concerned'. If MAGP objectives were not met, Member States were to ensure that from 2000 to 2001 the 'entry of new capacity with public aid is compensated by the withdrawal of capacity without public aid which is at least 30% greater than the new capacity introduced in the segment concerned'.[223] The capacity withdrawn could not be replaced with new capacity introduced under the provision.

Annex III required that newly built vessels had to comply with Community measures governing hygiene, safety, health, product quality, working conditions, measurement and monitoring requirements and also had to be registered in the appropriate fleet segment. Fleet renewals by transfer of ownership transactions were excluded from Community aid.

Art. 9(4)(a) stated that the scale of public expenditure for the construction of vessels could not exceed *twice* the scale of the following Table 1 in Annex IV.

Table 4.11

Fleet segment (GT)			Euro	
0	<	10	11,000/BRZ(GT)	+ 2,000
10	<	25	5,000 / BRZ(GT)	+ 62,000
25	<	100	4,200 / BRZ(GT)	+ 82,000
100	<	300	2,700 / BRZ(GT)	+ 232,000
300	<	500	2,200 / BRZ(GT)	+ 382,000
500 and over			1,200 / BRZ(GT)	+ 882,000

In relation to fleet modernisation, Art. 9 (1)(c) only allowed public aid to be granted if it did not lead to an increase in capacity. Regarding the scale of public expenditure for the modernisation of vessels, Art. 9 (4)(b) provided that it should not exceed the scale which is applicable in relation to the renewal of the fleet in Annex IV (not twice the scale).

[223] Art. 9(b) of Regulation 2792/99.

3.7.4 Joint enterprises

Measures concerning exploratory expeditions and joint ventures were no longer included in Regulation 2792/99, leaving private co-operation as the sole form of joint enterprise. To receive public funding the beneficiary must enter into an agreement with a commercial enterprise located in a third country in accordance with the third country's laws. The enterprise must be involved in commercial activities in the fishing sector of that country, obtain resources from waters under the jurisdiction or sovereignty of the country, and the beneficiary must own a significant proportion, generally between 25 and 75% of the share capital.[224] In addition, the ownership of the vessel must be transferred to the enterprise. The vessel may not be used for any other fishing activities other than those authorised by the authorities of the third country, and cannot be used by other ship owners. The premiums granted for vessels could not exceed 80% of the following rates:

Table 4.12.a

Fleet segment (GT)			Euro	
0	<	10	11,000/BRZ(GT)	+ 2,000
10	<	25	5,000 / BRZ(GT)	+ 62,000
25	<	100	4,200 / BRZ(GT)	+ 82,000
100	<	300	2,700 / BRZ(GT)	+ 232,000
300	<	500	2,200 / BRZ(GT)	+ 382,000
500 and over			1,200 / BRZ(GT)	+ 882,000

Table 4.12.b

Fleet segment (BRT)			Euro	
0	<	25	8,200 / BRT	
25	<	50	6,000 / BRT	+ 55,000
50	<	100	5,400 / BRT	+ 85,000
100	<	250	2,600 / BRT	+ 365,000

These premiums apply to vessels 10 to 15 years old. For vessels between 16 and 29 years old, the scales decreased by 1.5% per year, and for vessels 30 years old or more, scales are decreased by 22.5%.[225]

3.7.5 Subsidising temporary or permanent withdrawals

Regulation 2792/99 continued support for the scrapping, the permanent reassignment, and the transfer of vessels, or any combination

[224] Art. 8(2)(a) Regulation 2792/99.

[225] Art. 8(3) of Regulation 2792/99.

thereof.[226] This applied to vessels older than 10 years.[227] The premiums granted for the scrapping of vessels were as follows:[228]

Table 4.13.a

Fleet segment (GT)			Euro	
0	<	10	11,000	/ BRZ(GT) + 2,000
10	<	25	5,000	/ BRZ(GT) + 62,000
25	<	100	4,200	/ BRZ(GT) + 82,000
100	<	300	2,700	/ BRZ(GT) + 232,000
300	<	500	2,200	/ BRZ(GT) + 382,000
500 and over			1,200	/ BRZ(GT) + 882,000

Table 4.13.b

Fleet segment (GRT)			Euro	
0	<	25	8,200	/ BRT
25	<	50	6,000	/ BRT + 55,000
50	<	100	5,400	/ BRT + 85,000
100	<	250	2,600	/ BRT + 365,000

These premiums applied to vessels 10 to 15 years old. For vessels from 16 to 29 years old, the scales decreased by 1.5% per year, and for vessels of 30 years old or more, scales decreased by 22.5%.

Regarding the permanent transfer of vessels in support of joint enterprises, new conditions regarding the eligibility to public funding were introduced. According to Art. 7(3)(b), transfers to third countries were subject to the following conditions:

- the existence of appropriate guarantees that international law is not likely to be infringed, in particular with respect to the conservation and management of marine resources or other objectives of the CFP and with respect to working conditions of fishermen;
- the third country to which the vessel is transferred is not a candidate for accession;
- the transfer results in a reduction of fishing effort on the resources previously exploited by the vessel transferred (unnecessary in cases where the vessel transferred has lost fishing possibilities under a fisheries agreement with the Community or another agreement).

[226] Art. 7(1) of Regulation 2792/99.

[227] Art. 7(2) of Regulation 2792/99.

[228] Annex IV of Regulation 2792/99.

The amount of public aid for permanent transfers for joint enterprises equalled the amount for joint enterprises.[229]

Conditions on the eligibility for aid for the permanent reassignment of vessels for purposes other than fishing did not change: the maximum amounts for the scrapping premiums referred to in table 1 and 2 above, less 50%. However, no public aid for this purpose can be given to vessels with a tonnage less than 20 GRT or 20 GT, except where vessels were assigned to research or training activities.[230]

If Member States decided to adjust effort by restricting fishing days or days at sea, such measures, as under the two prior Regulations, did not give rise to any public funding.[231]

3.7.6 Encouraging capital investment for the protection and development of aquatic resources, aquaculture, fishing port facilities, processing and marketing and inland fishing

Projects were eligible for financial assistance for the protection and development of aquatic resources, aquaculture, fishing port facilities, processing and marketing and inland fishing if they met the following objectives:

- contribute to lasting economic benefits from the structural improvement in question;
- offer adequate guarantee of technical and economic viability;
- avoid counterproductive effects, particularly the risk of creating surplus production capacity.

According to Annex III(2), such projects could involve capital investments in production and management (construction, enlargement, equipping and modernisation of facilities) as well as investments to improve conditions of hygiene or human or animal health, to improve product quality, reduce environmental pollution and, where relevant, to increase production.

Projects targeting the protection and development of aquatic resources received aid from the FIFG only where they were concerned with the installation of fixed or moveable facilities to protect and develop aquatic resources. Such projects had to be in the collective interest, implemented by a public or publicly

---

229 Art. 7 (5)(b) of Regulation 2792/99. The premium for other permanent transfers to third countries was set at 50 percent of the maximum amounts for the scrapping of vessels, as described above. However, vessels with a tonnage less than 20 GRT or 22 GT as well as vessels over 30 years old were excluded from funding, except in cases where vessels were assigned for the preservation of historical heritage, research or training activities; see Art. 7(5)(c) of Regulation 2792/99.

230 Art. 7(5)(d) of Regulation 2792/99.

231 Art. 7(7) of Regulation 2792/99.

recognised body and could not negatively affect the aquatic environment. Aid for the scientific monitoring of such projects was available.

Public aid for aquaculture was granted only for projects concerned with environmental impact assessments and works concerned with the installation or improvement of water circulation in aquaculture enterprises and on service vessels.

Fishing port facilities, processing and marketing projects were financed under the same conditions as under the two prior Regulations, *i.e.*, Regulations 2468/98 and 3699/93.

Inland fishing was increasingly promoted. 'Inland fishing' is defined in Art. 11 of Regulation 2792/99 as 'fisheries carried out for commercial purposes by vessels operating exclusively in inland waters of the territory of a Member State [...].[232] The same conditions that applied to marine vessels under Annex III, also applied the construction or modernisation of inland fishing vessels. Investments could not jeopardise the equilibrium between fleet size and available resources. No funds were granted for scrapping premiums for inland fishing vessels. Member States had to ensure that vessels that had received financial aid continued to exclusively fish in inland waters.

3.7.7 Finding and promoting new market outlets

Under Regulation 2792/99 the Community began to support investments which aimed at increasing the quality for fishery and aquaculture products as well as the promotion of products obtained by using environmentally friendly methods.[233]

3.7.8 Socioeconomic measures

Regulation 2792/99 introduced initiatives to improve socioeconomic conditions. According to Art. 12(3)(c), financial assistance from the FIFG was available to fishermen who worked at least five years as fishermen, helping them to retrain or diversify their activities outside maritime fisheries. Member States had to determine the individual amount of aid in terms of the overall scale of the projects outside maritime fisheries. The highest possible amount of FIFG aid was set at €50,000.

A second measure aimed at supporting young fishermen under 35 years who recently acquired part or total ownership of a fishing vessel. Aid was only provided if the vessel was 10 to 20 years old, between 7 and 24 metres in length, registered in the Community's fishing vessel register, and as long as the transfer

[232] '[...] and which are *not* subject to the provisions of Title II', *i.e.*, small-scale coastal fishing being defined as fishing operations by vessels under 12 metres.

[233] Art. 14(2) of Regulation 2792/99.

ownership did not occur within the same family.[234] Aid could not exceed 10% of the costs of acquisition of ownership, up to a ceiling of €50,000.

3.7.9 Support for small-scale coastal fishing

Regarding capacity adjustments, the new 1999 structural regulation drew a distinction between small-scale coastal fishing and regular fishing operations.[235] Small-scale coastal fishing was defined in Art. 11(1) of Regulation 2792/99 as 'fishing carried on by vessels of an overall length of less than twelve metres'.

Member States were permitted to adopt measures to supplement Community measures, which seek to improve the conditions of small-scale coastal fishing. To this end, lump sum premiums could be granted to beneficiaries, which were partly financed by the FIFG. Grants were available to a group of owners of vessels, or families of fishermen involved in small-scale fishing activities for the following purposes:

- improvements to on-board safety equipment and sanitary and working conditions;
- technological innovation (more selective fishing techniques);
- organisation of the production, processing and marketing chain (promotion and added value of the products);
- professional re-qualification or training.

The competent authorities were to determine, with reference to the scale of the project and the financial commitments made by each participant, the amount they would grant and how it would be shared between the beneficiaries.

3.7.10 Temporary cessation of activities and other financial compensations

The Community continued to grant compensation for financial losses suffered by fishermen from unforeseen circumstances related to biological disturbances.[236] Aid was available for periods up to a maximum of two months per year or six months over the entire period of 2000 to 2006.

Assistance was also granted in cases where fisheries agreements with third countries were not renewed or suspended. Compensation was only granted for a six-month period, although it could by extended by additional six months with Commission approval.[237]

---

[234] Art. 12(3)(d) of Regulation 2792/99. According to Art. 12(3)(d)(i), the term 'family' included family up to the 'second degree'.

[235] Art. 2(3) of Regulation 1263/99 and Reason (7) of Regulation 2792/99.

[236] Art. 16 of Regulation 2792/99.

[237] Art. 16(1)(b) of Regulation 2792/99.

Financial aid could also have been paid to fishermen, vessel owners and processing industry members affected by recovery plans adopted under the management regime.[238] If recovery plans had been adopted, compensation could have been granted up to a three year maximum.

3.7.11 Innovative measures and technical assistance

When creating their development plans, Member States were required to provide the means necessary to undertake the following initiatives:[239]

- studies;
- pilot projects;
- demonstration projects;
- training measures;
- technical assistance;
- the exchange of experience and publicity connected with the preparation, implementation, monitoring, evaluation or adjustment of the operational programmes and the single programming documents.

Measures could concern topics listed in Art. 15(2), which include:

- improving management;
- working conditions;
- sanitary conditions concerning products on board or landed;
- aquaculture facilities;
- methods of fish farming;
- data collection;
- organisation of electronic trade;
- the creation of business incubators and centres for the collection of fishery and aquaculture products;
- access to training;
- knowledge and transparency in production and in the market;
- promoting equal employment opportunities for men and women working in the sector;
- designing systems that improve quality;
- creating added value in products.

Such measures had to be implemented by public, semi-public or publicly designated bodies.

[238] Art. 16(1)(c) of Regulation 2792/99; as described above in this chapter, recovery plans were foreseen under Regulation Art. 8(3)(i) of 3760/92. However, no multi-annual recovery plan had been adopted prior to the 2002 Reform.

[239] Art. 17 of Regulation 2792/99.

3.7.12 Operations by members of the trade

Support schemes for projects initiated by trade members were adjusted. One such change was the introduction of funding to facilitate the implementation of producer organisations' plans to improve their product quality. The amount of Community aid granted for such projects, in the first, second and third years could not exceed 60, 50 and 40%, respectively, of the costs incurred by the organisation in implementing its project.[240] It also established funding for short-term operations of collective interest, which could be undertaken by trade members and encouraged by Member States. Eligible operations had to relate to:[241]

- improving management, working and sanitary conditions concerning products on board or landed;
- aquaculture facilities and fish farming methods;
- data collection, the organisation of electronic trade;
- the creation of business incubators and centres for the collection of fishery and aquaculture products;
- access to training;
- design systems that improve quality and creation of added value in products;
- improvement of knowledge and transparency in production and in the market.

3.7.13 Reform in 2002 – The new common fisheries policy

The CFP reform in 2002 was particularly important to structural policies, and thus to promotion. Substantive changes were introduced by all regulations adopted in the course of the reform.[242] The new management regulation replaced the complicated and ineffective system of MAGPs and permanent arrangements for monitoring fleet renewals with a simpler entry-exit scheme. A ceiling on the size of Member States' fleets was introduced, mandatory reduction targets were abolished and Member States were required to adjust the fleet size to correspond to the amount of available resources. It also made substantial changes to its policies on fleet renewal, modernisation and the permanent transfer of Community vessels to third countries.

3.7.13.1 Adjustment of fishing capacity and the new entry-exit scheme

Arts. 11-14 of Regulation 2371/2002 replaced the MAGPs, prescribing a simple entry-exit scheme for different fishing fleet segments with

---

[240] Art. 15(2)(b) of Regulation 2792/99.

[241] Art. 15(2) and (3) of Regulation 2792/99.

[242] Regulation 2371/02, Regulation 2369/02, Regulation 2370/02.

specific reference levels. The reference level was the sum of the objectives of MAGP IV for each fleet segment as fixed by the Council.[243] According to Art. 13, the entry-exit scheme would operate as follows:

> 1. Member States shall manage entries into the fleet and exits from the fleet in such a way, from 1 January 2003:
> (a) the entry of new capacity into the fleet without public aid is compensated by the previous withdrawal without public aid of at least the same amount of capacity,
> (b) the entry of new capacity into the fleet with public aid granted after 1 January 2003 is compensated by the previous withdrawal without public aid of:
> (a) at least the same amount of capacity, for the entry of new vessels equal or less than 100 GT, or
> (ii) at least 1,35 times that amount of capacity, for the entry of new vessels of more than 100 GT.
> 2. From 1 January 2003 until 31 December 2004, each Member State which chooses to enter into new public aid commitments for fleet renewal after 31 December 2002 shall achieve a reduction in the overall capacity of its fleet of 3% for the whole period in comparison to the reference levels referred to in Article 12.

No exit was supported by public aid unless preceded by the fishing license.[244] Furthermore, where withdrawals exceed the compulsory reduction rates with the help of public aid, that amount of capacity will be automatically deducted from the reference level, and 'the reference level thus obtained shall become the new reference level'.[245]

Member States must adopt measures to adjust the fishing capacity of their fleets. This includes measures to ensure that capacity reference levels are not exceeded.[246] At the end of 2005, all Member States, except Greece, had complied with the entry-exit regime. Moreover, the Commission has stated that between 2003 and 2005, the fleet of the EU-15 Member States was reduced by 6.27% in terms of tonnage and 7.28% in engine power. The Commission as well as a wide range of fisheries experts considered this reduction as 'relatively small' or 'insuf-

---

[243] See Arts. 6 and 7 of Commission Regulation (EC) 1438/2003 laying down implementing rules on the Community Fleet Policy as defined in Chapter III of Council Regulation (EC) 2371/2002, OJ 2003 No. L204/21; Arts. 1(2) and 1(3) of Commission Regulation (EC) 916/2004 amending Regulation (EC) 1438/2003 laying down implementing rules on the Community Fleet Policy by reason of the accession of Czech Republic, Estonia, Cyprus, Latvia, Lithuania, Hungary, Malta, Poland, Slovenia and Slovakia to the European Union, OJ 2004 No. 163/81.

[244] Art. 11(3) of Regulation 2371/02.

[245] Art. 11(4) of Regulation 2371/02.

[246] Arts. 11(2) and 12(1) of Regulation 2371/02.

ficient' in relation to the high levels of fishing pressure on most Community fisheries.[247]

3.7.13.2 Fleet renewal and equipment or modernisation of vessels

According Art. 1(9) of Regulation 2369/02, aid for fleet renewals (*i.e.*, the construction of vessels) could only be granted until 31 December 2004 and only to vessels under 400 GT. After this time, financial transfers for new entries into the fleet had to be in line with the new entry-exit scheme in Arts. 11-14 of Regulation 2371/02.

Regarding the modernisation of fishing vessels, Art. 1(9)(d) referred to Art. 11(5) of Regulation 2371/02, which limited grants for improving on-board safety, working conditions, hygiene and product quality. Such funds could lead to an increase in tonnage as long as they did not increase the ability of vessels to catch fish.[248]

In regards to the modernisation and construction of vessels, the amount and participation rates for public aid remained unchanged for this period.[249]

3.7.13.3 Scrapping of vessels

In addition to the regular funds provided for the scrapping of vessels, between 2003 and 2006, supplementary assistance was given to vessel owners that were severely affected by recovery plans.[250] According to Art. 2, such owners are eligible for the increased scrapping premium, provided that they are also eligible for scrapping premiums under Regulation 2792/99 and that the effect of the recovery plan was to reduce their fishing effort by 25% or more. Art. 3 of Regulation 2370/02 increased the regular funding of the scrapping of vessels under Regulation 2792/99 by 20%. From 1 July 2001 onwards, regular premiums have been calculated in accordance with the more detailed rates set out in Table 3, Group 1 of Annex IV of Regulation 2792/99 (see Table 6).[251]

---

[247] Reductions of new Member States fleets lay at 18% both in power and tonnage; see Commission Com(2006) 872 final annual report from the Commission to the Council and the European Parliament on Member States' efforts to during 2005 to achieve a sustainable balance between capacity and fishing opportunities, p. 11. See also House of Lords, The Progress of the Common Fisheries Policy, pp. 29-30.

[248] See Arts 8 and 9 of Commission Regulation (EC) 1438/2003.

[249] The amount or twice the amount respectively provided in Table 1 of Annex IV, as presented in the previous sub-chapter.

[250] Arts. 1 and 2 of Regulation 2370/02; recovery plans are adopted by the Council under Art. 5 of Regulation 2371/02.

[251] Council Regulation (EC) 1451/2001 amending Regulation (EC) No. 2792/1999 laying down the detailed rules and arrangements regarding Community structural assistance in the fisheries sector, OJ 2001 No. L198/9.

3.7.13.4 The permanent transfer of vessels

According to Art. 1(7)(b) of Regulation 2369/02, the permanent transfers of vessels to third countries was promoted only until 31 December 2004. In addition, such transfers were permitted only when certain criteria were met:

- the Community and the third country must have entered into a fisheries agreement with each other;
- appropriate guarantees exist that international law is not likely to be infringed, particularly with respect to conservation and management of marine resources and other objectives of the CFP and with respect to working conditions of fishermen;
- a third country has not been a candidate for accession and in case it was not a member of a regional fisheries organisation;
- it had not been identified by the latter as a country which permits fishing in a manner that jeopardises international conservation measures.[252]

Where vessels were permanently transferred within the framework of joint ventures, public payments were set at a maximum of 80% of the maximum amount of the premium for scrapping vessels as referred to in Tables 1 and 2 in Annex IV of Regulation 2792/99 (presented above).[253] No public aid was available either for the transfer of vessels with a tonnage less than 20 GRT or 22 GT, or vessels older than 30 years.

3.7.13.5 Socioeconomic measures

Art. 1(12)(a) of Regulation 2369/2002 amended the funding schemes for socioeconomic measures. The new scheme placed a greater priority on retraining. Aid allocated to fishermen to diversify their activities outside marine fisheries was reduced from a maximum of €50,000 to €20,000 per individual, while retraining measures remained at €50,000.

3.7.13.6 Temporary cessation of fishing activities and other financial compensation

The Community continued to provide structural aid in the event of unforeseeable circumstances, particularly those with biological causes. Compensation was granted for three consecutive months or for six months over the entire period of 2000 to 2006. Prior to the amendment, aid was not granted

---

252 Exceptions were only made until 2006 for transfers to countries hit by the Tsunami in 2004, see Council Regulation (EC) 485/2005 amending Regulation 2792/99 as regards a specific action for transfers of vessels to countries hit by the tsunami in 2004, OJ 2005 No. L81/1.

253 Art. 1, paragraph 7(d)(b) of Regulation 2369/02 in combination with Art. 8(3) and Annex IV of Regulation 2792/99.

for more than 'two months per year or six months over the entire period from 2000 to 2006'.[254]

According to Art. 1(a)(ii) of Regulation 2369/02, aid was also granted to fishermen and owners of vessels in cases where the Council adopted recovery or management plans under Arts. 5 and 6 of Regulation 2371/02. Funds were also made available in instances where the Commission or the Member States adopted emergency measures under Arts. 7 and 8 of Regulation 2371/02. Compensation provided by Member States could last for one year and extended by an additional year. FIFG contributions could not exceed the higher of the two following limits: €1 million in total, or 4% of the Community financial assistance allocated to the sector in the Member State concerned from 2000 to 2006.[255] These thresholds could be exceeded where measures included a decommissioning scheme aimed at withdrawing, within two years of the adoption of the measure, a number of fishing vessels with a fishing effort at least equal to the effort of the fishing vessels suspended from fishing as a consequence of the plan or the emergency measure.[256]

Table 4.14: Shares of Community, Member States and beneficiary contributions to the total amount of co-financed projects in the fisheries sector from 2000 to 2006.[257]

Area	Legal base	Community share	National contribution	Beneficiary contribution	Total million Euro
Fleet renewal & modernisation	Regulations. 1260/99, 1263/99, 2792/99	15% 35% (ldr)	min. 5% min. 5% (ldr)	min. 60% min. 60% (ldr)	855.5
Aquaculture	Ibid.	15% 35% (ldr)	min. 5% min. 5% (ldr)	min. 60% min. 40% (ldr)	299.4
Adjustment of fishing effort/ capacity	Ibid.	25–50% 50–75% (ldr)	min. 50% min. 25%	remain remain	482.5
Facilities at fishing ports	Ibid.	25–50% 50–75% (ldr)	min. 50% min. 25%	remain remain	248.7
Search of new market outlets	Ibid.	25–50% 50–75% (ldr)	min. 50% min. 25%	remain remain	
Processing and marketing	Ibid.	15% 35% (ldr)	min. 5% min. 5% (ldr)	min. 60% min. 40% (ldr)	634

254 Art. 1(13)(a) of Regulation 2369/02.

255 Council Regulation (EC) 1421/2004 amending Regulation (EC) No. 2792/1999 laying down the detailed rules and arrangements regarding Community structural assistance in the fisheries sector, OJ 2004 No. L260/1 introduced financial compensation to shellfish farmers where the contamination of shellfish, due to the growth of toxin-producing plankton or the presence of plankton containing marine biotoxius, makes it necessary to suspend harvesting.

256 Art. 1(13) of Regulation 2369/02.

257 Regulation 2792/99; European Communities, Facts and Figures on the CFP (Luxembourg, 2006).

Socio-economic measures	Ibid.	25–50% 50–75% (ldr)	min. 50% min. 25%	remain remain	120.2
Other measures	Ibid.	varying	varying	varying	1,300.4

Table 4.15: Shares of Community, Member States and beneficiary contributions of co-financed projects in the fisheries sector from July 2001 to 2006[258]

Area	Legal base	Community share	National contribution	Beneficiary contribution
Adjustment of fishing effort/capacity (including rates for temporary cessation)	Regs. 2792/99 as amended by Regs. 1451/2001, 2369/02	•Objective 1 region: 50–75% •Objective 1 region: 50–80% covered by cohesion fund •Outermost regions: 50–80% •Outlying Greek islands: 50–85% •Other areas: 25–50%	•Min. 25% •Min. 20% •Min. 15% •Min. 15% •Min. 50%	(-) (-) (-) (-) (-)
Fleet renewal & modernisation	Regs. 2792/99 as amended by Regs. 1451/2001, 2369/02	•Objective 1 region: 35% •Objective 1 region: 35% covered by cohesion fund •Outermost regions: 40% •Outlying Greek islands: 35% •Other areas: 15%	•Min. 5% •Min. 5% •Min. 10% •Min. 5% •Min. 5%	• Min. 60% • Min. 40% • Min. 50% • Min. 60% • Min. 60%
Socio-economic measures	Regs. 2792/99 as amended by Regs. 1451/2001, 2369/02	•Objective 1 region: 50–75% •Objective 1 region: 50–80% covered by cohesion fund •Outermost regions: 50–80% •Outlying Greek islands: 50–85% •Other areas: 25–50%	•Min. 25% •Min. 20% •Min. 15% •Min. 15% •Min. 50%	(-) (-) (-) (-) (-)

3.8 Promotion from 2007 to 2013 – the European Fisheries Fund

In June 2006, the Council adopted a new central Regulation to govern the structural aspects of the CFP, entitled Council Regulation (EC) 1198/06 on the European Fisheries Fund (Regulation 1198/06).[259] The Regulation is expected to be in force from 1 January 2007 to 31 December 2013. It has

[258] Regulations 1451/2002; 2792/99; 2369/02.

[259] OJ 2006 No. L223/1.

replaced and simplified the existing FIFG system.[260] The Commission has also adopted Commission Regulation (EC) 498/2007 laying down detailed rules for the implementation for Council Regulation (EC) 1198/06 on the European Fisheries Fund.[261]

3.8.1 Changes and innovations

The EFF continues the most important reforms, which were initiated in 2002. It also introduces some novelties to the structural policies. First, it altered the FIFG's approach to structural planning and aid delivery. Each Member State is required now to draw up a 'national strategic plan' to serve as a basis for a dialogue between the Member States and the Commission regarding developments in the entire fisheries sector. Within this plan, Member States must lay out their priorities and suggest which projects should be eligible for aid. Strategic plans are drawn up in partnership with regional and local, private and public stakeholders. They form the basis of so-called 'operational programmes' which, after receiving Commission approval, constitute the sole programming instrument for the national fisheries sector. To receive approval, an operational programme must take into account the objectives and rules of the national strategic plan, as well as those of the CFP and EFF.

Another innovation is that projects eligible for assistance are categorised under so-called 'priority axes'. A priority axis is 'one of the priorities in an operational programme comprising a group of measures which are related and have specific measurable goals'.[262] Axes are listed as follows:

- priority Axis 1: Measures for the adaptation of the Community fishing fleet (€888 million, 26, 9% of total);
- priority Axis 2: Aquaculture, inland fishing, processing and marketing of fishery and aquaculture products (€1056 million, 32% of total);
- priority Axis 3: Measures of common interest (€929 million, 28, 2% of total);
- priority Axis 4: Sustainable development of fisheries areas (€314 million, 9,5% of total);
- priority Axis 5: Technical assistance (€112 million, 3,4% of total).[263]

Member States are required to include tables within their operational programmes, which set out how much aid each axis will receive.[264]

Under the different axes, the following promotional instruments can be found. 'Priority Axis 1' concerns measures directed at fleet adjustment. These measures have been adapted to comply with management and structural

---

[260] Regulations 1263/99; 2792/99; 366/2001.

[261] OJ 2007 No. L120/1.

[262] Art. 3(i) of Regulation 1198/06.

[263] See for the allocation of funds, House of Lords, *The Progress of the Common Fisheries Policy*, p. 23.

[264] Art. 20(1)(f)(ii) of Regulation 1198/06.

requirements introduced under the current central regulation, Regulation 2371/02. Accordingly, aid for fleet and gear adjustments is targeted at supporting operators affected by recovery and management plans, emergency measures, the non-renewal of fisheries agreements, national decommissioning schemes, and closures for reasons of public health or natural disasters.[265] In accordance with the 2002 Reform, the EFF continues to deny subsidies for the construction of vessels or the increase of fish holds. However, some aid will be granted for engine replacements, subject to a stringent entry-exit ratio and an increased contribution rate by beneficiaries.[266] With regards to fleet adjustment, the EFF provides some privileges, such as higher funding rates, to small-scale coastal fishers. Money is also available to improve socioeconomic conditions, such as support for the diversification, financing re-trainings, early retirements, etc. Financial aid will be available to help rescue or restructure firms in economic difficulty.[267] In order to balance the age structure in the fisheries sector, aid will be preferentially granted to young fishermen, on the condition they acquire a used vessel measuring 24 metres or less.

With regards to Priority Axis 2 concerning 'aquaculture, inland fishing, processing and marketing', the CFP has shifted its focus, now providing aid for micro, small and medium-sized enterprises. In some cases, however, larger enterprises may also receive funding. The EFF introduced funds for the eradication of animal diseases, as well as support for mollusc farmers for temporary harvest suspensions due to contaminations of molluscs by toxin-producing plankton. The processing and marketing segment increasingly receives aid for improving the use of underutilized species, by-products and waste.[268] The EFF has also introduced compensatory funds for the temporary cessation of inland fisheries affected by recovery plans for species found in inland waters.[269]

The new Priority Axis 3 on 'measures of common interest' introduces aid for the removal of lost fishing gear, the set up of management plans for local stocks, pilot research projects, the promotion of partnerships between scientists and operators, and funds for the promotion of equal opportunities. Such measures may be undertaken by private operators of the fisheries and aquaculture sector.

Priority Axis 4 on the 'sustainable development of fisheries areas' establishes a new set of comprehensive measures to CFP' structural policy. As a response to decreasing fish stocks, aid is given for the sustainable development and improvement of socioeconomic conditions in areas dependant on fisheries. Assistance is not limited to the fisheries sector: funds are available for environmental protection, enhancing inter-regional and transnational co-operation as well as capacity building for the preparation of local development strategies. To increase stake-

---

[265] As required by Arts. 5, 6, 7, 8, 11 to 16 of Regulation 2371/02.

[266] Art. 25 of Regulation 1198/06.

[267] Art. 21(f) of Regulation 1198/06.

[268] Art. 35(1)(e) of Regulation 1198/06.

[269] Art. 33(4) of Regulation 1198/06.

holder involvement, aid is distributed on a co-operative basis, so that local public and private partners from different sectors are also eligible for funding.[270]

Priority Axis 5 on 'technical assistance' provides aid for the preparation and implementation of EFF programmes. Member States can opt to use up to 5% of their total EEF funding for the implementation of the operational programmes or the improvement of administrative capacities in fisheries areas, subject to the convergence objective. Such measures can include evaluations, expert reports, statistics and studies, information measures, build-up networks, promote co-operation throughout the Community, etc. The Commission may use up to 0.8% of its annual EFF funds towards implementing Regulation 1198/06.

For all Priority Axes, EFF measures increasingly aim at environmental protection. For example, under Priority Axis 1, support is granted for management measures such as recovery plans, management plans and emergency measures which increasingly take into consideration environmental protection. The EFF also finances equipment which has a smaller impact on non-commercial species, marine ecosystems, and the sea bed.[271] Priority Axis 2 supports aqua-environmental measures to reduce the negative impact of the processing sector.[272] Under Priority Axis 3, the EFF supports measures that contribute to collective actions contributing to improved resource management, promote selective fishing methods or gears, the recovery of lost fishing gear from the sea beds as well as measures to protect aquatic fauna and flora.[273] The EFF also provides funding for fisheries products obtained using low-impact methods and eco-labelling schemes.[274] Priority Axis 4 supports eco-tourism and environmental protection.[275] Where the EEF finances environmentally friendly aquaculture projects or measures which protect the aquatic fauna and flora, it must take into account the requirements of NATURA 2000 areas.[276]

The EFF also provides funding for networking and information exchange among organisations promoting equal opportunities for men and women and other stakeholders.[277] Women are also eligible for special aid towards business training.[278]

In order to improve the quality and the effectiveness of the Community assistance, all operational programmes are subject to an *ex ante*, interim and *ex post* evaluation. The Member States or Commission carry out these evaluations, as appropriate. Member States are now required to designate a managing, a

---

[270] Art. 45 of Regulation 1198/06.

[271] Art. 25(6)(c)-(d), (7)-(8) of Regulation 1198/06.

[272] Arts. 30(2)(a-c) and 35(1)(d) of Regulation 1198/06.

[273] Arts. 37 and 38 of Regulation 1198/06.

[274] Arts. 40(3)(d),(f) of Regulation 1198/06.

[275] Arts. 43(2)(c), 44 (1)(b) and (f) of Regulation 1198/06.

[276] Arts. 30(2)(d), (4)(d) and 38 (2)(c) of Regulation 1198/06.

[277] Art. 37(k) of Regulation 1198/06. The promotion of equality between men and women is introduced in Art. 11, and mentioned in Art. 8 of Regulation 1198/06; see also reason 13, *ibid.*

[278] Art. 44(2) Regulation 1198/06.

certifying and an audit authority responsible for monitoring and control. These authorities are responsible for implementing the EFF, submitting statements of expenditure and ensuring that audits are carried out to ensure the proper implementation of structural aid.

Another new element under the EFF is that the 2007-2013 structural policy is its reliance on the 'principle of partnership', stressing close co-operation between the Commission, Member States, competent regional, local and other public authorities, economic and social partners and other appropriate bodies. The reliance on joint action relates in particular to the set-up of the national strategic plan and the preparation, implementation, monitoring and evaluation of the operational programme.[279]

Participation rates for EFF initiatives are listed in Annex II of Regulation 1198/2007 (see Table 4.16 below). Annex II records which operation is categorised in which group and provides the rates of public (A) and private (B) contributions. Numbers express a percentage of total eligible costs being the sum of all public and private contributions (*i.e.*, A + B).

Table 4.16

	Group 1	Group 2	Group 3	Group 4
Regions covered by the Convergence objective and outlying Greek islands	•A < 100% •B > 0%	•A < 40% •B > 60%	•A < 80% •B > 20%	•A < 60% •B > 40%
Regions not covered by the Convergence objective	•A < 100% •B > 0%	•A < 40% •B > 60%	•A < 60% •B > 40%	•A < 40% •B > 60%
Outermost region	•A < 100% •B > 0%	•A < 50% •B > 50%	•A < 80% •B > 20%	•A < 75% •B > 25%

3.8.2 National Strategic Plans and National Programmes

Art. 15(1) requires Member States to adopt and submit to the Commission, following an appropriate consultation with stakeholders, a national strategic plan on its national fisheries industry. These plans have served as the basis for dialogue between the Commission and Member States in regards to structural support schemes over 2007 to 2013. They contain a description of the national sectors in the light of *all aspects of the CFP*, and set out national priorities, objectives, estimates financial resources required, and implementation deadlines.[280] Before the end of 2011, the Commission will organise a strategic follow-up debate on the content and the progress of the implementation of the national strategic plans.[281]

---

[279] Art. 8 Regulation 1198/06.

[280] Art. 15(2) of Regulation 1198/06.

[281] Art. 16 Regulation 1198/06.

According to Art. 17(1) and (2), each Member State must also draw up an operational programme to implement the policies and priorities to be co-financed by the EFF. The programme must be in line with national strategic plan objectives. Member States are required to involve regional, local, economic and social partners in the fisheries sector as well as all other appropriate bodies in the development of its national plan. According to Art. 20, the programme must include, for example:

- a synthesis of the situation of policy areas eligible for support in terms of strength and weaknesses;
- a description and justification of the priority axes chosen, having regard to the relevant part of the national strategic plan and the guiding principles set out in Article 19, as well as the expected impact resulting from the *ex ante* evaluation referred to in Article 48;
- specific targets set for each priority axis;
- a summary description of the principal measures envisaged for implementing the priority axes;
- information on complementarity with other funds;
- tables breaking down in detail the financial allocations from the EFF and the Member States;
- detailed information on the competence of involved authorities, transparency, monitoring, etc.

The Commission evaluates whether the programmes contribute to EFF objectives in Art. 4, the guiding principles for the operational programmes in Art. 19, and whether they take into account the national strategic plans. Art. 4 basically states that the EFF must support the CFP so as to ensure sustainable exploitation of marine capture, aquaculture and inland resources as well as protect the marine environment. Art. 19 requires that the preparation and the implementation of the operational programme by Member States shall take into account a variety of guiding principles such as:

- consistency with the principles of the CFP and the national strategic plan in order to achieve, in particular, a stable and enduring balance between fishing capacity and fishing opportunities;
- enhancement of a harmonious, balanced and sustainable development of economic activities, jobs and human resources, as well as protection and the improvement of the environment;
- an appropriate allocation of the available financial resources between the priority axes and, in particular, where relevant, an appropriate level for the adaptation of the Community fleet;
- promotion of operations contributing to the Lisbon and Gothenburg strategy (sustainable level of employment; reduce impact of fishing on the environment);[282]

[282] Lisbon European Council: Presidency Conclusions Nr. 100/1/00; Göteborg European Council: Presidency Conclusion Nr. 200/1/01.

- improving human resources in the sector (upgrading and diversifying professional skills, developing lifelong learning and improving working conditions and safety);
- encouraging operations which increase the added value of products;
- contribute to better supply of Community markets;
- promote gender balance in the fisheries sector;
- promotion of an integrated sustainable development of the fisheries areas;
- where relevant, improvement of the institutional and administrative capacities aiming at good governance of the CFP and the implementation of the operational programmes.

When appraising the operational programme, the Commission must also consider the *ex ante* evaluations required under Art. 48. These assessments are carried out by independent assessors supervised by the Member States or the Commission, and aim to ensure the coherence between the guiding principles, the strategic plans and operational programmes.[283] If the Commission considers the programme as incongruent with CFP and EFF objectives and guidelines, it may request the Member State to amend its programme accordingly. If the proposal is approved, the Commission adopts an approving decision.[284] The operational programmes cover the period from 1 January 2007 to 31 December 2013. They may be re-examined and even revised where difficulties arise in their implementation, a significant strategic change is necessary, or for reasons of sound management.[285]

3.8.3 Priority Axis 1: Measures for the adaptation of the Community fishing fleet

To achieve a stable and enduring balance between fishing capacities and fishing opportunities, Member States must establish a policy for adjusting fishing effort within so called 'fishing effort adjustment plans'.[286] These plans are integrated in the national strategic plans. According to Art. 21, public aid may be provided for owners of fishing vessels and fishers affected by recovery plans, emergency measures, the non-renewal of fisheries agreements, management plans, Member States measures requiring higher environmental protection standards within their 12 nm zone, national decommissioning schemes, the temporary cessation of fishing. Additionally, support may be granted for investments on board of fishing vessels and selectivity, for small-scale coastal fishing, socioeconomic purposes, and for restructuring and

[283] Arts. 47(3), (6) and 48 of Regulation 1198/06.

[284] Art. 17(6) of Regulation 1198/06.

[285] Art. 18 of Regulation 1198/06.

[286] Art. 22 of Regulation 1198/06.

rescuing firms in difficulty. Member States must give priority to those industry members affected by recovery plans under Art. 5 of Regulation 2371/02.[287]

Where stocks under recovery plans do recover, fishing effort adjustment plans shall be adopted (or adapted) by Member States within six months of the date of the decision of the Council or Commission. In case of a non-renewal of a fisheries agreement, Member States must adopt the effort adjustment plans for the fishing vessels and fishers affected within six months of the notification by the Commission.[288] Within the annual and final reports on implementation, Member States must provide the results achieved in implementing their fishing effort adjustment plans.

3.8.3.1 Public aid for the permanent cessation of fishing activities

The EFF contributes to financing the permanent cessation of fishing activities where this is achieved by the scrapping of vessels, reassignment of vessels for non-fishing activities, and use for the creation of artificial reefs. Public aid paid to the owners of fishing vessels shall apply to the vessel's fishing capacity and, where appropriate, to the fishing licenses associated with it.[289] Member States may set the level of public aid taking the best cost-effectiveness ratio on the basis of objective criteria, such as the price of vessels, its turnover, or its fishing power.[290] According to Annex II, participation rates for the permanent cessation of fishing belong to 'group 1'; see Table 4.16 above.

3.8.3.2 Public aid for the temporary cessation of fishing activities

The EFF will finance aid measures for the temporary cessation of fishing activities. Aid is provided for fishers who are affected by management measures that stall fishing activities.[291]

Where recovery plans under Art. 5 of Regulation 2371/02 require a stop in fishing activities, aid can be granted for up to a 24-month period.[292]

Where *Member States* adopt emergency measures under Art. 8 of Regulation 2371/02, the EFF may grant financial support for up to three months.[293] Where the *Commission* adopts emergency measures under Art. 7 of Regulation 2371/02, aid may be granted for up to six months.[294]

---

[287] Art. 22(1) of Regulation 1198/06.

[288] Art. 22(3) of Regulation 1198/06.

[289] Art. 23(1) of Regulation 1198/06.

[290] Art. 23(3) of Regulation 1198/06.

[291] Art. 24 of Regulation 1198/06.

[292] Arts. 24(1)(i), 22(3) and 21(a)(i)of Regulation 1198/06.

[293] Arts. 24(1)(ii) and Art. 21(a)(ii) of Regulation 1198/06.

[294] Arts. 24(1)(iii) and 21 (a)(ii) of Regulation 1198/06.

In cases where fisheries agreements between the Community and third countries have not been renewed, the EFF may provide aid for up to 12 months.[295]

Where the Community adopts effort adjustment plans in the course of management plans under Art. 6(4) and 4(2)(f) of Regulation 2371/02 or the Member States adopt management plans at the national level, the EFF may provide aid for up to 8 months.[296]

Aid may also be granted for three months to firms facing economic difficulties during the period of replacement of engines.[297]

Finally, aid may be provided for up to six months in the event of natural disasters or where fisheries are closed by Member States for reasons of public health or exceptional occurrences.[298]

Where financial contributions are made for measures under Art. 24(1)(i)-(v), payments have to be integrated into Member States' fishing effort adjustment plans.[299] The financial contribution for measures mentioned in Arts. 24(1)(i)-(vi) may not exceed €1 million or 6% of the Community assistance allocated to the sector in the Member States concerned.[300] These thresholds may be exceeded by decisions found in compliance with the Comitology management procedure under Art. 101(3).

According to Annex II, participation rates equal those for public aid for the permanent cessation of fishing activities in 'group 1', see Table 4.16, above.

3.8.3.3 Investments on board fishing vessels and selectivity

According to Art. 25(1), the EFF may contribute to the financing of equipment and the modernisation of vessels of five years or older. Investments may concern the improvement of safety on board, working conditions, hygiene, product quality, energy efficiency and selectivity. Aid is granted subject to the condition that the ability of vessels to catch fish is not increased, particularly not above levels provided in Art. 12 of Regulation 2371/02. No aid will be granted for the construction of fishing vessels or for the increase of fish holds.[301]

The EFF may contribute to financing the replacement of one engine per vessel, subject to the following scheme[302]:

[295] Arts. 24(1)(iv) and 22 (a)(iii) of Regulation 1198/06.

[296] Art. 24(1)(i) of Regulation 1198/06.

[297] Art. 24(1)(vi) and Art. 21(f) of Regulation 1198/06.

[298] Art. 24(1)(vii) of Regulation 1198/06.

[299] Art. 22(3) of Regulation 1198/06.

[300] Art. 24(2) of Regulation 1198/06.

[301] Art. 25(2) of Regulation 1198/06.

[302] Art. 25(3) of Regulation 1198/06.

Vessels less than 12 metres in overall length, which do not use towed gears,[303] may receive aid, provided that the new engine has the same power or less than the old one.

Vessels between 12 and 24 metres in overall length may receive aid, provided that the new engine has at least 20% less power than the old one.

Trawlers of more than 24 metres in overall length may receive aid, provided that the new engine has at least 20% less power than the old one, and that the vessel is subject to a rescue and restructuring plan for firms in difficulty and uses less fuel intensive fishing methods.

Member States may agree, in accordance with the Comitology management procedure, that the reductions mentioned in the second and third categories may be achieved by a group of vessels of that particular category.[304]

Second, according to Art. 26(6), the EFF may finance equipment and modernisation works which allow keeping catches on board which may no longer be discarded, cover the preparation or trial of new technical measures, reduce the impact of fishing on non-commercial species, reduce the impact on ecosystems and the sea bottom, and protect catches and gear from wild predators.

Third, the EFF may support investments to achieve the selectivity of fishing gear, including up to two replacements between 2007 and 2013, provided that the vessel concerned is affected by a fishing effort adjustment plan, changes its fishing methods and is leaving the fishery concerned to go to another fishery, or the new gear meets recognised environmental criteria and practices which go beyond the existing regulatory obligations under Community law. The first of the two contributions may be granted for the purpose of guaranteeing compliance with new technical requirements on selectivity under Community law, or for reducing the impact of fishing on non-commercial species. According to Annex II, participation rates for all measures, except those concerning the replacement of engines, belong to 'group 2'. For engine replacements under Art. 25(3), B-rates are increased by 20% and A-rates are reduced accordingly, see Table 4.16 above.[305]

3.8.3.4 Small-scale coastal fishing

The EFF increasingly promotes small-scale coastal fisheries by modifying existing promotional measures of general applicability and participation rates to benefit small-scale coastal fisheries. Art. 26(1) defines small-scale coastal fishing as 'fishing carried out by fishing vessels of an overall length of

303 As defined in Table 3 of Annex I of Commission Regulation 26/2004 regarding the fishing vessels register of the Community, OJ 2004 No. L5/25.

304 Art. 25(3) and (4) of Regulation 1998/06.

305 Annex II of Regulation 1998/06.

less than 12 metres and not using towed gear [...]'.[306] Art. 26(2) sets out special participation rates for this sector, stating that where the EFF provides financial aid for investments on-board fishing vessels and selectivity under Art. 25, private participation rates may be reduced by 20%. Art. 26(3) allows the EFF to finance socioeconomic measures established in Art. 27 to benefit small-scale coastal fishers. According to Art. 26(4), the EFF may contribute to the payment of premiums for fishers and owners of fishing vessels involved in small-scale coastal fishing in order to:

- improve management and control of access conditions to certain areas;
- promote the organisation of the production, processing and marketing chain of fisheries products;
- encourage voluntary steps to reduce fishing effort for the conservation of resources;
- encourage the use of technological innovations [...] that do not increase fishing effort; [and]
- improve professional skills and safety training.

Operations mentioned in Arts. 26(3) and (4) are subject to 'group 1' participation rates, while 26(2) are subjects to 'group 2' rates; see Table 4.16 above.

3.8.3.5 Socioeconomic compensation for management of the Community fishing fleet

According to Art. 27, the EFF may contribute to the financing of socioeconomic measures, including economic diversification, the upgrading of professional skills in particular for young fishers, retraining in occupations outside sea fishing, early departure and retirement. Compensation must be refunded if the beneficiaries return to their work as fishermen within a period of less than one year after receiving it.

To counterbalance a demographic trend showing an increased average age of the Community fleet, young fishermen can receive funds for the purchase of second-hand vessels. Art. 23(2) requires that fishermen be younger than 40 years with a minimum of five years experience as fishers or equivalent professional training to receive aid. The vessel must not exceed 24 metres in overall length and must be between 5 and 30 years old. The premium must not exceed 15% of the cost of acquisition of ownership nor exceed €50,000.

Conditions for non-renewable compensation for vessel acquisitions for young fishermen may be modified in accordance with the Comitology management procedure as referred to in Art. 101(3). Operations mentioned in Art. 27 are subject to 'group 1' participation rates, see Table 4.16 above.

[306] Not using towed gear as listed in Table 3 of Annex I of Commission Regulation 26/2004 regarding the fishing vessels register of the Community, OJ 2004 No. L5/25.

3.8.4 Priority Axis 2: Aquaculture, inland fishing, processing and marketing of fishery and aquaculture products

The EFF also intervenes in the aquaculture sector, providing support for aqua-environmental, public health, and animal health measures as well as productive investment, see Art. 28. Member States are also required to avoid counterproductive effects such as surplus production capacity, which adversely affects CFP conservation policy.[307] EFF initiatives are also linked to environmental Community measures. Projects falling under Annex II of Council Directive (EEC) 337/85 on the assessment of the effects of certain public and private projects on the environment[308] will only be supported where the information required in Annex IV of that Directive has been provided.[309]

3.8.4.1 Productive investments in aquaculture

The EFF also supports investments for the construction, extension, equipment and modernisation of production installations.[310] Projects eligible for aid must contribute to one or more of the following objectives:

- diversification towards new species and production of species with good market prospects;
- implementation of environmentally friendly aquaculture methods;
- support of traditional aquaculture;
- purchase of equipment that protects the farms from wild predators;
- improvement of working and safety conditions.

Funding is limited to enterprises not exceeding a specific size, *i.e.*, micro, small and medium sized enterprises as well as enterprises with less than 740 employees or with a turnover of less than €200 million. 'Micro, small and medium sized enterprises' are defined in Art. 3(f) as 'enterprises which employ fewer than 250 persons and which have an annual turnover not exceeding €50 million, and/or an annual balance sheet total not exceeding €43 million'.[311] Aid is also granted to all enterprises in the outermost regions, as well as small Greek islands. Member States are required to ensure that priority is given to micro- and small-sized enterprises, *i.e.*, enterprises employing less than 50 persons and

[307] Art. 2(5) of Regulation 1198/06.

[308] OJ 1985 No. L175/40.

[309] Art. 28(6) of Regulation 1198/06.

[310] Art. 29 of Regulation 1198/06.

[311] Art. 3(f) of Regulation 1198/06 in combination with Annex Art. 2(a) of Commission Recommendation (EC) 361/2003 concerning the definition of micro, small and medium sized enterprises, OJ 2003 No. L124/36.

having an annual turnover not exceeding €10 million.[312] Productive investments in aquaculture subject to 'group 4' participation rates';[313] see Table 4.16 above.

3.8.4.2 Aqua-environmental measures

According to Art. 30, the EFF may grant financial aid for the use of environmentally friendly aquaculture production methods. The EFF shall promote:

- environmentally friendly farming;
- participation in the Community eco-management and audit scheme (EMAS);[314]
- organic aquaculture;[315]
- creating compatibility of aquaculture with environmental constraints resulting from the designation of NATURA 2000 areas.[316]

Art. 30(3) provides that 'in order to receive compensation [...], beneficiaries must commit themselves for a minimum of five years to aqua-environmental requirements which go beyond the mere application of normal good aquaculture practice'. To receive aid for environmentally friendly farming, environmental benefits must be demonstrated to competent bodies designated by a Member State.[317] Compensation is calculated on the basis of different criteria, such as the loss of revenue incurred and/or additional costs due to the use of aqua-environmental measures, etc. Compensation for environmentally friendly farming, organic aquaculture and improving compatibility with NATURA 2000 sites will be allocated in a single payment.[318] Aqua-environmental measures subject to 'group 1' participation rates; see Table 4.16 above.

3.8.4.3 Public health measures

The EFF grants aid to mollusc farmers whose harvest is suspended due to the contamination of molluscs owing to the proliferation of

[312] *Ibid.*, Annex Art. 2 (b) and (c).

[313] Annex II of Regulation 1198/06.

[314] Regulation (EC) 761/2001 of the European Parliament and of the Council allowing voluntary participation by organisations in a Community eco-management and audit scheme (EMAS), OJ 2001 No. L114/1 as last amended by Commission Regulation (EC) 196/2006, OJ 2006 No. L32/4.

[315] Within the meaning of Council Regulation (EEC) 2092/91 on organic production of agricultural products and indications referring thereto on agricultural products and foodstuff, OJ 1991 No. L198/1, as last amended by Commission Regulation (EC) 780/2006, OJ 2006 No. L137/9.

[316] Council Directive (EEC) 43/92 on the conservation of natural habitats and wild fauna and flora, see OJ 1992 No. L206/7, as last amended by Regulation of the European Parliament and of the Council (EC) 1882/2003, OJ 2003 No. L284/1.

[317] Art. 30(3) of Regulation 1198/06.

[318] Art. 30(4) and (5) of Regulation 1198/06.

toxin-producing plankton or the presence of plankton containing biotoxins. Aid may be granted where the suspension of the harvest lasts for more than four consecutive months or where the loss suffered as the result of the suspension amounts to more than 35% of the average annual turnover from the last three years. The maximum duration for which aid shall be granted is 12 months over the entire programming period (2007 to 2013).[319] Public health measures are subject to 'group 1' participation rates;[320] see Table 4.16 above.

3.8.4.4 Animal health measures

The EFF may allocate funds to control and eradicate aquaculture disease, pursuant with Council Decisions (EEC) 424/90 on expenditures in the veterinary field. Animal health measures are subject to 'group 1' participation rates;[321] see Table 4.16 above.

3.8.4.5 Inland fishing

Inland fishing is defined in Art. 33 as 'fishing carried out for commercial purposes by vessels operating exclusively in inland waters or by other devices used for ice fishing'. Support may be granted for the following activities:

- construction, extension, equipment and modernisation of inland fishing facilities;
- investments on board in accordance with the criteria applicable to sea fishing vessels;
- the reassignment of vessels operating in inland fishing to other activities outside fishing;
- temporary cessation which is foreseen in a legal management act.

No support will be given for the transfer of the ownership of a business.[322] Moreover, without prejudice to the above-mentioned funding opportunities, the permanent or temporary cessation of fishing as well as the construction of vessels operating in inland waters are not eligible for assistance.[323] Third, aid is not granted where it is likely to jeopardize the equilibrium between fleet size and resources available.[324] Fourth, Member States are required to ensure that inland vessels which have received aid under Art. 33 continue to fish in inland waters.[325]

---

[319] Art. 31 of Regulation 1198/06.

[320] Annex II of Regulation 1198/06.

[321] Annex II of Regulation 1198/06.

[322] Art. 33(5) Annex II of Regulation 1198/06.

[323] Art. 33(6) of Regulation 1198/06.

[324] Art. 33(7) of Regulation 1198/06.

[325] Art. 33(8) of Regulation 1198/06.

Modernisation of inland fishing facilities under Art. 33(2) is subject to the participation rates listed in 'group 1' of Annex II. Depending on the decision of the management authorities appointed by the Member States, investments on board vessels are subject either to 'group 2' or 'group 4' rates. The reassignment in Art. 33(3) corresponds with the rates provided in 'group 1', see Table 4.16 above.

3.8.4.6 Processing and marketing of fisheries and aquaculture products

The EFF may grant support to process and market fisheries and aquaculture products.[326] Investments are not eligible for aid if they concern products intended to be used and processed for purposes other than human consumption, 'with the exception of investments for the treatment, processing and marketing of fisheries and aquaculture product waste'.[327] In general, investment should promote sustainable employment in the fisheries sector.[328] Assistance for lifelong learning can also be granted.[329] According to Art. 35, the EFF may also support the 'construction, extension, equipment and modernisation of enterprises, concentrating on achieving the following objectives:

- improving working conditions;
- improving and monitoring public health and hygiene conditions or product quality;
- producing high quality products for niche markets;
- reducing negative impacts on the environment and improving the use of little used species, by-products and waste;
- producing of marketing new products, applying new technologies, or developing innovative production methods;
- marketing products mainly originating from local landings and aquaculture.

Investments in this area are also channelled to micro and small enterprises.[330] Aid is granted under the same conditions which apply to productive investments in aquaculture. No aid is granted to the retail trade sector.[331] Measures under Art. 35 are financed in accordance with participation rates laid down in group 4 of Annex II, see Table 4.16 above.

[326] Art. 34(1) of Regulation 1198/06.

[327] Art. 34(4) of Regulation 1198/06.

[328] Art. 35(2) of Regulation 1198/06.

[329] Art. 34(3) of Regulation 1198/06.

[330] Arts. 35(3) to (5) of Regulation 1198/06.

[331] Art. 35(6) of Regulation 1198/06.

3.8.5 Priority Axis 3: measures of common interest

EFF provisions allow for the support of measures of common interest, described as measures having a 'broader scope than measures normally undertaken by private enterprises and which help to meet the objectives of the CFP'.[332] Measures of common interest concern collective actions, protection and development of aquatic fauna and flora, fishing ports, landing sites and shelters, development of new markets and promotional campaigns, pilot projects, modification for reassignment of fishing vessels.

3.8.5.1 Collective actions

According to Art. 37, the EFF can fund initiatives which are implemented with the active support of operators themselves or by organisations acting on behalf of the producers or other organisations recognised by the Member States, which target the following aims:

- contribute to a better management of resources;
- promote selective fishing methods and gears;
- remove lost fishing gears;
- improve working conditions and safety;
- contribute to transparency of markets;
- improve quality and food safety;
- develop, restructure or improve aquaculture sites;
- promote investments concerning production, processing or marketing equipment and infrastructure including for waste treatment;
- upgrade professional skills, or develop new training methods;
- promote partnerships between scientists and operators;
- promote equal opportunities;
- improve management and control of access to fishing areas;
- create or restructure producer organisations and implement their plans;
- initiate feasibility studies relating to the promotion of partnerships with third countries.

Aid for the creation or restructuring of producer organisations shall be granted only for a maximum of three years following the date of recognition or following the date of the decision on restructuring of the producer organisation. Aid shall be digressive over the three years. Collective actions are financed, depending on the decision of the management authority, in accordance with rates provided in 'group 1' or 'group 3' of Annex II, see Table 4.16 above.

---

[332] Art. 36 of Regulation 1198/06.

3.8.5.2 Measures intended to protect and develop aquatic fauna and flora

Art. 38 lists three types of projects which can receive EFF support for the protection of aquatic fauna and flora: (1) the construction or installation of static or moveable facilities; (2) measures on the rehabilitation of inland waters; and (3) initiatives promoting the protection and enhancement of the environment in accordance with the NATURA 2000 guidelines on fishing activities. Such initiatives must be implemented by public or semi-public bodies, recognised trade organisations or other bodies appointed for that purpose by the Member States.[333] Measures in Art. 38 are financed, depending on the decision of the managing authority, either in accordance with participation rates laid down in 'group 1' or 'group 3' of Annex II; see Table 4.16 above.

3.8.5.3 Fishing ports, landing sites and shelters

The EFF may support investments in existing public or private fishing ports or landing sites. The investments must relate to the following list of activities:

- improving the conditions under which fisheries and aquaculture products are landed, processed, stored in the ports and auctioned;
- provision of fuel, ice, water and electricity;
- repairs to equipment and the maintenance of fishing vessels;
- construction, modernisation and extension of quays;
- computerised management of fishing activities;
- improving safety and working conditions;
- storage and treatment of waste;
- measures to reduce discards.

The EFF can grant funds for safety related investments for the construction or modernisation of small fishing shelters. The managing authority may choose to finance measures under Art. 39 in accordance with participation rates laid down in 'group 1' or 'group 3' of Annex II, see Table 4.16 above.

3.8.5.4 Development of new markets and promotional campaigns

The EFF may support measures to implement a policy of quality and value enhancement, development of new markets or promotional campaign for fisheries and aquaculture products.[334] Measures shall, in particular, relate to:

- initiation of promotional campaigns for fisheries and aquaculture products;

[333] Art. 38(3) of Regulation 1198/06.

[334] Art. 40 of Regulation 1198/06

- marketing of surplus or underexploited species;
- implementation of a quality policy for fisheries and aquaculture products;
- promotion of products obtained using methods with low impact on the environment;
- promotion of products recognised under the terms of Council Regulation (EC) 510/2006 on the protection of geographical indications and designations of origin for agricultural products and foodstuff[335];
- quality certification (including eco-labelling);
- campaign to improve the image of the fisheries and aquaculture products and the image of the fisheries sector;
- implementation of market surveys.

The managing authority may decide whether the measure under Art. 40 is subject to participation rates under 'group 1' or 'group 4'; see Table 4.16 above.

3.8.5.5 Pilot projects

The EFF may also support pilot projects, which aim at acquiring and disseminating new technical knowledge, carried out in partnership between industry members and a scientific or technical body.[336] Pilot projects may involve the following:

- testing the technical or economic viability of an innovative technology with the aim of disseminating the acquired knowledge;
- enable tests to be carried out on management plans and fishing effort plans, including, if necessary, the establishment of no fishing zones, in order to evaluate the biological and financial consequences, and experimental restocking;
- developing more selective and environmentally less harmful gear;
- testing alternative types of fishing management techniques.

All projects need to include adequate scientific follow-up and results must be made available to the public.[337] Measures in Art. 41 are financed in accordance with participation rates laid down in 'group 1' of Annex II; see Table 4.16 above.

3.8.5.6 Modification for reassignment of fishing vessels

The EFF may support the modification of fishing vessels for training and research purposes in the fisheries sector or for other activities outside fishing.[338] Operations are limited to public or semi-public bodies. Modifica-

---

[335] OJ 2006 No. L93/12.

[336] Art. 41 of Regulation 1198/06.

[337] Art. 41(2) and (3) of Regulation 1198/06.

[338] Art. 42 of Regulation 1198/06.

tions made under Art. 42 are financed in accordance with participation rates laid down in 'group 1' of Annex II, see Table 4.16 above.

3.8.6 Priority Axis 4: Sustainable development of fisheries areas

Due to reduced catches, the Community has attempted to reduce the economic dependence of coastal fishing areas on catching activities. To this end, support is granted to measures which accomplish the following:

- maintain the economic and social prosperity in these areas and add value to fisheries and aquaculture products;
- support diversification or the economic and social restructuring of areas facing socioeconomic difficulties as a result of changes in the fisheries sector;
- promote the quality of the coastal environment;
- promote national and trans-national co-operation between fisheries areas.

Areas eligible for aid must have a sea or lakeshore, and include ponds or river estuaries.[339] Assistance should target regions with a low population density, fishing areas in decline, or small fisheries communities.[340] Areas are selected by Member States.[341] Support under Priority Axis 4 is granted for:

- all measures under Axes 1, 2, and 3, with the exception of permanent and temporary cessation of fishing activities[342];
- strengthening competitiveness of fisheries areas;
- support for workers in or persons with a job linked to the fisheries sector to restructure, redirect and diversify their economic activities towards jobs outside the fisheries sector[343];
- adding value to fisheries products;
- supporting small fisheries and tourism related infrastructure and services;
- protecting the environment;
- restoring production potential damaged by natural or industrial disasters;
- promoting inter-regional and trans-national co-operation of actors;
- capacity building to prepare and develop local development strategies;
- running costs of groups (representing local public and private partners

---

339 Commission 'Fisheries and Aquaculture in Europe – European Fisheries Fund: driving sustainable development' (Brussels: 2006), p. 8

340 Art. 44(4) of Regulation 1198/06; areas shall be limited in size, *i.e.*, they shall be smaller than NUTS Level 3 within the meaning of Regulation (EC) 1059/2003 of the European Parliament and of the Council on the establishment of a common classification of territorial units for statistics (NUTS), OJ 2003 No. L154/1.

341 Art. 44(5) of Regulation 1198/06.

342 See Art. 44(3) of Regulation 1198/2007.

343 See Arts. 44(1)(b), (c) and (2) and (4) of Regulation 1198/2007.

from the various local relevant socioeconomic sectors).[344]

According to Art. 44(2), the EFF may finance up to '15 percent of the priority axis involved, measures such as the promotion and improvement of professional skills, worker adaptability and access to employment, particularly in favour of women', provided that these measures are part of a sustainable management strategy and have direct link to the above-mentioned measures in Art. 44. If the measures adopted are eligible for aid under Priority Axis 4 as well as under another Community support fund, Member States must clearly indicate which instrument has been applied.[345] Depending on the decision of the Member State's managing authority, measures pursued under Art. 44 are financed in accordance with participation rates laid down in groups 1- 4 of Annex II, see Table 4.16 above.

An important innovation is the implementation of support measures by local entities (the groups).[346] A group proposes and implements an integrated local development strategy based on a bottom-up approach in agreement with the relevant management authority.[347] According to Art. 45(4), the group itself determines how operations under the local development strategy are to be conducted and correspond to measures in Art. 44. It also requires that the greater part of the operations be led by the private sector. Territories covered by one group should be coherent and have sufficient critical mass in terms of human, financial and economic resources to support a viable local development strategy.[348]

3.8.7 Priority Axis 5: Technical assistance

Under the EFF regime, the Commission and the Member States can propose that aid for technical assistance to further the implementation of the EFF can be provided. On the initiative of the Commission and subject to a 0.8% ceiling of its annual allocation, the EFF may finance measures such as:

- evaluations, expert reports, studies concerning the EFF;
- measures aimed at the partner, the beneficiaries and the general public, including information measures;
- measures to disseminate information, networking and co-operation;
- the installation, operation and interconnection of computerised systems for management, monitoring, inspection and evaluation;

---

[344] Running costs for groups may not exceed, as a general rule, 10% of the total budget allocated to a fisheries area. Member States may decide to increase this threshold; see Art. 44 (5) of Regulation 1198/2007.

[345] Art. 44(6) of Regulation 1198/2007.

[346] Art. 45(1) of Regulation 1198/2007.

[347] Art. 45(2) of Regulation 1198/2007.

[348] Art. 43(3) of Regulation 1198/2007. As a general rule, areas shall be smaller than NUTS level 3 of the common classification of territorial units for statistics within the meaning of Regulation (EC) 1059/03 on the establishment of a common of territorial units for statistics (NUTS), OJ 2003 No. L154/1.

- improvement of evaluation methods and exchange of information on practices in this field;
- the establishment of trans-national and Community networks of actors in relation to sustainable development of fisheries areas.

Measures must be adopted in accordance with the advisory procedure under the Comitology Decision.[349]

At the initiative of the Member States, the EFF may finance actions relating to the preparation, management, monitoring, evaluation, publicity, control and audit of the operational programme, as well as networking up to 5% of its total amount.[350] The EFF can also finance actions under the operational programme relating to the improvement of administrative capacities of Member States whose regions fall under the convergence objective.[351]

4 Instruments promoting fisheries in the market organisation

As explained above, the CFP is also concerned with the organisation of a common market in fisheries products. Promotional interventions under the market organisation, however, only constitute a minor fraction of structural interventions (see Table below). The market policy's objectives are: stabilising the markets, guaranteeing and widening supplies as well as increasing profitability of production.[352] Currently, the market organisation regime is laid down in Regulation 104/00 on the common organisation of the markets in fishery and aquaculture products.[353] According to this regulation, Community market policy in fisheries builds on four basic mechanisms: common marketing standards,[354] producer organisations, a common price system, and a system of trade with third countries. To make clear the promotional aspects of the common market policy, the structure of the common price system as well as the producer organisations must be examined in greater detail.

To stabilise prices, the Community will intervene in the market. To achieve this, it does not institute price guarantees, but instead modifies the supply and demand mechanism by increasing prices by creating artificial scarcity.[355] This

349 Art. 46(4) of Regulation 1198/2007.

350 Art. 46(2) of Regulation 1198/2007.

351 Art. 46(4) of Regulation 1198/2007.

352 Regarding market stability see Reasons (4), (11) and (27); regarding increasing profitability see Reason (6) and Arts. 2 and 3; regarding increasing the variety of supply, see Reason (8) of Regulation 104/2000.

353 OJ 2000 No. L17/22.

354 Recently, rules on marketing have been complemented by rules on consumer information, see Commission Regulation (EC) 2065/01 laying down detailed rules for the application of Council Regulation (EC) 104/00 as regards informing consumers about fishery and aquaculture products, OJ 2001 No. L 287/6.

355 Churchill, *EEC Fisheries Law*, p. 233.

system was established in 1970 and has not been changed substantially.[356] In instituting such measures, the Council will first, by qualified majority, determine guide prices for different species before the beginning of each year.[357] Prices are set for products at the first stage of marketing, *i.e.*, the sale of fish from fishers to wholesalers or retailers.[358] Guide prices are based on the 'average of prices recorded for a significant proportion of Community output on wholesale markets or in ports during the three fishing years immediately preceding the year for which the price is fixed'.[359] They also take into account trends in production, in demand, the need to stabilise markets, to help support producers' income and consumer interests. Next, the Commission fixes Community 'withdrawal prices', which cannot exceed 90% of the guide prices set by the Community.[360] Producer organisations also have the option of fixing withdrawal prices.[361] When fish prices fall below withdrawal prices, fish can be removed from the market by the producer organisations.[362] Producer organisations receive remuneration for certain amounts of their expenditures, only where they apply Community withdrawal prices (or a price close to the Community withdrawal price).[363] They are also required to withdraw products that do not meet market standards.[364] Withdrawn products have to be 'disposed of for purposes other than human consumption or in such a way as not to interfere with normal marketing of other products' as well as with the products in question.[365] According to Art. 21(3), the scale of remuneration ranges up to 85% of the withdrawal price. When 4% of annual quantities of the product put up for sale are withdrawn, the producer organisation receives 85% of the withdrawal price. Where 5 to 10% are withdrawn, remuneration rates range between 55 and 75% for different species. For withdrawn products that exceed 10% of annual quantities of the product put up for sale, no aid will be granted. Thus, increasing withdrawals results in decreasing remuneration. This digressive mechanism is intended to motivate producer organisations to match supply and demand. This system was introduced in response to a European Parliament resolution objecting to

---

[356] See Arts. 7-16 of Regulation 2142/70 on the common organisation of the market in fishery (and aquaculture) products, OJ 1970, No. L236/5; Arts. 8-17 of Regulation 100/76, OJ, 1976, No. L020/1; Arts. 9-18 of Regulation 3796/81, OJ 1981, No. L379/1; Arts. 9-21 Regulation 3687/91, OJ 1991, No. L354/1; Arts. 8-18 of Regulation 3759/92, OJ 1999, No. L388/1.

[357] Art. 18(3) of Regulation 104/00.

[358] Churchill, *EEC Fisheries Law*, p. 233.

[359] Art. 18(2) of Regulation 104/00.

[360] Art. 20(3) of Regulation 104/00, referring to Art. 38(2) which again refers to the management procedure under Arts. 4 and 7 of the Comitology Decision, *i.e.*, Council Decision 468/1999, OJ 1999 No. L184/23.

[361] Art. 17(3) of Regulation 104/00; R.C. Fischer, 'Die gemeinsame Fischereipolitik', in Hilf (ed.), *Kommentar zur Europäischen Union*, (München: 1996), p. 28.

[362] Arts. 17 and 21 of Regulation 104/00.

[363] Art. 21(1)(a) of Regulation 104/00.

[364] Art. 21(1)(b) of Regulation 104/00.

[365] Art. 21(2) and 17(2) of Regulation 104/00.

the fact that in 1980 more than 100,000 tonnes of fish had been withdrawn and destroyed.[366] Where producer organisations apply their own withdrawal prices, they may finance these measures by creating their own intervention funds.[367] To complement the digressive remuneration system, keeping waste to an absolute minimum, the Community adopted a system of so-called 'carry-over-aid', *i.e.*, according to Art. 23, processing and storing withdrawn products can be eligible for aid. Products have to meet certain quality, size and presentation requirements. In addition, they have to be supplied by a member producer. Remuneration will, in principle, only be granted for withdrawals which do not exceed 18% of the annual amount put up for sale. The amount of aid may not exceed the amount of costs for stabilisation or storage.[368] The Member States remunerate producer organisations, and are later refunded by the Community's agricultural funds.[369]

Market interventions are linked to the fisheries management system, particularly the quota system. Community aid for the organisation of the markets shall be granted in respect of products from a stock or group of stocks only up to the limit of any quantities allocated to the Member States in question from the total volume of allowable catches for the stocks in question.[370]

Table 4.17: Community assistance under the common organisation of markets between 1973 and 1998 (in total mil)[371]

1973	1974	1975	1976	1977	1978	1979	1980	1981	1982	1983	1984	1985	1986	1987
1.19	1.18	9.43	10.61	8.09	14.11	17.21	23.04	28.02	?	25.45	14.59	18.49	17.23	17.45

1988	1989	1990	1991	1992	1993	1994	1995	1996	1997	1998
46.9	23.99	23.59	26.15	32.07	30.39	33.04	28.1	25.31	21.8	10.9

5 Other sources of assistance to the Community's fisheries sector

The Community's fisheries sector receives further assistance from other public funds and sources as well as for purposes other than promot-

[366] Churchill, *EEC Fisheries Law*, pp. 235-236.

[367] Art. 17(3) of Regulation 104/00.

[368] Art. 23(3) of Regulation 104/00.

[369] Since 2005, the 'European Agricultural Guidance and Guarantee Fund (EAGGF)' has been replaced by the 'European Agricultural Guarantee Fund (EAGF)' and the 'European Agricultural Fund for Rural Development (EAFRD)'. See Art. 35(1) of Regulation 104/00 in combination with Arts. 1(2)(b), 2 (2) and 5 of Regulation 1258/99, OJ 1999 No. L171/19.

[370] Art. 35(2) of Regulation 104/00.

[371] See M. Holden, *The Common Fisheries Policy*, p. 36; C. Lequesne, *The politics of fisheries in the European Union*, p. 95.

ing or adjusting its structures or stabilising markets. It would go beyond the scope of this Report to examine these promotional activities in depth; however, a brief overview will be provided below.

Two additional Community structural funds provide aid to the fisheries sector: The European Regional Development Fund ('ERDF') and the European Social Fund ('ESF'). However, the funding and its effects can be characterised as indirect. The ERDF aims at reducing the disparities in socioeconomic development throughout the Community. Thus, financial transfers are made to less developed regions. Only a small share of the total budget is allocated to poor areas, which are highly dependant on fishing. Between 1994 and 1999, aid amounted to 96.89 million ecus. The ESF aims at adapting people's occupational skills to the changing demands of the Community's labour markets. Between 1994 and 1999, the ESF gave about 63.13 million ecus to the fisheries sector.

As mentioned many times throughout this chapter, the Community finances access to the fishing zones of third countries in benefit of its distant-water fleet. Payments made directly to the third countries amounted to over one billion euros between 1993 and 1997.[372] While the Community financed about 82.8% of the agreement's costs, the fisheries sector paid the remaining share.[373]

Community funds are given for research activities within the fisheries sector under its 'research and technological development framework programmes'. Each programme focuses on a specific research area. The objective of the framework programme, covering the period from 1991-1994, for example, was to 'establish harmony between the production of biological resources and the needs of consumers and the processing industry.' Between 1991 and 1998, under the third and fourth programmes, about 163 million ecus were made available for marine and aquaculture projects, and an additional 2.9 million ecus were spent on accompanying measures, as well as 7.9 million ecus for scholarships.[374]

Community aid is also paid to Member States to support their control and monitoring efforts. Aid is granted for buying or modernising ships, aircrafts, and vehicles as well as other technical systems. The budget for the period from 1996 to 2000 was limited to 205 million ecus.[375]

Finally, Member States also give financial aid to the fisheries sector. National transfers are comprised of three kinds of payments.

First, Member States contribute their share to all Community projects that are co-financed. For example, national contributions to Community's structural

---

372 IFREMER, '*Evaluation of the Fisheries Agreements Concluded by the European Community, European Contract No. 97/S 240-152919*', (1999) available online at: http://ec.europa.eu/fisheries/publications/studies/, p. 15.

373 *Ibid.*, p. 15.

374 OECD, *Transition to Responsible Fisheries – Government Financial Transfers and Resources Sustainability: Case Studies*, (Paris: 2000), pp. 50-52.

375 *Ibid.*, pp. 52-54.

measures between 1994 and 1999 amounted to the large sum of 1 298 million ecus.[376]

Second, Member States may also grant additional aid in compliance with the narrow requirements established in the Community rules on state aid. The Commission must receive notice of all national measures providing aid to the fisheries sector prior to their entering into force. The Commission considers, against the background of Art. 92 of the Treaty and its own 'Guidelines for the Examination of State Aid to Fisheries and Aquaculture',[377] whether the proposed measures have the potential to distort competition in the common market. The Commission may dismiss or approve the proposals. To speed up this procedure, the Commission adopted block exemptions.[378] As a result thereof, Member States can currently give companies a certain amount of aid over a given period of time without requiring advance clearance from the Commission. In 2004, the Commission adopted a *de minimis* regulation which established a ceiling of €3,000 of aid per beneficiary per three-year period. In 2007, however, this ceiling was upgraded to €30,000.[379] For example, in 1997, state aid (excluding co-financing) amounted to about 64.2 million ecus.[380]

The third category relates to Member States paying the costs of general services, such as management, control and research costs. In 1997, these costs were about 472 million ecus.[381]

6 The future of promotion under the World Trade Organization (WTO)

In November 2007, the WTO group on 'Rules' issued a draft proposal for a revised version of the Subsidies and Countervailing Measures Agreement. The draft agreement largely prohibits subsidies that contribute to maintaining or building overcapacities.[382] Annex VIII of the revised SCM-Agreement i.a. provides that subsidies for the construction or modernisation of fleets, vessel transfers, operating costs, port infrastructure, income or price supports,

376 *Ibid.*, p. 56.

377 OJ 2004, No. C229/5.

378 Commission Regulation (EC) 1595/04 on the application of Articles 87 and 88 of the EC Treaty to State aid to small and medium-sized enterprises active in the production, processing and marketing of fisheries products, OJ 2004, No. L291/3.

379 See Art. 3(2) of Commission Regulation (EC) 875/2007 on the application of Articles 87 and 88 of the EC Treaty to *de minimis* aid in the fisheries sector and amending Regulation 1860/2004, OJ 2007, No. L193/6.

380 OECD, *Transition to Responsible Fisheries – Government Financial Transfers and Resources Sustainability: Case Studies*, p. 57.

381 *Ibid.*, p. 58.

382 The draft is available online at: http://www.wto.org/english/tratop_e/rulesneg_e/ rules_chair_text_nov07_e.htm.

as well as access agreements, are forbidden. General exceptions are made for the funding of safety improvements, gear selectivity, as well as for environmental and compliance measures. While prohibitions generally do not apply to least-developed countries, developing countries will be exempted under certain conditions.

CHAPTER 5

Towards an Integrated Approach to Promotion and Management of Fisheries

The following chapter proposes an integrative approach to promoting and managing fisheries. This approach will serve as a benchmark to assess the state of integration of the Community's fisheries management and its promotional activities from 1970 to 2008. Subsequently, it will try to draw conclusions from the CFP experiences which complement the integrative approach provided here.

1 The logic and limits of promotion and management

Traditionally, policy makers treat the promotion and management of fisheries as separate categories, drawing up their policy approaches in relative isolation from each other.[1] This can lead to a lack of coherence in fisheries policy as a whole.[2] In this author's opinion, in order to integrate these policy areas, promotion should aim first and foremost at adjusting consumption and production capacity to the size commensurate with sustainable exploitation limits. Pursuing other objectives, such as increasing the relative size and competitiveness of the fleet or increasing supplies, threatens fish stocks by increasing overall production power. The approach is based on two observations:

First, the exploitation of marine capture resources ultimately depends on the level of available fish stocks. To maintain sustainable exploitation, fish stocks cannot be continuously fished beyond sustainable limits. Thus, this level of exploitation must be the initial point of reference for all other policy segments, promotion in particular.

Second, as long as there is no rights-based management system, the open access conflict reoccurs within the promotion area. Economic theory assumes investment in the fishing industry up to the point at which revenues equal production costs (the point of Maximum Economic Yield). This means that, in general, exploitation would be below the MSY-level.[3] However, economics also teach that open access not only fuels the 'race to fish', but it also encourages investment in the sector, creating overcapacity.[4] This dynamic, again, is aggravated by the granting of subsidies that alter cost-revenue structures, by reduc-

1 See, with regards to investments in fishing capacities, D. Greboval, 'The International Plan of Action for the Management of Fishing Capacity: Retrospect and Prospect' p. 563.

2 Y.H. Song, 'The Common Fisheries Policy of the European Union: Restructuring of the Fishing Fleet and the Financial Instrument for Fisheries Guidance', pp. 537-577; Court of Auditors Special Report No. 3/93 concerning the implementation of the measures for the restructuring, modernisation and adaptation of the capacities of fishing fleets in the Community together with the Commission's replies, OJ 1994 No. C2/1, paras. 1.18.-1.23; W.E. Schrank, *Introducing Fisheries Subsidies – FAO Technical Paper*, pp. 32 et seq.; OECD, Transition to Responsible Fisheries – Government Financial Transfers and Resource Sustainability, pp. 6-226.

3 Iudicello *et al.*, *Fish, Markets, and Fishermen – the Economic of Overfishing*, pp. 47-49.

4 Hans Frost and Peder Andersen, 'The Common Fisheries Policy of the European Union and fisheries economics' (2006) 30 *Marine Policy*, p. 738; Gordon R. Munro, 'The economics of overcapitalisation and fishery resource management: a review', in Hatcher and Robinson (ed.), *Overcapacity, overcapitalisation*

ing production costs and investment risks.[5] Such overcapacity creates economic waste[6] and poses a major obstacle to the successful implementation of fisheries management objectives.[7]

To guarantee the long term economic and ecologic viability of the fishing industry, both *management and promotion* policies must aim at maintaining fisheries resources at sustainable levels. *Management* policies must ensure that fish stocks are not continuously exploited beyond MSY-levels as qualified by ecosystem requirements. *Promotion* measures must reduce production capacity in the industry to levels commensurate with stocks available (under the sustainable management regime). Ideally, the production sector should be of a size where it can produce efficiently and where demand for fish products does not contribute to overexploitation.[8]

Several components of an integrated approach to promotion and management of fisheries can be deduced from these premises. An integrated approach requires the following actions of coastal states:

First step:

- (scientifically) assess which quantity of marine capture resources they can take from the seas (including their own as well as third countries' waters) without jeopardising the long-term existence of fish stocks and the ecosystem;
- adopt and effectively implement sustainable exploitation limits which consider ecosystem requirements.

Second step:

- assess the impact of the production sector on fish stocks, identifying which segment of the production sector pressures which stocks.

---

and subsidies in European fisheries. Proceedings of the first Concerted Action workshop on economics and the Common Fisheries Policy, (Portsmouth: 1999), pp. 7-23 (see particularly, p. 20).

5 Lindebo, 'Role of Subsidies in EU Fleet Capacity Management' (2005) 20 *Marine Resources Economics*, pp. 448-449; A.S. Khan *et al.* 'The nature and magnitude of global non-fuel subsidies', pp. 7-8.

6 Erik Lindebo, 'Role of Susbidies in EU Fleet Capacitiy Management', p. 447; OECD, *Towards Sustainable Fisheries* (Paris: 1997), pp. 9-10.

7 J. Sutinen, A. Rieser and J.R. Gauvin, 'Measuring and Explaining Noncompliance in Federally Managed Fisheries' (1990) 21 *Ocean Development and International Law*, pp. 335-372; on the CFP experiences, see Report 1991 from the Commission to the Council and the European Parliament on the Common Fisheries Policy, SEC (91) 2288 of 18 December 1991; Commission Report Com (2007) 167 final on the monitoring of the member states' implementation of the common fisheries policy 2003-2005; see also Introduction Section of the International Plan of Action for the Management of Fishing Capacity, available online at www.fao.org, para. 1.

8 A similar argument with regards to the adjustment of the fleet size is proposed in the Court of Auditors Special Report No. 3/93 concerning the implementation of the measures for the restructuring, modernisation and adaptation of the capacities of fishing fleets in the Community together with the Commission's replies, OJ 1994 No. C2/1, para. 1.22; see also D. Greboval, 'The International Plan of Action for the Management of fishing Capacity: Retrospect and Prospect', p. 569.

- restructure the sector to reduce production overcapacities to levels commensurate with the production potential of fish stocks, *i.e.*, sustainable exploitation limits.

Third step:
- assess whether public promotion schemes actually adjust production capacity so that it is more in line with management objectives and conservation measures;
- change national promotion policies so that they accord with management objectives by progressively eliminating subsidies that contribute to the maintenance or build-up of unsustainable production powers.

Fourth step:
- repeat steps one to three periodically.
- adjust policies, where management and promotion do not contribute to the attainment of sustainable management objectives.

The efficient allocation of public funds and compatibility with environmental objectives heavily depends on public participation and transparent procedures.[9] *Transparency* and *participation* are crucial elements of an integrated approach to promotion and management of fisheries.

An assessment of the history of the CFP provides valuable information on the effects of varying degrees of integration of the promotion and management of fisheries. This approach also exposes the difficulties associated with improving integration.

2 The state of integration of fisheries promotion and management

The following section will investigate the state of integration of promotion and management measures over the 38 years of the CFP. For practical reasons, the history of the CFP will be subdivided into six phases. Each phase will be assessed according to the following structure:
- describe the pertinent developments of the management regime;
- describe the pertinent development of the promotion regime;
- assess the state of integration;
- review the effects of the management and promotional measures on *fleet developments* (fleet size is a pertinent indicator of a fisheries sector's overall production power).[10]

[9] C. Coffey, 'Reforming the European Union Fisheries Subsidies – A briefing paper by the Institute for European Environmental Policy (IEEP), available online www.panda.org., p. 3.

[10] It must, however, be kept in mind that in a globalising market, it is not only the fleet size of an individual fisheries nation that determines the overall demand for fish products.

2.1 The first phase, 1970 – 1982[11]

In the 1970s and the early 1980s, the CFP's prevailing goal was to increase production to guarantee Europe's food supplies.[12] This motivation arose from an acute awareness of the suffering experienced by Europeans as a result of starvation during and after World War II. Accordingly, the management regime was, at that time, underdeveloped. Only a few laws were in place that gave the Community the power to adopt conservation measures.[13] From 1978 to 1982, the Commission only proposed 'non-binding' TACs. The competence to adopt management rules still largely lay with the Member States. Besides, until 1978, there was no scientific expertise in the Commission.[14]

Public funding, at that time, basically aimed at expanding the capture fisheries and aquaculture sectors. From 1980, due to a largely overcapitalised long-distance water fleet, the Community also began to 'externalise' its overcapacity problem by financing the conclusion of third country access agreements with the Faeroe Islands, Norway, Senegal, Guinea Bissau.[15]

Integration was low during the first phase. Member States did not adopt and implement sustainable TACs as recommended by scientists, nor did they consider the impact of the production sector on the state of the resources. Accordingly, promotional measures did not aim at reducing production capacity so that it was in line with sustainable management objectives; instead, they had the effect of expanding production power.

The combination of an underdeveloped management regime and promotion measures lead to an enormous increase in fleet overcapacities. In 1970, the Community fleet's size was 794 000 gross registered tonnage (GRT). By 1983, it had increased by 64% to 1 303 000 GRT; and engine power was estimated at 1 983 000 kilowatts (kW). This figure had almost doubled by 1977 to 3 691 000 kW.[16] One must also take into account that from 1977, these increased fishing capacities were primarily concentrated in the Community EEZ.

---

11 Information for the time prior to this phase, see H. -Y. Song, EC's Common Fisheries Policy in the 1990s', *Ocean Development and International Law* (1995) 26, pp. 36-37; Holden, *The Common Fisheries Policy*, pp. 17-18. See also the 'first' proposal of the Commission for a CFP, i.e. 'Report on the Situation in the Fisheries Sector of the EEC Member States and the Basic Principles for a Common Policy', (67/196/EEC), OJ 1967, p. 862.

12 M. Holden, *The Common Fisheries Policy*, pp. 21 and 39 et seq.

13 For example, Art. 5 of Regulation 2141/70.

14 M. Holden, *The Common Fisheries Policy*, p. 21.

15 OJ 1980 No. L226/12; OJ 1980 No. L 226/47; OJ 1980 No. L226/16; OJ 1980 No. L226/33.

16 These numbers are all taken from M. Holden, *The Common Fisheries Policy*, p. 22. They do not include Italian data and the tonnage of the Dutch fleet.

2.2 The second phase, 1983 – 1986

The Community adopted a resource management regime on 25 January 1983.[17] The basic management regulation provided a legal framework for the adoption of Community conservation measures, such as TAC, technical measures, minimum fish sizes.[18] A control regulation had been adopted in 1982.[19] However, Member States were unwilling to accept cutbacks in their fishing opportunities. In order to prevent destabilization of the new policy (referred to by *Holden* as the 'the house of cards'), the Commission proposed high TACs, which basically referred to existing (and biologically unsound) exploitation rates.[20] As *Hegland* and *Raakjær* have put it, the first years served to 'institutionalise' the yet fragile CFP package.[21]

The Community continued existing promotional schemes and also introduced new types of subsidies. It promoted access to the fishing grounds of Guinea, Equatorial Guinea, Sao Tomé and Principe, and Madagascar,[22] the exploitation of under- or unexploited fish species, and the temporary lay-up of vessels. It introduced support for the marketing for fisheries products and assistance for exploratory expeditions and joint ventures. Special financial grants were made available to poorer regions on the Mediterranean coasts building-up production capacities. Some financial assistance was granted for the protection of coastal areas and for the biological protection of marine areas and the creation of marine parks.[23]

Integration of the promotional and management regimes remained low throughout the second phase. The basic idea of integration had found its way in CFP legislation: management policy purportedly aimed at 'conservation of the biological resources of the sea and their balanced exploitation on a lasting basis and in appropriate economic and social conditions'.[24] However, the Council interpreted 'balanced exploitation' and 'rational use' in a way that was biologically unsustainable, setting TACs above healthy levels. Nevertheless, during this time the Community had begun to consider the connection between the produc-

---

[17] Regulations 170/83 and 171/83. According to Art. 102 of the 1972 Accession Treaty, the management policy should have been adopted no later than 1978. A control regulation had already been adopted in 1982, Regulation 2057/82.

[18] The decision to adopt TACs and not impose fishing limits by effort regulation was influenced by the fact that policy makers were accustomed to this instrument through their work in NEAFC, see Hegland and Raakjær, 'Recovery Plans and the Balancing of Fishing Capacity and Fishing Possibilities', p. 142.

[19] Council Regulation (EEC) 2057/82 establishing certain control measures for fishing activities by vessels of the Member States, OJ 1982 No. 220/1.

[20] Holden, *The Common Fisheries Policy*, p. 57.

[21] Hegland and Raakjær, 'Recovery Plans and the Balancing of Fishing of Fishing Capacity and Fishing Possibilities', p. 142.

[22] OJ 1983 No. L111/2; OJ 1984 No. L 188/2; OJ 1984 No. L54/2; OJ 1986 No. L73/26.

[23] Regulation 2908/83, OJ 1983 No. L290/1 and OJ 1985 No. L197/1.

[24] See Art. 1 of Regulation 170/83, OJ 1983 No. L24/1.

tion sector and the state of fish stocks. As a first step, the Community attempted to *link the promotion of the fleet* to the available resources. The Member States' MAGPs required Commission approval, which was to be determined on the basis of whether they were a 'suitable framework for projects which may qualify for financial support from the Community'.[25] Throughout the MAGP I period, Member States were required to maintain capacity at 1983 levels. Ultimately, the effort to adjust promotion to reduce fleet capacity failed for various reasons:

- insufficient classification of the fleet into categories related to the species caught, fishery zones and methods of fishing;
- monitoring of the fleet based on a limited number of physical capacity parameters only, without any consideration of the remaining parameters and fleet activity (fishing effort);
- absence of short- and long-term objectives based on the actual situation of particular stocks;
- lack of statistical data and inadequate measures to control fishing effort;
- non-obligatory status of the programme.[26]

The Community, however, failed to take this third step of integration, meaning that it had not assessed whether its promotion schemes really contributed to adjustments in production capacity in accordance with management objectives. From 1983 to 1986, those subsidies that tended to increase production greatly exceeded those that decreased fishing pressure.

There were a few major consequences arising from this contradictory and disintegrated approach. From 1983 to 1987, the Community fleet had increased by 315 519 GRT to 1 618 519 GRT, reaching approximately twice the size of the 1970 fleet. Engine power had almost tripled in comparison to 1970 levels.[27] In 1970, the Community engine power lay at estimated 1 983 000 kW and in 1977 at 3 691 000 kW; by 1987 it had increased to 6 476 471 kW. These numbers still excluded the Spanish and Portuguese fleets.[28]

---

[25] See Art. 5 (2) of Regulation 2908/83.

[26] See Com SEC(91) 2288 final, Report 1991 from the Commission to the Council and the European Parliament on the Common Fisheries Policy ('Commission Report – 1991'); Court of Auditors, Special Report No. 3/93 concerning the implementation of the measures for the restructuring, modernisation and adaptation of the capacities of fishing fleets in the Community together with Commission's replies ('Court of Auditors Report – 1993'), OJ 1994 No. C2/1, pp. 18. *et seq.*

[27] M. Holden, *The Common Fisheries Policy*, p. 22. These numbers deviate a little from data available in the 'Court of Auditors Report – 1993', see pp. 15-16 (footnote 54).

[28] *Ibid*, p. 22.

2.3 The third phase, 1987 – 1993

In response to widespread criticisms by environmental NGOs, the Commission, fisheries experts and stakeholders,[29] the Community reformed its management regime in 1992. The Council adopted new management and control regulations (Regulations 3760/92 and 2847/93).[30] Under the new regime, the Community has been able to limit the amount of time vessels could spend at sea (*i.e.*, limit fishing effort), adopt multi-annual management plans, and introduce a licensing system.[31] The control and enforcement system was not just limited to technical aspects, but extended to structural and marketing aspects of the CFP. The Commission also began to propose TAC rates closer to those recommended by scientists.[32] However, in the Council, Member States continued to oppose any reductions in their fishing opportunities.[33]

On the promotional side, the contradictory spending practices became increasingly visible. On one hand, promotion schemes were altered to reduce the total production capacities. MAGP II aimed at reducing the fleet by 3% in terms of tonnage and 2% for engine power. On the other hand, the total amount of aid for the renewal and modernisation of vessels was raised substantially, and included the financing of construction and modernisation projects in Spain and Portugal. Furthermore, the Community took over Spanish and Portuguese access agreements with third countries.[34] In addition, support for processing and marketing as well as investments in fishing ports was bolstered.[35] Overall,

---

29 See Com SEC(91) 2288 final, Report 1991 from the Commission to the Council and the European Parliament on the Common Fisheries Policy ('Commission Report – 1991'); Court of Auditors, Special Report No. 3/93 concerning the implementation of the measures for the restructuring, modernisation and adaptation of the capacities of fishing fleets in the Community together with Commission's replies ('Court of Auditors Report – 1993'), OJ 1994 No. C2/1, pp. 18. *et seq*; The first edition of Mike Holden's book, *The Common Fisheries Policy*, was published in 1992. Another independent expert group published a report, which supported most of the Commission's views, the so-called 'Gulland Report' (no longer available).

30 Council Regulation (EEC) 2847/93 establishing a control system applicable to the common fisheries policy, OJ 1993 No. L261/1.

31 See Art. 4(2)(d), Arts. 5 and 7(1) and Art. 8(3) of Regulation 3760/92.

32 *Ibid.*, p. 67.

33 M. Holden, *The Common Fisheries Policy*, p. 67.

34 U.S., Seychelles, Mozambique, Gambia, Angola, Dominican Republic, Comoros, Mauritius, Sierra Leone, Cap Verde, Ivory Coast and Tanzania, see OJ 1984 No. L272/1; OJ 1987 No. L160/1; OJ 1987 No. L201/1; OJ 1987 No. L146/1; OJ 1987 No. L341/1; OJ 1993 No. L299/1; OJ 1988 No. L137/24; OJ 1989 No. L159/1; OJ 1990 No. L125/27; OJ 1990 No. 212/1; OJ 1990 No. L379/1; OJ 1990 No. L379/24.

35 Regarding the processing and marketing segments, new measures had been introduced in 1989; Regulation 355/77 had been replaced by Regulation 4042/89 on the improvement of the conditions under which fisheries and aquaculture products are processed and marketed, OJ 1989 No. L388/1.

aid granted for purposes that negatively impacted fish stocks still exceeded those amounts earmarked for conservation purposes.[36]

The integration of promotional and management strategies was still insufficient. The situation remained unchanged in the second phase. Management had not earnestly committed to limiting exploitation; and even though it increasingly tried to adjust fleets to correspond with available exploitation opportunities, overall promotion actually lead to an increase in production capacity.[37]

With regards to the effects of such a policy, a scientific report from 1992 estimated that fishing power would have to be reduced by approximately 40% in order to adjust fishing mortality to reflect available resources.[38]

2.4 The fourth phase, 1994 – 1999

At the end of the preceding phase, the Community had introduced new management strategies: multi-annual TACs, effort limitations and a licensing system. In addition, new management requirements had developed in international law. It would be expected that the CFP would be re-orientated, taking a more sustainable approach. In actual fact, however, this did not occur. The Council continued to set exploitation limits high above those recommended by scientists[39] and it did not make use of the possibility to adopt more effective management instruments, *i.e.*, multi-annual TACs[40] and effort limitations.[41] Nevertheless, a licensing system was adopted. All Community vessels were required to have Community fishing licenses by the end of 1993.[42] Additionally, special fishing permits were mandatory for fishing in the waters of third countries, and in sensitive areas, such as the so-called Western Waters.[43]

---

36 See Table 3 a. See also, Report 1991, p. 29.

37 M. Holden, *The Common Fisheries Policy*, p. 67.

38 Erik Lindebo, 'Role of Subsidies in EU Fleet Capacity Management', p. 446 (citing the 'Gulland Report').

39 Commission Communication Com(2001) 135 final Green Paper on the Future of the Common Fisheries Policy vol. I, p. 4.

40 *Ibid.* pp. 6-7.

41 Commission Communication Com(2001) 135 final Green Paper on the Future of the Common Fisheries Policy, vol. II, pp. 5-6.

42 Art. 5(1) of Regulation 3760/93. Subsequent Regulation 3690/93 OJ 1993 No. L341/93; Regulation 1627/94 OJ 1994 No. L 171 and Regulation 3317/94 OJ 1994 No. L350 elaborate on the obligation in Art. 5(1) and establish further requirements concerning licensing and permits, and fishing operations in third country waters.

43 Art. 7(1) of Regulation 3760/93. Subsequent Regulation 3690/93 establishing a Community System laying down rules for the minimum information to be contained in fishing licenses, OJ 1993 No. L341/93; Regulation 1627/94 laying down general provisions concerning special fishing permits, OJ 1994 No. L 171/7 and Council Regulation (EC) 3317/94 setting out general provisions concerning the authorisation in the waters of a third country under a fisheries agreement, OJ 1994 No. L350/13.

On the promotional side, administrative structures and spending schemes had been marginally improved. All public aid had to be consistent with the newly introduced 'community programmes'. These programmes aimed at strengthening co-ordination between the Member States and the Community with regards to the development and adjustment of the fisheries sectors of the Member States. First, each Member State had to provide detailed and comprehensive information on their fishing industry and current funding activities. They also had to present a strategy for the adjustment of this sector – the so-called 'sectoral plan'. Next, the Commission examined whether these plans were consistent with the aims and objectives of the CFP in general, the FIFG, and MAGPs. From 1991 to 1993, the Community increased its efforts to reduce fishing pressure under MAGP III and MAGP IV, setting more ambitious goals for the reduction of fishing efforts. Accordingly, aid for the adjustment of the fleet was raised. Grants were provided for permanent reductions – scrapping vessels, the permanent re-assignment of vessels for other non-fishing purposes, and permanent transfer to third countries. The Community also became increasingly aware of the need to offset the social consequences of the restructuring process. Assistance was made available, for example, for early retirement schemes or compensatory payments to fishers who withdrew from fishing permanently. With regards to the construction and modernisation of the fleet, the Community began channelling its financial assistance to different fleet segments, reducing support for the powerful large-scale vessels, in particular. Moreover, a special programme called the PESCA-Programme was adopted to support the fisheries sector during the restructuring process. This programme aimed at promoting the transformation, diversification and the redeployment of the sector's work force and also provided funding for coastal regions affected by the continuing decline of the fisheries.

Integration of promotional and management policy improved to a marginal extent during the fourth phase. Improvements were largely the result of improved understanding of the connection between production sector and the state of fisheries resources. To improve the development and adjustment of the fisheries sectors, Member States now had to assess their sectors thoroughly under the so-called Community Programmes. In addition, reduction goals under MAGP III and IV were more ambitious than under the two prior programmes. However, under the management regime, the Council had continued to set TAC levels above those recommended by scientists and new managing instruments were not used. In addition, the Community continued to advance subsidies for investments to increase fishing and production power. Aid, which tends to increase pressure on marine resources, still greatly exceeded the amounts granted for sectoral adjustments. For example, the Community provided 640.71 million ecus for fleet modernisation and construction, 190.72 million ecus for facilities at fishing ports, and 601.61 million ecus for the processing and marketing sectors. These numbers do not even include ancillary expenditures, such as payments for temporarily laying-up vessels, support for the search of market outlets or payments for access to foreign waters.

In 2001 and 2002, the Commission described the effects of this policy on the status of the Community fleet:

> '[...] the current fleet is much too large. Technological progress is increasing the efficiency of fishing vessels and it undermines the efforts of capacity reduction programmes. The Multi-Annual Guidance Programmes (MAGPs) were set by the Council at levels that were not ambitious enough to address the problem of excess capacity effectively and have often not been enforced. They were also complex to administer. Subsidies for the construction/modernisation and running costs may have aggravated this situation.'[44]
>
> '[...] the fishing capacity of the Community fleet far exceeds that required to harvest the available fishery resources in a sustainable manner. The most recent scientific advice from ICES suggests that the level of fishing mortality of the main Community fish stocks need to be reduced by between one-third and one-half, depending on the type of fishery [...] and area concerned, in order to ensure sustainable fishing.'[45]

2.5 The fifth phase, 2000 – 2006

In the recent years, the Council's overall approach to fisheries conservation seems to have changed to a limited extent. *Hegland* and *Raakjær* have observed a 'crisis driven change' in the Council's decision making, increasingly emphasising conservation.[46] The Council adopted a new basic management regulation in 2002.[47] The regulation tries to move away from the destructive end-of-year negotiation pattern by taking more seriously the possibility to adopt multi-annual management and recovery plans.[48] The second important step is to move towards a combination of management strategies, including the combination of TACs and effort restrictions.[49] A third step is to equip the Commission with new control and sanction competences. Further-

---

44 Commission Communication Com(2001) 135 final Green Paper on the Future of the Common Fisheries Policy, vol. I, p. 10.

45 Commission Communication Com(2002) 181 final on the Reform of the Common Fisheries Policy ("Roadmap"), p. 3.

46 Hegland and J. Raakjær, 'Recovery Plans and the Balancing of Fishing of Fishing Capacity and Fishing Possibilities', p. 152-157.

47 Regulation (EC) No. 2371/2002 on the Conservation and Sustainable Exploitation of Fisheries Resources under Common Fisheries Policy (Regulation 2371/2002), OJ 2002 No. L358/59. It repeals Regulation 3760/92, see Art. 34(1) of Regulation 2371/02.

48 Arts. 5 and 6 of Regulation 2371/02.

49 See Art. 4(2)(f), 20 (1) of Regulation 2371/02. 'Recovery plans' under Art. 5 in particular require the use of effort restrictions; see Art. 5(4), paragraph 2 of Regulation 2371/02; see also Council regulation (EC) 1954/2003 on the management of fishing effort relating to certain Community fishing areas and

more, stakeholders are increasingly consulted within the newly established Regional Advisory Councils.[50] Finally, the new regulation leaves room for taking into account scientific insecurities and the environmental effects of fishing operations, adopting an *'ecosystem approach'* and a *'precautionary approach'*.[51]

Substantive changes were also introduced into the promotional regime. Throughout the reform process, the Community has tried to establish a system which better links investments in capacity to fishing effort limitations. Procedural hindrances arising from the rather complex system of MAGPs, fleet segmentations and reduction targets were abolished in 2002, and replaced by a simpler entry-exit regime. The Community simply stopped its attempts to make Member States reduce their fishing capacity through mandatory reductions; instead, it has established ceilings on the size of Member States fleets and attempted to manage fishing effort. Aid is now particularly allocated to fishers affected by conservatory management and structural measures, such as recovery plans and fleet reductions. There was also a push to phase out subsidies that were most harmful to resources, and end the most obvious contradictory spending policies. Additionally, inland fisheries and environmentally friendly fishing methods have been increasingly promoted. Premiums for the permanent transfer of vessels to third countries were phased out in 2004. However, not all harmful subsidisations have been eliminated. For example, from 2000 to 2006, the processing and marketing sector received €634 million of aid, while €248.8 million were granted to build facilities at fishing ports. In fact, even where subsidies for the modernisation of vessels do not immediately lead to an increase in fishing power, they do create incentives for fishers to remain in an overcapitalised industry, and thus, in actual terms, promote the 'race to fish'. It must also be recognised that, even though aid for fleet renewal and modernisation ended in 2004, the total amount paid in the fifth phase added up to €855.5 million.

The integration of promotional activities and management efforts has been substantially enhanced by the 2002 Reforms. Even though the Council continued to set fishing rates above sustainable levels, it seems inclined to *gradually* accept cut-backs to exploitation opportunities. Several multi-annual plans have been adopted throughout the fifth phase and effort limits are increasingly used to manage fisheries. The Member States (in the Council) have also approved broader sanctioning powers for the Commission. With regards to the second step of the integration concept, assessing and adjusting production capacities to the available resources, the entry-exit scheme provides an improved and relatively simple means to link promotion to management objectives. Finally, subsidisation most harmful to the management system has been eliminated.

---

resources and modifying Regulation (EC) 2847/93 and repealing Regulation (EC) 685/95 and (EC) 2027/95, OJ 2003 No. L289/1.

50 J.B. Ingerowski and M. Salomon, 'Ein kritischer Blick auf die aktuellen Entwicklungen in der Gemeinsamen Fischereipolitik unter Einbeziehung der neu geschaffenen regionalen Fischereibeiräte', pp. 540-541.

51 See Arts. 2(1), second paragraph, 4(1) and (2)(g)(iv), 5(2), 6(2), 7(1), 8(1) of Regulation 2371/02.

Money has even been made available to decrease capacity and to buffer the socio-economic effects of management measures and fleet reductions suffered by fishermen. By granting aid to fishers who leave the business due to the adoption of management or recovery plans, aid is channelled to the reduction of fleet segments that target the most exploited fish stocks. In addition, public support is increasingly channelled to more environmentally friendly fishing and inland fishing.

Some of the effects of an increasing integration of promotion and management have the following results: From 2003 to 2006, the fleet of the EU-15 Member States was reduced by 217,000 GT or 772,000 kW, giving a net reduction of approximately 10%.[52] By that, Member States have met their targets under Art. 12(1) of Regulation 2371/02. The Commission, however, considers these reductions to be 'insufficient' and 'too modest' with a view to the fishing pressures, the poor economic performance and the constant technological improvements of large parts of the Community fleet.[53] In addition, the Court of Auditors (as well as the Commission) criticised that the entry-exit system and the reference ceilings 'do not oblige, or even encourage, Member States to reduce their overcapacity.'[54] It also stated that given that the Community restricted itself to merely managing fishing effort (by limiting fishing activity and power), existing overcapacities have a direct bearing on the profitability of the fishing industry. By reducing days at sea, the Community ultimately creates incentives for the fishing industry to increase fishing productivity, which *de facto* increases fishing capacity.[55]

2.6 The sixth phase, 2007 – 2008

During the sixth phase, the management regime has hardly been modified. Only several multi-annual recovery plans have been adopted. Overall, the Council's practice to set catch limits at unsustainable levels seems

---

[52] Commission Com(2007) 828 final annual report from the Commission to the Council and the European Parliament on Member States' efforts to during 2006 to achieve a sustainable balance between fishing capacity and fishing opportunities, pp. 9-10; see also Commission Com(2006) 872 final annual report from the Commission to the Council and the European Parliament on Member States' efforts to during 2005 to achieve a sustainable balance between capacity and fishing opportunities, p. 11.

[53] Commission Com(2006) 872 final annual report from the Commission to the Council and the European Parliament on Member States' efforts to during 2005 to achieve a sustainable balance between capacity and fishing opportunities, p. 11; Commission Staff Working Document SEC(2008) 2760 accompanying the Proposal for a Council Regulation establishing a Community control system for ensuring compliance with the rules of the Common fisheries Policy, p. 6; see also House of Lords, *The Progress of the Common Fisheries Policy*, pp. 29-30.

[54] Court of Auditors, *Special Report 7/2007*, para. 114.

[55] Court of Auditors, *Special Report 7/2007*, para. 115.

to have prevailed. As a result, many stocks remain outside safe biological limits and no stock has yet recovered through the intervention of a recovery plan.[56]

Nevertheless, promotional policies have undergone substantial changes in 2007. Since the beginning of 2007, the European Fisheries Fund (EFF)[57] replaced the FIFG, attempting to implement the 2002 reforms.[58] One way it does this is by supporting the new fleet adjustment strategy included in the basic management regulation. To this end, it supports fleet exits and continues to refuse aid for the construction of vessels and the increase of fish holds on fishing vessels. The EFF also provides for the adjustment of fleet parts that target the most exploited stocks by giving aid to fishers affected by recovery and management plans or other fishing limitations. In addition, the EFF increasingly integrates environmental objectives into its plan.[59] For example, the NATURA 2000 requirements have to be taken into account in the grant of EFF support for aqua-environmental measures.[60] Another important development is its promotion of local development strategies in areas heavily dependent on fisheries. It is particularly noteworthy that these strategies are set up and implemented by local stakeholders themselves. Despite these advances, some problems remain. During the difficult EFF negotiations, a 'pro subsidy group' of Member States called for a re-uptake of subsidy schemes that harm resources.[61] They argued in favour of subsidisation, despite the fact that this type of aid had been phased out with the 2002 reforms.[62] In the end, Member States agreed to grant some aid for engine replacement, subject to the condition that vessels' ability to catch fish is not enhanced. It must be noted, however, that even where these subsidies do not increase fishing capacity, they do create an incentive to remain in the business. Besides, given the ineffective control mechanisms under the CFP, such subsidies could, in fact, lead to an increase in fishing capacity.[63] Furthermore, the Community continues to subsidise fishing ports, landing sites and shelters and develop markets and promotional campaigns. The

---

56 WWF, *WWF Mid-Term Review of the EU Common Fisheries Policy*, (Brussels: 2007), p. 4; Communication Com(2007) 295 final from the Commission to the Council on Fishing Opportunities for 2008 – Policy Statement from the European Commission, p. 5.; European Environment Agency, *Europe's environment – The fourth assessment*, (Luxembourg: 2007), pp. 210-250; J.B. Ingerowski and M Salomon, Ein kritischer Blick auf die aktuellen Entwicklungen in der Gemeinsamen Fischereipolitik unter Einbeziehung der neu geschaffenen regionalen Fischereibeiräte', pp. 540-541.

57 Regulation 1198/2006, OJ 2006 No. L223/1.

58 See Reason (1) and Art. 6 of Regulation 1198/06.

59 See Art. 25(6)(c)-(d), (7)-(8); Arts. 30(2)(a-c) and 35(1)(d); Arts. 37 and 38; Arts. 40(3)(d),(f); Arts. 43(2)(c), 44 (1)(b) and (f); Arts. 30(2)(d), (4)(d) and 38 (2)(c) of Regulation 1198/06.

60 Arts. 30(2)(d), (4)(d) and 38 (2)(c) of Regulation 1198/06.

61 The Council had failed to reach an agreement twice (June 2005 and May 2006) until finally a regulation was adopted on 19 June 2006; see Council of the European Union Press Release, 19 June 2006, No. 10302/06 (Presse 178), p. 17.

62 Proposal Com(2004) 497 final on the European Fisheries Fund, pp. 1-86.

63 Court of Auditors, *Special Report 7/2007*, para. 119.

processing industry also receives support. All these subsidies reduce the costs of fishing as well as processing costs, resulting in increased fishing and overall demand for fish. Finally, the Commission has concluded that the entry-exit system neither obliges nor encourages Member States to reduce their fleets.

Despite apparent weaknesses in the management and structural regimes, the 2006 amendments still do a better job of integrating the approaches to promotion and management. The EFF does not provide funding for the most harmful subsidies and aid is made available to support fishers where they are affected by the reduction of fishing under the management regime. In addition, participation and transparency have been improved by including local stakeholders in the strategic planning of sector developments. It also increasingly supports environmentally sustainable fishing and aquaculture methods.

3 Practical lessons to be learned from CFP experiences

Important practical lessons on how to integrate promotion and management in fisheries policies can be gathered by tracking the development of the CFP. These practical lessons can supplement the suggested four-step integration concept.

First, to achieve a stable balance between production capacities and fishing opportunities, policy makers must establish sustainable exploitation limits for Community and third country waters, as well as for the high seas. This will ensure that policy makers can effectively plan for the grant of public aid. Throughout the process of adopting sustainable exploitation rates and promotional measures, policy makers must keep in mind that a major reason why unsustainable quotas were adopted in the Community's past and present, was the strong fisheries' lobby which aimed at maintaining their investments in an overcapitalised sector. While it may be rather simple to agree politically on the build-up of a fishing fleet, the creation of jobs and the distribution of financial grants, it appears far more difficult to reduce existing fleets, create unemployment and reduce existing benefit schemes.

A second insight is that the effects of fisheries promotion largely depend on the state of the stocks and the effectiveness of a management regime. Theoretically speaking, promotional activities will not pressure fisheries resources if fish stocks are underexploited or exploitation limits are set at sustainable levels and implemented effectively. The CFP is an apt example of how subsidies should not be spent. Since the late 1970s, the Community has faced dwindling resources, large overcapacity, and no effective management regime. Nevertheless, subsidies have been made available for many purposes, increasing or maintaining overall production power and leading to production overcapacity in the fisheries sector. It is only recently that the Community has begun to change its subsidisation practices, offering grants for the most harmful types of aid. Subsidies increasingly aim at improving the effectiveness of the management regime, for example, by supporting fishers who leave the sector or promoting environmentally friendly fishing methods.

Another lesson to be gleaned from an examination of the development of the CFP is that an effective integration of promotion and management approaches requires appropriate scientific assessment and input. Scientific input is not only necessary for establishing safe exploitation limits, but must increasingly take into focus the effects of promotional measures on production sectors and the fisheries resources. For a long time, the CFP has not really accounted for the effects of its subsidies schemes. Promotional policies grew out of the interest of the Member States in supporting their fishing industries, securing food supplies, stabilising markets and increasing competitiveness. From its onset in 1970, the promotion of fisheries rose continuously, reaching its height in the period between 2000 and 2006. This policy has resulted in huge overcapacity. Two concrete lessons can be learned with regards to the interaction between promotional measures and fisheries production: In an ideal world, when determining which segments of the production sector should be eligible for aid, it is important to understand the pressure each segment places on each fish stock. It is important, for example, to classify the fleet into categories relating to the exploited species, fishery zones and methods of fishing.[64] Furthermore, policy makers must always take into account the technological progress of catching techniques, which is estimated at about 2% per year.[65]

It is important that any new programme to reduce exploitation, so that it is more in line with the available resources, has to be simple, precise and binding. The Community since 2002 manages fleet entries and exits by means of a simple and binding system that uses fleet reference levels and a clear and simple reduction formula. Any aid granted to the industry must comply with this system. This approach, even though regarded by the Commission as being insufficient, seems, nevertheless, more effective than earlier approaches. Under MAGP I and II, the Commission had only few available means to ensure that fleet reduction objectives were met. MAGP III and IV were regarded as costly and difficult to administer and not ambitious enough in terms of their reduction objectives.[66]

The experience of the CFP also teaches that the competitiveness of the production sector should not be increased by so-called 'contrary spending'. Contrary spending grew out of the desire to pursue two opposing objectives: increasing competitiveness and efficiency of the production sector, on one hand, and reducing the fleet size, on the other. As a result, for a long time the Community was financing both production increases and decreases. Aid for fleet renewals and fleet modernisation, for example, have neutralised aid for

---

64 See critique on MAGP I in Com SEC(91) 2288 final, Report 1991 from the Commission to the Council and the European Parliament on the Common Fisheries Policy ('Commission Report – 1991').

65 Lindebo, 'Role of Subsidies in EU Fleet Capacity Management', pp. 236-237.

66 Commission Communication Com(2001) 135 final Green Paper on the Future of the Common Fisheries Policy, vol. I, p. 11.

the adjustment of the sector and contributed to maintaining overcapacities. Contrary spending has been an enormous waste of public money and increased pressure on fisheries resources. Any gains in efficiency and competitiveness must be achieved through other means. Gains may be increased, for example, by managing stocks at MSY (or even the MSE) levels or through the introduction of ITQs.

An integrated approach would include grants for social and environmental purposes. The Community, for example, has provided financial aid to buffer the social consequences of limiting fishing opportunities or unforeseen natural occurrences.[67] It has also provided funds to reduce the overall economic dependence of coastal fishing areas on catching activities. This supports the diversification and the economic and social restructuring of regional economies. The Community has adopted measures for early retirement schemes, retraining, and improving tourism infrastructure. The CFP has also increasingly recognised the need for providing funds to support environmentally friendly production methods, both for catching and aquaculture. As long as markets do not fully remunerate fish caught and processed using environmentally friendly and sustainable methods, sound production should be rewarded with public money.

Promotion schemes can negatively affect third country fisheries. As explained above, for a long time, the Community 'externalised' its overcapacity problem by financing the conclusion of international access-agreements, joint ventures and the transfer of old vessels.[68] This externalisation has been criticised on various grounds, particularly from a developing country perspective. Access-agreements, for example, are deemed to be not based on fair benefit-sharing ratios. The first generation of agreements included financial compensation which only amounted to 2% to 17% of the market value of the catch.[69] In addition, fishing abroad was heavily subsidised. However, as in Community waters, subsidies tend to increase the fishing activities in third country waters. This has contributed to overfishing, particularly in West African countries waters.[70] Another criticism is that public aid tends to distort competition between the Community's and the hosting states' fishers and industries.[71] More recently, the Community and third countries have re-negotiated the existing agreements.[72]

---

67 However, public funding, (*e.g.*, for the temporary lay up of vessels) should not help to maintain existing overcapitalisation.

68 For the current status of the Community's 'external fleet', see Commission Study (Contract FISH/2006/02) on the European External Fleet, available online at http://ec.europa.eu/fisheries/publications/studies/external_fleet_2008_en.pdf.

69 IFREMER, 'Evaluation of the Fisheries Agreements Concluded by the European Community', p. 35.

70 Mwikya, Fisheries Access Agreements: Trade and Development Issues, pp. 25-26, B. Gorez, *Policy Study: EU-ACP Fisheries Agreements*, (Brussels: 2005); Kaczinsky, V.M. and Fluharty, D.L., 'European policies in West Africa: who benefits from fisheries agreements? (2002) 26 *Marine Policy*, pp. 75-93.

71 Mwikya, Fisheries Access Agreements: Trade and Development Issues, pp. X, and 20-24.

72 Commission Communication Com(2002) 637 final on an integrated framework for fisheries partnership agreements with third countries, p. 3; Commission Communication Com(2001) 135 final Green

The Community claims that the old treaties have now transformed into genuine partnerships. Agreements officially aim at accounting for the varied and often complex circumstances faced by third countries. They also promise that financial contributions will increasingly go towards covering expenses linked to management, scientific assessment, control of the fisheries of third countries and to the build-up of the fisheries sectors of third countries. Although it is too early to assess the success of these new agreements, some general observations can be made in this vein. First, the Community continues to substantially subsidise the activities of its fleet. Second, it has been argued that multilateral access-agreements, and not bilateral agreements, would be more effective at achieving sustainable exploitation and management of highly migratory species in third country waters.[73] Third, the Community's (and most third countries') weak control systems continue to threaten the proper implementation of control requirements laid down in the agreements. Finally, given that Community fishers increasingly compete with local and Asian fleets, particularly in West-African waters,[74] it is suggested here that the Community ought to consider whether it should cut back subsidies to maintain its 'external fleet' and instead focus on helping third countries to build up their own fleets, manage their resources properly and to protect them against illegal, unreported and unregulated fishing by foreign fleets.

An effective management regime as well as an efficient allocation of public funds depends on transparency and public participation. Since 2002, the Community started to include interested non-governmental parties in the political process leading up to the adoption of CFP measures. On the management side, RACs have (arguably) increased stakeholder involvement.[75] An inclusive approach has also been taken under the EFF. To achieve the best possible results, local public and private stakeholders are included in the planning and implementation of promotional funding restructuring fishing economies.

Finally, an integrated approach must recognise that it is not only the catching segment of the fisheries industry which creates pressure on the marine resources. It is well accepted that subsidies that reduce production costs allow the fisheries sectors to produce even beyond the point where the real production costs equal revenues. However, policymakers and scientists have failed to acknowledge that *most* subsidies to reduce production costs may also lower retail prices. Lower prices create demand for a product,[76] which, in the long-term,

Paper on the Future of the Common Fisheries Policy, vol. I, pp. 17-19.

73 Mwikya, Fisheries Access Agreements: Trade and Development Issues, pp. 6, 10-11.

74 Commission Study (Contract FISH/2006/02) on the European External Fleet, pp. 8-10.

75 Though it can be criticised how modest inclusion really is, see Hatchard and Gray, 'The 2002 Reform of the Common Fisheries Policy's System of Governance – Rhetoric or Reality?', pp. 545-554; Ingerowski and Salomon, 'Ein kritischer Blick auf die aktuellen Entwicklungen in der Gemeinsamen Fischereipolitik unter Einbeziehung der neu geschaffenen regionalen Fischereibeiräte', pp. 539-542.

76 Ludicello *et al.*, *Fish, Markets and Fishermen – The Economics of Overfishing*, p. 60; see also G.A. Goeffrey and P. J. Remy, *Advanced Microeconomic Theory*, (Boston: 2001), p. 54. There are, of course, exceptions to

increases pressure on marine resources. Thus, any evaluation of the production sector must take into consideration the relationship between *all factors* that contribute to rising exploitation. This may include the production sectors' *processing* powers as well as *consumption patterns*. Where the demand for fish is more than can be exploited on a sustainable basis, consumption and production sectors must be downsized so that they accord with the amount of exploitable resources. The adjustment process can be supported through public aid.

To date, the Community has failed to adopt measures which try to steer consumers towards sustainable products. Current Community measures in this regard have merely refrained from subsidising fishing and aquaculture production for purposes other than human consumption.[77] Further steps, however, seem necessary.[78] The Community's first step should be to provide support for the establishment of ambitious labelling schemes that give credible product information. Information should cover the entire 'life cycle' of the product, including details on catching, processing and transportation, and the label should state whether the product is produced sustainably or not.[79] Such a system would help consumers distinguish between sustainably- and unsustainably-produced fisheries' products and thus empower them to link their purchase decisions to the ecological impacts of fishing. The Community, on the other hand, could also redirect its promotional measures. For example, it could make public aid partially conditional upon sustainable production. This would create a competitive advantage for sustainably-produced fish and reduce the price of such products. The Community could also decide to launch marketing campaigns to foster the idea that sustainable products are value-added.

---

this demand pattern, *e.g.*, luxury goods.

77 See, for example, Art. 11 and Annex III of Regulation 3699/93 or Art. 34(4) of Regulation 1198/06.

78 Tavis Potts and Marcus Haward, 'International Trade, Eco-Labelling, and Sustainable Fisheries – Recent Issues, Concepts and Practices', pp. 95-97.

79 J.T. Sproul, 'Green fisheries: certification as a management toll', in Pitcher, Hart and Pauly (eds.), *Reinventing Fisheries Management*, (London: 1998), p. 140.

CHAPTER 6

Access to Justice

The following chapter analyses the question of who has standing in Community courts to institute proceedings against CFP measures.

1 Access to judicial review

Community law aims at providing an effective regime of legal protection, which in principle grants Community courts the jurisdiction to review every Community act.[1] The Court of Justice (ECJ) and Court of First Instance (CFI) are to ensure that 'the law is observed' in the interpretation and the application of the Treaty.[2] In addition, the Courts have exclusive jurisdiction to declare void any act taken by a Community institution.[3] Within their competence, Community courts may *i.a.* review Community acts, interpret Community law, or decide about the legality of Member States acts or omissions which concern the application of Community law.[4]

This chapter will assess the extent to which the Community courts can review the legality of CFP measures. According to Arts. 230(2), (4) and 234 of the Treaty, the Member States, the Commission, the Council, Parliament, and under very restricted conditions, individuals, can all file a court action. It has been described above, however, that the Council and the Commission adopt fisheries legislation by mutual agreement. The plenary body provides ineffective oversight of Parliament's Committee on Fisheries which is mainly composed of intermediaries for the short-term interests of the fishing industries. As a result, these organs are unlikely candidates to challenge CFP laws before the Court, particularly where there is concern about whether specific measures or actions comply with the conservation requirements laid down in Community law. It is therefore important that individuals and organisations that represent this public interest have standing before the courts. The following section will outline the prospects for such parties to institute proceedings before the Community courts and thereby challenge CFP legislation.

2 Treaty provisions on access

The Treaty provides for two forms of standing to challenge Community legislation: *direct* and *indirect*. The Council, Commission, Parliament, Member States and private individuals all have *direct* access to the Court of Justice under the conditions laid down in Art. 230(2) and (4) of the Treaty.

---

1 Bernhard Wegener, 'Art. 230 EGV', in Calliess and Ruffert (eds.) *EUV – EGV Kommentar* (München: 2007), p. 1954.

2 Arts. 220, para. 1; 230(4) and 234 of the Treaty.

3 See Art. 230 of the Treaty as interpreted in Case 314/85, Firma Foto-Frost v. Hauptzollamt Lübeck-Ost [1987] ECT 4199, para. 17.

4 Arts. 220, 226, 228 230(2) and 230(4) of the Treaty.

Indirect access is permitted under Art. 234 of the Treaty for private parties that initiate proceedings in national courts which then may request a preliminary ruling from the ECJ.

2.1 Standing of privileged applicants

Art. 230, para. 2 of the Treaty provides that the ECJ shall have jurisdiction over actions brought by the *Member States*, the *European Parliament*, the *Council* or the *Commission* 'on grounds of lack of competence, infringement of an essential procedural requirement, infringement of the Treaty or of any rule of law relating to its application, or misuse of powers'. In comparison to natural or private entities, the Community institutions and the Member States have been granted privileged access: they do not have to prove any individual concern or particular interest to have a Community legal act reviewed by the Court of Justice.[5] This special privilege is based on the assumption that Community institutions and the Member States act as representatives of the public interest, thereby improving the efficacy of the Community's legal system.[6]

2.2 Standing for non-privileged private parties and interests groups

Direct standing is granted to natural and legal persons only under certain conditions and is, in practice, difficult to obtain. Art. 230, para. 4 of the Treaty lays down the conditions under which non-privileged parties entities have access to Community judicature. It states:

'Any natural or legal person may, under the same condition, institute proceedings against a decision addressed to that person or against a decision which, although in the form of a regulation or a decision addressed to another person, is of direct and individual concern to the former.'

To clearly assess the potential for private parties such as fishermen or environmental groups to obtain judicial review, it is important to distinguish between the different causes of action and outline the European Court's interpretation of the terms '*direct* and *individual* concern'.

2.2.1 Causes of action

Art. 230, para. 4 provides that any natural or legal person may institute proceedings against (a) decisions addressed to them, (b) decisions

[5] Only in one case has the ECJ denied standing to Germany because it was not 'adversely affected' by a Commission decision in question; see Case 242/2002, Germany vs. Commission, 2002 [ECR] I-5603, para. 46. However, given the very special circumstances in that case, it cannot be concluded that the ECJ, in general, requires a Member States to be 'adversely affected'; see Wolfram Cremer, 'Art. 230', in Calliess and Ruffert (ed.), *EUV/EGV*, (München: 2007), para. 21.

[6] Alexander Thiele, *Europäisches Prozessrecht*, (München: 2007), p. 64; Booß, 'Art. 230 EGV', para. 40.

addressed to others, and (c) decisions in the form of regulations. Decisions addressed to others and decisions in the form of regulations will only be subject to judicial review instituted by natural or legal persons where they *directly* and *individually* concern the litigant. With a view to the possibility to challenge 'decisions in the form of regulations', Art. 230, para. 4, in principle, does not refer to 'true regulations'. It merely aims at preventing that Community institutions, by choosing the form of a regulation, preclude an individual from bringing an action against a decision which concerns him or her directly and individually.[7] With a view to 'true regulations', *i.e.*, measures of general application, the Court will hear the claim of non-privileged persons only in exceptional cases and only where they can prove that they are *directly* and *individually* concerned by them.[8]

2.2.2 The 'direct concern' criteria

Community measures are of *direct concern* where they directly affect the legal situation of the applicant. These include laws that are self-executing, *i.e.*, where implementation results from Community rules without Member States' application of other intermediate rules, actions, or discretion.[9] For example, CFP management measures, such as the yearly regulations on TACs and TAEs, are not implemented directly by the Community; the Member States decide how to allocate their national quota among their fishermen. The same is true for those CFP measures adopted under the structural and market policy, which are implemented by Member States and producer organisations. In these cases, it is difficult to argue that Community rules are of direct concern.[10] Thus, the most important CFP measures of relevance to fisheries management and conservation are not of direct concern. However, technical restrictions such as net rules, no-take areas or no-take periods adopted by the Council or the Com-

---

7 Joined Cases 789/79 and 790/79 Calpak and Società Emiliana Lavorazione Frutta v Commission [1980] ECR 1949, para. 7; Case T-447/05 SPM v Commission [2007] ECR II-0000, para. 61; Case T-91/07, WWF-UK Ltd. v. Council, not yet in the ECR, para. 61.

8 Cremer, 'Art. 230', p. 2027. In the earlier days of the Community, the ECJ strictly denied that 'true' regulations could be of individual concern and subject to judicial review under Art. 230, para. 4; see Case 16-17/62, Conféderation Nationale des Producteurs de fruits et Légumes v. Council, [1962] ECR I-471 (English special edition), p. 478 et seq.

9 Case C-386/96, Société Louis Dreyfus & Cie v. Commission [1998] ECR I-2309, paras. 40-56; Case C-15/06 P, Regione Siciliana v. Commission, [2007] ECR I-2591, para. 31; see also Jonas Ebbesson, 'European Community', in Ebbesson (ed.), *Access to Justice in Environmental Matters in the EU*, (The Hague: 2002), p. 76.

10 With regards to Commission TAC deductions and TAC regulations in general, see Case T-172/07, Atlantic Dawn Ltd. et. al v. Commission, see online at www. http://curia.europa.eu, or OJ 2007 No. C170/29, para. 41.

mission do not require further Member States actions to be applicable.[11] Thus, they directly concern fishers.

2.2.3 The 'individual concern' criteria

The criteria of *individual concern* is much more controversial than that of *direct concern*.[12] The meaning of 'individually concerned' was defined by the ECJ in its seminal *Plaumann* decision. In that case, the Court stated that:

Persons other than those to whom a decision is addressed may only claim to be individually concerned if that decision affects them by reason of certain attributes which are peculiar to them or by reason of circumstances in which they are differentiated from all other persons and by virtue of these factors distinguishes them individually just as in the case of the person addressed.[13]

In later decisions, the Court has stated that it will not question the general application of a measure by the fact that it is:

possible to determine the number or even the identity of the persons to whom it applies at any given time, so long as it is established that such application takes effect by virtue of an objective legal or factual situation defined by the measure in question in relation to its purpose.[14]

It is only under exceptional circumstances that individuals are granted direct access to Community judicature. Different authors have tried to categorise these exceptions, proposing three main groups.[15]

---

[11] In Case T-177/01, Jégo Quéré & Cie Sa v Commission, [2002] ECR II-02365, a fishing company claimed that a Commission regulation which established minimum mesh sizes was invalid, because they were adopted in breach of the principle of proportionality and equality, see para. 18. Neither the Commission nor the CFI denied the 'direct concern' but concentrated their arguing on the criteria of 'individual concern'.

[12] See, for example, Ebbesson, 'European Community', pp. 49-94; Craig and de Burcá, *EU Law*, pp. 515-527; Craig, *EU Administrative Law*, pp. 340-344; Berend Lindner, 'Zur Klagebefugnis natürlicher und juristischer Personen für Nichtigkeitsklagen gem. Art. 230 IV EG gegen EG-Verordnungen' (2003) *Neue Zeitschrift für Verwaltungsrecht*, pp. 569-572; C. Calliess, 'Individualrechtsschutz im europäischen Verfassungsbund' (2004) 1 *Göttinger Online-Beiträge zum Europarecht*, pp. 1-17; Cremer, 'Art. 230', p. 2030-2037; Cornelia Koch, ‚International Decisions – Commission of the European Community v. Jégo Quéré & Cie SA' (2004) 98 *The American Journal of International Law, 4, pp.* 814-819; Carsten Nowak, ‚Das Verhältnis zwischen zentralem und dezentralem Individualrechtsschutz im Europäischen Gemeinschaftsrecht' (2000) Europarecht, 5, pp. 724-742; more compromising: see Chalmers *et al.*, *European Union Law*, pp. 424-436 (at 425-436), Peter Baumeister, 'Effektiver Rechtsschutz im Gemeinschaftsrecht' (2005) 40 *Europarecht*, 1,, pp. 1-35.

[13] Case 25/62, Plaumann & Co v. Commission, [1963] ECR 95, page 107.

[14] Case T-472/93, Campo Ebro and other v. Council, [1995] ECR II-421, para. 32; Case C-209/94, Buralux and others, [1996] ECR I-615, para. 24.

[15] Thiele, *Europäisches Prozessrecht*, p. 105; Cremer, 'Art. 230', pp. 2031-2033; Booß, 'Art. 230 EGV', paras. 55-61. The categories, however, vary from author to author.

First, the Court has found claimants are individually concerned where they were affected by a Commission regulation which was directed at a fixed number of identifiable individuals.[16]

Second, the threshold of individual concern was met where the complainants played a certain role within a procedure that led to the adoption of the measure in question. These cases were often in the areas of antidumping and competition law.[17] In such cases, the Court did not even require that the participation of the claimant in the process preceding the adoption of the disputed legislation be based on a guarantee provided by law.[18] Nevertheless, the Court stated that it is not sufficient for the applicant to obtain *locus standi* by merely submitting a complaint and corresponding subsequently with the Commission.[19] In the area of state aid, the Court required that the complaining party have a significant role in the procedure under Art. 93(2) of the Treaty and that their 'position on the market is significantly affected by the aid which is subject to the decision at issue'.[20] Two further decisions may be included in this group.[21] In both cases, the applicants had little involvement in the procedure, which led up to the adoption of antidumping duties or to the adoption of an import quota.[22]

Third, the Court has held that the test of individual concern is met where applicants can prove that a contested measure of general application affects their business activities '*seriously*'[23] or affects their established individual rights.[24] However, the claimant[s] must be individually and exclusively entitled to that right.[25]

---

16 Case 41/70, International Fruit Company and others v. Commission, [1971] ECR p. 111, paras. 15-22; C-73/97, France v. Comafrica and others [1999] ECR I-185, paras. 40, 41.

17 Case 26/76, Metro v. Commission, [1977] ECR p. 1875, para. 13; Case T-585/93, Greenpeace and others v. Commission [1995] ECR II-2205, para. 59 (and the cases cited therein); see also cases cited in Craig and de Burcá, *EU Law*, p. 520; Thiele, *Europäisches Prozessrecht*, p. 105; Cremer, 'Art. 230', p. 2031.

18 Case T-435/93, ASPEC and others v. Commission, [1995] ECR II-1281, para. 63; Case T-442/93, AAC and others v. Commission, [1995] ECR II-1329, para. 48.

19 Case T-585/93, Greenpeace and others v. Commission [1995] ECR II-2205, para. 56.

20 Case C-106/98, Comité d'entreprise de la Société française de production v. Commission, [2000] ECR I-3659, paras. 40-41.

21 Booß, 'Art. 230 EGV', para. 60. Booß refers to Case C-385/89, Extramet Industries SA v. Council [1991] I-2501 as well as to Case 11/82, Piraiki-Patraiki v. Commission [1985] ECR 207.

22 Booß, 'Art. 230 EGV', para. 60.

23 Case C-358/89, Extrament Industrie SA v. Council, [1991] ECR I-2501, para. 17.

24 Case C-309/89, Codorniu SA v. Council, [1994] ECR I-1853, para. 18-22; Case T-482, Weber v. Commission [1996] ECR II-609, paras. 67-70.

25 Booß, 'Art. 230 EGV', para 61.

2.2.4 Standing of interest groups and environmental organisations

The rights of interest groups or environmental organisations with legal personality to institute proceedings before the Community courts are similar to those of natural persons. Accordingly, if associations want to challenge a Community measure, they must show that they are directly addressed and individually concerned by the law in question.

The Court has confirmed that interest groups are 'directly and individually concerned' in four situations: (a) legislative provisions expressly confer a range of procedural powers on associations; (b) associations represent the interests of undertakings or individuals that themselves have *locus standi* to seek the annulment of the provision in question; (c) associations represent the (collective) interest of individuals or undertakings and have been, on request of the Commission, actively involved in a procedure which led to the adoption of the contested law;[26] and (d) where the association is distinguishable because its own interests as an association are concerned, in particular, because its negotiation position has been affected by the measure whose annulment is being sought.[27]

However, when assessing whether environmental interest groups can institute proceedings before the Community Courts, regard must be had to the fact that the groups are agents of environmental protection; this means that they represent a public or general and not just an individual interest.[28] In the *Greenpeace* case, the environmental organisation Greenpeace asked the Court to adopt a liberal approach to the individual concern criteria. *Greenpeace* argued as follows: First, it contended that according to the established case law associations are granted access where their members are themselves individually concerned.[29] *Greenpeace* reasoned that since some of the individuals it represented would actually be concerned by the contested decision, it should have standing.[30] Second, *Greenpeace* argued in the alternative that it should be regarded as 'individually concerned by reason of the particularly important role [it] play[s] in the process of legal control by representing the general interest shared by a number of individuals in a focused and coordinated manner [...]'.[31]

---

26 Case 67/85, Van der Kooy v. Commission [1988] ECR 219, paras. 18-25; Case C-313/90, CIFRS and others v. Commission, [1991] ECR I-2257, para. 29. See Rehbinder, 'Rechtsschutz gegen Handlungen und Unterlassungen der Organe und Einrichtungen der Europäischen Gemeinschaft im Lichte der Aarhus-Konvention', p. 662.

27 Case T-38/98 Associazione Nazionale Bieticolltori v. Council [1998] ECR II-44191, para. 25. Booß, 'Art. 230 EGV', paras. 70-75; Chalmers *et al.*, *European Union Law*, p. 434.

28 See, for example, J.H. Jans and Hans H.B. Vedder, *European Environmental Law*, (Groningen: 2008), pp. 211-214; Ebbesson, 'European Community', pp. 80-81.

29 Case T-585/93, Greenpeace and others v. Commission [1995] ECR II-2205, para. 37.

30 *Ibid.*, para 38.

31 *Ibid.*, para. 39.

The Court of First Instance (CFI) disagreed with Greenpeace's arguments, declaring the proceeding inadmissible. In response to the first argument, the CFI stated that the claimants 'do not rely on any attribute substantially distinct from those of all people who live or pursue an activity in the areas concerned [...]'.[32] The CFI held that:

> an association formed for the protection of the collective interest of a category of persons cannot be considered to be directly and individually concerned [...] by a measure affecting the general interests of that category, and is therefore not entitled to bring an action for annulment where its members may not do so individually [...].[33]

This decision was upheld by the ECJ on appeal.

The approach adopted by the CFI and the ECJ in this case was widely criticised,[34] as it bars direct action suits initiated by environmental associations against Community institutions before the European Courts. The CFI did indicate, however, that under certain circumstances it would grant standing to environmental associations such as *Greenpeace*, and referred to past case law (as outlined above). It can be inferred from its decision that associations formed for the protection of the collective interests of a category of persons may bring an action for annulment where its members are directly and individually concerned.[35] It also explicitly stated that:

> 'special circumstances such as the role played by an association in a procedure which led to the adoption of an act within the meaning of Art. 173 of the Treaty may justify holding admissible an action brought by an association whose members are not directly and individually concerned by the contested measure.'[36]

In addition, the Court stated that participation must be based on 'qualified procedural guarantees provided within the law'.[37]

---

32 *Ibid.*, para. 54.

33 *Ibid.*, para. 59.

34 Jans and Vedder, *European Environmental Law*, pp. 212-214; Ebbesson, 'European Community', pp. 78-81; A. Ward, 'Judicial, Review of Environmental Misconduct in the European Community: Problems, Prospects and Strategies' (2000) Yearbook of European Environmental Law, pp. 154-156.

35 Case T-585/93, Greenpeace and others v. Commission [1995] ECR II-2205, para. 59.

36 *Ibid.*, para. 59. See Manfred Rehbinder, 'Rechtsschutz gegen Handlungen und Unterlassungen der Organe und Einrichtungen der Europäischen Gemeinschaft im Lichte der Aarhus-Konvention', in Gaitanides, Kadelbach, Rodriguez and Gil (ed.), *Europa und seine Verfassung – Festschrift für Manfred Zuleeg* (Baden-Baden: 2005), pp. 661-662; Ebbesson, 'European Community', p. 82.

37 Case T-585/93, Greenpeace and others v. Commission [1995] ECR II-2205, para. 63; see also Case T-264/03, Schmoldt and others v. Commission, [2004] ECR II-1515, paras. 100, 131-145; Case T-47/00, Rica Foods v. Commission, [2000] ECR II-113, para. 55.

2.2.5 Indirect access

Art. 234(b) of the Treaty grants the national courts of the Member States the right to request the Court of Justice 'to give preliminary rulings concerning [...] the validity and interpretation of acts of the institutions of the Community [...]'. The standard case with regards to CFP legislation is that a Member State applies a CFP regulation to its national fishermen. Fishermen may then turn to the national courts and claim either that part or all of the regulation is invalid. The national court may then decide to refer the case to the ECJ.[38]

3 Recent critique

The restrictive interpretation of Art. 230(4) of the Treaty has been highly criticised. One ramification of the Court's approach is that the more people that are affected by a Community act (and the court decision), the less the likelihood there is that they will meet the threshold of 'individually concerned'.[39] Further criticism has been offered by *Advocate General Jacobs* and the *CFI* in the cases of *Extramet Industries, Unión de Pequeños Agiculturos v. Council* and *Jégo-Quéré SA v. Commission.*[40] Their arguments provide a good representation of the critique that exists in the literature: First, where Community measures are self-executing (such as certain technical restrictions), there must first be a violation of these measures by an addressee before they can challenge the legislation. Second, as it is difficult for natural and legal persons as well as interest groups to meet the test for direct access, pursuant to Art. 230, para. 4 of the Treaty, the main approach used by private entities to challenge CFP legislation is through requests for preliminary rulings by national courts under Art. 234 of the Treaty. However, indirect access does not provide a party with full access to the Community courts. For example, claimants cannot decide themselves whether a reference will be made to the ECJ. Instead, they must convince the national court to refer the case. Procedures under Art. 234 of the Treaty also present a number of disadvantages in comparison to those under Art. 230(4) of the Treaty. Indirect access involves delays, increased costs, precludes third party interventions as

---

[38] Craig and de Búrca, *EU Law*, p. 525.

[39] Krämer and Winter, 'Umweltrecht', para. 75; see also Jans and Vedder, '*European Environmental Law*', p. 212.

[40] See Case C-358/89, Extramet Industrie SA v. Council [1991] ECR I-2501, Jacobs AG, paras. 70-74; Case C-50/00 P, Unión de Pequeños Agiculturos v. Council [2002] ECR I-6677, Jacobs AG, paras. 60-66, 102; Case T-177/01 Jégo-Quéré SA v. Commission [2002] ECR II-2365, para. 51; Case T-177/01 Jégo-Quéré SA v. Commission [2002] ECR II-2365, para. 51.

well as the award of interim measures.[41] This notwithstanding, the ECJ refused to follow the advice of AG Jacobs and rejected the CFI's decision on appeal.[42]

4 Recent developments

The following section will assess the extent to which recent developments in international and Community law affect the right of fishermen, interest groups and environmental groups to access European courts. In particular, it will look at specific provisions of the international Aarhus Convention as well as the implementing Aarhus Regulation. It is important to remember in this connection that neither international agreements nor secondary legislation can amend Art. 230, para. 4 of the Treaty.[43] Nevertheless, Community institutions and the Member States are free to improve access to the courts, which could be done in a number of ways. First, the Member States may choose to amend Art. 230, para. 4 of the Treaty. Second, it has been argued that the ECJ could reconsider its restrictive interpretation of the terms 'directly and individually concerned' under Art. 230, para. 4.[44] Third, given that the Council is granted a certain margin of discretion to implement Art. 230, para. 4,[45] it may grant individuals and their representing organisations 'qualified procedural guarantees' within CFP legislation as defined in the *Greenpeace* decision.

After describing the relevant provisions of the Aarhus Convention and the Aarhus Regulation, the focus will turn to a recent CFI order on the admissibility of a proceeding instituted by the WWF against a Council TAC regulation. This decision reflects the CFI's current understanding of the terms 'directly and individually concerned' in the face of the Aarhus Convention and the Aarhus Regulation.

---

41 Case C-50/00 P, Unión de Pequeños Agiculturos v. Council [2002] ECR I-6677, Jacobs AG, para. 102.

42 Case C-50/00 P, Unión de Pequeños Agiculturos v. Council [2002] ECR I-6677, paras. 37-45; Case C-263/02 P, Commission v. Jégo Quéré SA [2004] ECR I-3425, paras. 29-39.

43 In Case T-94/04 EEB and others v. Commission [2005] ECR II-4919, the CFI noted 'that the principles governing the hierarchy of norms [...] preclude secondary legislation from conferring standing on individuals who do not meet the requirements of the fourth paragraph of Art. 230 EC.' See also Kirsten Schmalenbach, 'EGV Art. 300', on Calliess and Ruffert (eds.), *EUV/EGV*, (München: 2007), para. 82.

44 Chalmers *et al.*, *European Union Law*, p. 433. However, the Court states that to amend the *Plaumann* formula the Treaty would have to be amended by the Member States, Case-50/00 P, Unión de Pequeños Agiculturos v. Council [2002] ECR I-6677, paras. 44-45.

45 Chalmers *et al.*, *European Union Law*, p. 433; Annette Guckelberger, 'Die EG-Verordnung zur Umsetzung der Aarhus-Konvention auf Gemeinschaftsebene' (2008) Natur und Recht, 30, p. 86; Eckhard Rehbinder, 'Rechtsschutz gegen Handlungen und Unterlassungen der Organe und Einrichtungen der Europäischen Gemeinschaft im Lichte der Aarhus-Konvention', pp. 658-664.

4.1 The Aarhus Convention

The Community signed the *Aarhus Convention*[46] on 25 June 1998 and approved it on 17 February 2005.[47] The Convention lays down specific conditions under which state parties have to provide individuals and environmental associations with access to judicial review.

The Court of Justice has recognised, subject to the abovementioned conditions, that environmental associations have standing, and can meet the 'individually concerned' test if they were involved in a procedure preceding the adoption of a legal provision. Since Arts. 6(2) and 2(5) of the Aarhus Convention allow environmental associations to be involved in process for adopting environmental legislation, one could argue that this grants them 'qualified procedural guarantees' in the sense of the *Greenpeace* decision.[48] According to Art. 6(2):

The public concerned shall be informed, either by public notice or individually as appropriate, early in an environmental decision-making procedure, and in an adequate, timely and effective manner [...].

Art. 2(5) defines the 'public concerned' as:

> the "public concerned" means that public affected or likely to be affected by, or having an interest in, the environmental decision-making; for the purpose of this definition, non-governmental organisations promoting environmental protection and meeting any requirements under national law shall be deemed to have an interest.

Art. 9(3) and (4) of the Aarhus Convention also require state parties to grant members of the public as well as environmental organisations the right to institute proceedings before competent courts where they deem private persons or public authorities to contravene environmental law.[49] Art. 9(3) specifically states:

> In addition and without prejudice to review procedures referred to in paragraphs 1 and 2 above, each party shall ensure that, where they meet the criteria, if any, laid down in its national law, members of the public have access to administrative or judicial procedures to challenge acts and omissions by private persons and public

46 United Nations Economic Commission for Europe (UNECE) Convention on Access to Information, Public Participation in Decision-making and Access to Justice in Environmental Matters. Available online at www.unece.org/env/pp/.

47 See Council Decision (EC) 370/05, on the conclusion, on behalf of the European Community, of the Convention on access to information, public participation in decision-making and access to justice in environmental matters, OJ 2005 No. L124/1.

48 See arguments of the WWF in Case T-91/07, WWF UK Ltd. v. Council, Order of the Court not yet in the ECR, para. 55.

49 See, for example, Rat von Sachverständigen für Umweltfragen (SRU), *Rechtsschutz für die Umwelt – die altruistische Verbandsklage ist unverzichtbar, Stellungnahme*, (Berlin: 2005), pp. 9-10; arguing similarly, Krämer, *EC Environmental Law*, pp. 161-162 (see below).

> authorities which contravene provisions of its national law relating to the environment.

According to Art. 9(4):

> In addition and without prejudice to paragraph 1 above, the procedures referred to in paragraphs 1, 2 and 3 above shall provide adequate and effective remedies, including injunctive relief as appropriate, and be fair, equitable, timely and not prohibitively expensive. Decision of courts, and whenever possible of other bodies, shall be publicly accessible.

However, this interpretation is widely contested. Others have argued that, given the express reference to national law to Art. 9(3), state parties have a wide margin of discretion and are thus not required to establish comprehensive access to judicial review.[50]

The author holds the first opinion. Almost without exception, Community law bars environmental organisations from instituting proceedings against measures of general application. Given that the rules on access to justice existed prior to the adoption of the Convention, it is difficult to see why the Community entered into the agreement, yet it never intended to amend its restrictive approach to judicial review. In this case, Arts. 9(3) and 9(4) of the Aarhus Convention would simply be a restatement of the existing law. This idea is supported by Reason 18 of the Aarhus Convention which states the objective of the Arts. 9(3) and (4) of the same Convention, namely, that 'effective judicial mechanisms should be accessible to the public, including organisations, so that its legitimate interests are protected and the law is enforced'. The *Aarhus Compliance Committee* and others have correctly concluded that parties to the Convention are not entitled to 'introduc[e] or maintain so strict criteria that they effectively bar all or almost all environmental organisations from challenging act or omissions that contravene national law relating to the environment'.[51]

4.2 The Aarhus Regulation 1367/06

The Community adopted Regulation (EC) 1367/06 on the application of the provisions of the Aarhus Convention on Access to Information,

[50] See Council of the European Union, common Position OJ 2005 No. C264E/18; A. Epiney, 'Zu den Anforderungen der Aarhus-Konvention an das europäische Gemeinschaftsrecht', in Falke and Schlacke (ed.), *Information – Beteiligung – Rechtsschutz, Neue Entwicklungen im, Umwelt- und Verbraucherrecht*, (Berlin: 2004), p. 23; Annette Guckelberger, 'Die EG-Verordnung zur Umsetzung der Aarhus-Konvention auf Gemeinschaftsebene', p. 86; Manfred Rehbinder, 'Rechtsschutz gegen Handlungen und Unterlassungen der Organe und Einrichtungen der Europäischen Gemeinschaft im Lichte der Aarhus-Konvention', pp. 665-666; Thomas Danwitz, 'Aarhus-Konvention: Umweltinformation, Öffentlichkeitsbeteiligung, Zugang zu den Gerichten' (2004) *Neue Zeitschrift für Verwaltungsrecht*, 3, p. 276. See also Sabine Schlacke, *Überindividueller Rechtsschutz* (Tübingen: 2008), pp. 242 et seq.

[51] See www.unence.org; also cited in Krämer, *EC Environmental Law*, pp. 161-162.

Public Participation in Decision-making and Access to Justice in Environmental Matters to Community institutions and bodies which entered into force on the 28 July 2007 ('Regulation 1367/06').[52]

Arts. 10 to 12 of Regulation 1367/02 concern 'internal review and access to justice'. The internal review procedure is a pre-condition for instituting judicial proceedings. Art. 10 provides that any non-governmental organisation is entitled to make requests for 'internal review to the Community institution or body that has adopted an administrative act under environmental law or, in case of an alleged administrative omission should have adopted such an act'. Several of the terms used in Art. 10 are defined more precisely in Art. 2 and Art. 11 of Regulation 1367/02. The term 'environmental law' is given a broad definition in Art. 2(f):

> Community legislation which, irrespective of its legal basis, contributes to the pursuit of the objectives of Community policy on the environment as set out in the Treaty: preserving, protecting and improving the quality of the environment, protecting human health, the prudent and rational use of natural resources, and promoting measures at international level to deal with regional or worldwide environmental problems.

According to this definition, administrative acts connected to CFP conservation measures could be subject to internal review. However, the scope of the internal review procedure is severely restricted by Art. 2(c) and 2(g). Art. 2(c) defines a 'Community institution or body as any public institution, body, office or agency established by, or on the basis of, the Treaty *except when acting in a judicial or legislative* capacity [...]'.[53] Art. 2(g) states that an 'administrative act means any measure of *individual scope* under environmental law, taken by a Community institution or body, and having legally binding effects'.[54]

Art. 12 provides that the non-governmental organisation which has requested an internal review pursuant to Art. 10, may institute proceedings before the ECJ

---

[52] See OJ 2006 No. L264/13. Prior to Regulation 1367/06, the Community has adopted a body of legislation which ought to contribute to attaining of the objectives of the Aarhus Convention: European Parliament and Council Directive (EC) 4/03 on public access to environmental information and repealing Council Directive 90/313/EEC, OJ 2003 No. L41/26; European Parliament and Council Directive (EC) 35/03 providing for public participation in respect of the drawing up of certain plans and programmes relating to the environment and amending with regard to public participation and access to justice Council Directives 85/337/EEC and 96/61/EC – Statement by the Commission, OJ 2003 No. L156/17; European Parliament and Council Directive (EC) 42/01 on the assessment of the effects of certain plans and programmes on the environment, OJ 2001 L 197/30.

[53] Italics are inserted by the author. Art. 2(2) of Regulation 1367/06 also provides that the terms 'administrative acts and administrative omissions shall not include measures taken or omissions by a Community institution or body in its capacity as an administrative review body, such as under [...] Arts. 81, 82, 86, 87, 226, 228, 195, 280 of the Treaty.

[54] Italics inserted by the author.

'in accordance with the relevant provisions of the Treaty'. Art. 12(2) is a restatement of the same clause with regards to omissions by a Community institution or body.[55]

The scope of the internal review procedure is rather limited and so is its use with respect to the CFP. First, the definitions of 'Community institution' and 'administrative act' clearly intend to exclude the internal review of the type of legal instrument most commonly used under the CFP, *i.e.*, regulations. Art. 2(c) of the Aarhus Regulation explicitly excludes the review of judicial or legislative acts; and the definition of 'administrative acts' in Art. 2(g) only refers to measures of individual scope. Although fisheries regulations contain detailed provisions, they regulate in an abstract and generally applicable way the use of specific fish stocks in a specific time period. As a result, they are not of an individual concern and thus should be regarded as legislative acts. Second, Art. 10(1) only allows non-governmental organisations to request an internal review; individual fishermen are excluded. Third, *Jans* and *Vedder* argue that Art. 10 does not clearly define the legal standard for the internal review. According to them, the level of proof required under the legislation to substantiate a request for internal review is unclear.[56] Ultimately, Art. 12(1) and (2) both explicitly subject access to the Court of Justice to 'the relevant Treaty provisions', *i.e.*, it refers to the strict conditions laid down in Art. 230, para 4 of the Treaty. The Aarhus Regulation therefore has had little impact on improving access to judicial review as provided under Art. 230, para. 4 of the Treaty.[57]

4.3 Case T-91/07 WWF UK Ltd. v. Council

A recent order by the CFI makes clear how it currently interprets the procedural rights of environmental groups under the CFP.

[55] The Commission's proposal foresaw that the environmental organisation would have the right to require an internal review from the relevant Community institution. The latter was then required to review the disputed act and address a reply to the organisation. Where the organisation still considered the act to be inconsistent with Community law, it should gain access to judicial review. This approach aimed at qualifying the challenging environmental organisation 'directly and individually concerned' in the sense of Art. 230, para. 4. See Commission Proposal COM(2003) 622 final for a Regulation of the European Parliament and of the Council on the application of the provisions of the Aarhus Convention on Access to Information, Public Participation in Decision-making and Access to Justice in Environmental Matters to EC institutions and bodies, page 16. See also Josef Falke, 'Die Aarhus-Konvention und der Zugang zu Gerichten in Umweltangelegenheiten', in Falke and Schlacke (ed.), *Information – Beteiligung – Rechtsschutz, Neue Entwicklungen im Umwelt- und Verbraucherrecht*, (Berlin: 2004), pp. 116-117; Krämer and Winter 'Umweltrecht', p. 1153.

[56] Jans and Vedder, *'European Environmental Law'*, pp. 216-217.

[57] Krämer, *EC Environmental Law*, p. 161.

4.3.1 Order of the CFI

In March 2007, the WWF UK instituted an annulment procedure before the CFI under Art. 230, para. 4 of the Treaty.[58] It challenged the legality of specific cod-TACs for 2007 adopted by the Council for 2007 within its annual TAC regulation.[59] The disputed TACs concerned stocks located in zones covered by the Council's Cod Recovery Plan.[60]

The WWF stated in its application to the Court that the Council had violated existing Community law in adopting TACs which were not in conformity with scientifically recommended limits to return cod stocks outside back to within safe biological limits.[61] First, it claimed that the TACs were not in accord with the cod recovery plan requirements adopted in Regulation 423/04. Second, they did not comply with the precautionary principle as laid down in international law as well as other provisions of Community law.[62] Third, it argued that 'should the Cod Recovery Plan not allow for the setting of a zero TAC, the plan in itself would be unlawful since it would not comply with the precautionary principle'.[63]

In addition, the WWF went to great lengths to explain why the Court should hold the action admissible under Art. 230, para. 4 of the Treaty. First, it argued that, in exceptional cases, measures of general application could be of direct and individual concern within the meaning of Art. 230, para. 4.[64] This exception would apply in cases 'when a person is entitled to submit observations and does lodge observations in order to attempt to influence the relevant decision'.[65] It then argued that, given its involvement in the adoption of the disputed TAC through its work in the Regional Advisory Council, it is 'individually concerned' within the meaning of Art. 230, para. 4 of the Treaty. It emphasised that its involvement was based on Arts. 4(2)(d) and (f), 5(4) and 31(5) of Regulation 2371/02. Art. 4(2)(d) and (f) provide that 'the limiting of catches and fishing effort shall be established by the Council taking into account available scientific, technical and economic advice and in particular [of] the reports drawn up by the STECF [...] and that advice provided by RACs established under Art. 31 of

58 Case T-91/07, WWF UK Ltd. v. Council, Order of the CFI, not yet in the ECR. The WWF appealed the case to the ECJ, see Case C-355/08, WWF UK Ltd. v Council, pending case.

59 Council Regulation (EC) 41/07 fixing for 2007 the fishing opportunities and associated conditions for certain fish stocks and groups of fish stocks, applicable in Community waters and, for Community vessels, in waters where catch limitations are required, OJ 2007 No. L15/1.

60 Council Regulation (EC) 423/04 establishing measures for the recovery of cod stocks, OJ 2004 No. L70/8.

61 Application of Case T/91/07 WWF UK v. Council in OJ 2007 No. C117/30.

62 The WWF claimed that the Council regulation would violate the requirements of the precautionary principle as laid down in the Cod Recovery Plan, in Art. 5(3), 4 and 2(1) of Regulation 2371/02, in Art. 174 of the Treaty and in Arts. 5 and 6 of the Fish Stocks Agreement.

63 The WWF brought two additional arguments which shall be neglected here.

64 Case T-91/07, WWF UK Ltd. v. Council, Order of the CFI, not yet in the ECR, para. 42.

65 *Ibid.*, para. 42

Regulation 2371/02 should be taken into account. Art. 31(5) states that RACs may 'submit recommendations and suggestions, of their own accord or at the request of the Commission or a Member State, on matters relating to fisheries management [...]'. The WWF also pointed out that the members of the North Sea RAC are a closed and identifiable group of individuals 'recognised by the Member States as having a personal and identifiable interest in a measure such as the contested regulation'.[66]

The WWF also argued that the Aarhus Convention establishes entitlements which individualise itself for the purpose of Art. 230, para. 4 of the Treaty. Its contention was that the Community had entered into the Aarhus Convention, which forms an integral part of the Community legal order, and thus is directly applicable.[67] Arts. 2(5) and 6(2) of the Aarhus Convention and Art. 9(2) of Regulation 1367/06 require the Community to involve the WWF early in the decision-making procedure that leads to the adoption of the TACs. This entitlement would confer a particular status on the WWF, making it individually distinguishable within the meaning of Art. 230, para. 4 of the Treaty.[68]

Moreover, the WWF argued that it should have standing to seek judicial review under Art. 230, para. 4 because the Commission and the Council have both *misused their power* when establishing the disputed TACs. According to the WWF, neither the Commission nor the Council had complied with the Cod Recovery Plan.[69] In making this argument, it also relied on the right to effective judicial protection. Its contention was that if it was not allowed to challenge the illegal TACs in Court, no institution, body or individual would do so.[70]

It would be abhorrent to the rule of law for there to be no person, body, institution or government able or willing in practice to take steps to ensure that emergency measures to ensure the survival of an environmental resource, whose destruction will be likely to have ecological and environmental ramifications for many individuals across the Community, are properly implemented by the Council in accordance with the law.[71]

Finally, it refers to the irreparable environmental damage caused by the disputed TACs.[72]

The Court, however, held the claim inadmissible. One-by-one, it rejected the WWF's arguments on admissibility and did not rule on any of the material legal questions, *i.e.*, the non-compliance of the disputed TACs with the provisions of the Cod Recovery Plan and the TACs' consistency with the precautionary principle as laid down in international and Community law.

---

66 *Ibid.*, para. 51. It basically refers to the functions of the RACs as described above. See also Art. 31(2) of Regulation 2371/02 and Art. 5(1) and (2) of Decision 585/04.

67 *Ibid.*, para. 53.

68 *Ibid.*, paras. 53-55.

69 *Ibid.*, para. 57.

70 *Ibid.*, para. 58.

71 *Ibid.*, para. 59.

72 *Ibid.*, para. 60

In principle, the Court accepted that the fact that a person is 'involved in a procedure leading up to a disputed measure is capable of distinguishing that person individually in relation to the measure in question only if the applicable Community legislation grants him certain procedural guarantees'.[73] It also recognised that the WWF is a member to the RAC that was involved in the procedure leading to the adoption of the disputed TAC. Furthermore, it did not explicitly deny that Arts. 4(2)(d) and (f) and 31(1), (4) and (5) of Regulation 2371/02 provide procedural guarantees which involve the claimant in the procedure leading up to the disputed measure. However, it then stated that *only the RACs* themselves could claim that these provisions could distinguish them individually for the purpose of Art. 230, para. 4 of the Treaty. The Court added that even if the WWF enjoyed procedural guarantees under Regulation 2371/02 and Decision 585/04, the WWF's action before the Court 'would not be aimed at safeguarding those guarantees'.[74] Finally, the CFI rejected the argument that the members of the RAC or the RAC's Executive Committee constitute a closed and identifiable group. It held that it 'is not sufficient to differentiate the members of the North Sea RAC or the members of its executive Committee from any other person and to distinguish them individually in the same way as an addressee. The contested regulation is to apply by virtue of an objectively determined situation in the context of which possible membership of a RAC or a RAC'S Executive Committee is irrelevant.'[75]

The CFI also held that the Aarhus Convention and its implementing measure (Regulation 1367/06) do not confer a particular status on the WWF in the sense of Art. 230, para. 4 of the Treaty. It provided that any entitlements under the Aarhus Convention or Regulation 1367/06 are granted to the WWF as a member of 'the public'. Such entitlements do not make the WWF's position distinguishable from all other persons within the meaning of the *Plaumann Doctrine*.[76]

Finally, the CFI found that the applicant was not individually distinguishable on the basis of the misuse of power or the irreparable environmental damage in accordance with Art. 230, para. 4. Addressing the argument on effective judicial protection, it stated that a bar to an action before the national courts cannot set aside conditions governing the admissibility of actions under Art. 230, para. 4.[77]

---

73 *Ibid.*, para. 69.

74 *Ibid.*, para. 77.

75 *Ibid.*, para. 79.

76 *Ibid.*, paras. 82 and 66.

77 *Ibid.*, paras. 87-88.

4.3.2 Comments and criticism of the decision in Case T-91/07 WWF v. Council

The CFI's decision in WWF v. Council leaves ample grounds for criticism. First, the Court rejected the argument that any of the procedural rights laid down in Council Regulation 2371/02 and Council Decision 585/04 can be regarded as granting the WWF 'qualified procedural guarantees provided within the law' which involve the WWF in the procedure leading up to the adoption of the disputed TACs. However, in the *Greenpeace* decision, the Court provides little guidance on how 'procedural guarantees provided within the law' should be designed. It remains therefore arguable whether procedural rights granted to the members of RACs qualify as 'procedural guarantees provided in the law'. The articles governing RACs must be interpreted as granting certain procedural guarantees for the following reasons: These provisions explicitly recognise the Council's intent to increase participation and represent general interests (particularly, the environment) in the process leading up to the adoption CFP legislation. The purpose of RACs is to draw on the knowledge and experience of fishermen and other stakeholders when creating policy[78] and to provide for new forms of participation when designing CFP measures.[79] Art. 31(2) of Regulation 2371/02 explicitly states which interests shall be represented. Art. 31(2) states that RACs shall consist, *inter alia*, of 'representatives [...] of the environment and consumer interests [...].' Accordingly, Arts. 1(2), 5(1), 7(3) of Decision 585/04 grant environmental organisations seats within the general assembly and on the executive committee of the RAC. Moreover, 7(3) of Decision 585/04 allows parties to RACs, which have been overruled in the process of adopting a recommendation, to express their dissenting opinions. Given that the Community legislator explicitly recognises environmental conservation as a general interest that needs representation, it is a contradiction for the Court to deny parties that represent that interest access to legal review on the grounds that it acts on behalf of that general (and not an individual) interest.

This view is supported by other Court of Justice decisions. The Court has on different occasions interpreted less strictly the requirements of Art. 230, para. 4 of the Treaty with regards to the involvement of groups that represent private or commercial interests in legislative procedures.[80] For example, in the *Van der Kooy* decision, a public body, which represented agricultural businesses,[81] was

---

78 Reasoning (27) of Regulation 2371/02.

79 Reasoning (1) of Decision 585/04. Reasoning (4) emphasises that RACs shall include all the interests affected by the CFP; see also Art. 2(2)(c) of Regulation 2371/02, naming 'a broad involvement of stakeholders at all stages of the policy from conception to implementation' as guiding principle of the CFP.

80 Case 67/85, Van der Kooy v. Commission [1988] ECR 219, paras., 18-25; Case C-313/90, CIFRS and others v. Commission, [1991] ECR I-2257, para. 29.

81 The body instituting proceeding in *Van der Kooy* was a 'body established under public law to protect the common interests of agricultural undertakings, taking into account the public interest'; Case 67/85, Van der Kooy v. Commission [1988] ECR 219, para. 1(3).

granted standing and allowed to institute proceedings under Art. 230, para. 4 of the Treaty. In that case, the ECJ recognised the active role taken by the association in the procedure under Art. 88(2)[82] of the Treaty through its submission of written comments and by maintaining close contact with officials throughout the procedure. However, the decisions *Greenpeace* and *WWF UK* do not provide a reasonable explanation of why the Court is willing to grant associations that represent a few private interests *locus standi*, while at the same it disallows associations representing environmental interests, *i.e.*, the greater public interest, to institute proceedings under Art. 230, para. 4 of the Treaty. It has been stated that applying consequently, Art. 230, para. 4 leads to a remarkable situation: the more people are affected by a court decision, the less is the likelihood that claimants can meet the threshold of being individually concerned.[83]

The second criticism regards the CFI's argument that possible procedural guarantees accorded by Regulation 2371/02 and Decision 585/04 would exist solely for the benefit of the RACs and not for the benefit of its members.[84] This is a misinterpretation of the purpose of the RACs, which function as mere institutional mechanisms that allow representatives of certain (general) interests to express their views to the Commission. It is not the RACs genuine will or interest which shall be expressed to the Commission but the views of those who represent the interests recognised within the law establishing RACs (including the environment).

The third point to be criticised is that CFI has not interpreted the case in the light of Arts. 2(5), 6(2) and 9(3) and (4) of the Aarhus Convention. It remains true that international law cannot amend Treaty provisions. However, according to Art. 300(7) of the Treaty, from the day of its entering into force the Convention forms an integral part of the Community's legal order and immediately binds Community institutions.[85] The ECJ must then decide whether the provisions of the Aarhus Convention affect its body of case law on the interpretation of Art. 230, para. 4 of the Treaty.[86] Given that Art. 2(5) and 6(2) of the Aarhus Convention 'merely' require state parties' legislators to inform the public, the CFI may have been right to not interpret these provisions as qualified procedural rights in the sense of the requirements established in *Greenpeace*. However, as stated above, it is arguable whether Art. 9(3) and (4) of the Convention actually requires the Community to grant access to judicial review to the public. The provisions should be interpreted in a way that provides state parties with a very wide margin of discretion. However, a central objective of the Convention is to make judicial mechanisms accessible to the public; this should include its repre-

---

82 Ex Art. 93(2) of the Treaty.

83 Krämer and Winter, 'Umweltrecht', para. 75; see also Jans and Vedder, '*European Environmental Law*', p. 212.

84 Case T-91/07, WWF UK Ltd. v. Council, Order of the CFI, not yet in the ECR, para. 75.

85 Case 181/73, Haegemann [1974] ECR 449, paras. 2 to 6; Case T-115/94, Opel Austria/Council, [1997] ECR II-39, paras. 90, 101, 102 Case 12/86, Demirel, [1987] ECRI-3719, para.7.

86 Schmalenbach, 'EGV Art. 300', para. 85.

sentative organisations.[87] If environmental organisations, such as the WWF, are denied the right to the judicial review, Art. 9(3) and (4) of the Aarhus Convention would clearly be rendered meaningless.

Even if the current interpretation of Art. 230, para. 4 does not violate the Aarhus Convention, the adoption of the Convention should provide sufficient impetus for the Community and the European Court to amend its current system of access to judicial review on environmental matters.

5 Conclusion

It is clear that there is an institutional imbalance within the CFP legal framework. The Commission and the Council may violate existing Community law by adopting flawed CFP legislation. Potential claimants that have the right to file a claim in Community courts are simply not interested in doing so, and those who are interested are simply not allowed to do it. The WWF has concluded correctly that this situation poses a threat to the rule of law. The Court's interpretation of Art. 230 para. 4 of the Treaty has resulted in very strict conditions being placed on access to judicial review for individuals and environmental groups. In general, this right is rarely available. Access to the judicial branch is primarily gained indirectly, through national courts requesting a preliminary ruling from the ECJ. The problems with this procedures have been described above. Accordingly, these individuals and groups may not be afforded judicial review.

To uphold the rule of law and to give real effect to the Community's legal obligations under the CFP and the Aarhus Convention, Community institutions should consider the following options. First, the Member States could amend the Treaty and establish procedural rights for individuals and interest groups in fisheries conservation and management matters.[88] However, the text of the Lisbon Treaty, which has not yet been ratified, indicates that such an amendment will not occur in the near future.[89] Second, the Court should reconsider its restrictive interpretation of both the terms 'directly and individually concerned' under Art. 230, para. 4.[90] Third, given that the Council is, in fact, granted a

87 Recital 18 and Art. 2 Nr. 4, 5 of the Aarhus Convention.

88 Ebbesson, 'European Community', p. 95. For an overview of the literature on the rights of environmental groups under (Member States) national law, see, for example, Hans-Joachim Koch, 'Die Verbandsklage im Umweltrecht' (2007) *Neue Zeitschrift für Verwaltungsrecht*, pp. 369-379; see Nicolas de Sadeleer, Gerhard Roller and Miriam Dross, *Access to Justice in Environmental Matters and the Role of NGOs – Empirical Findings and Legal Appraisal* (Groningen: 2005).

89 See words of Art. 263 of the Consolidated Version of the Treaty on the Functioning of the European Union.

90 Chalmers *et al.*, *European Union Law*, p. 433; Manfred Rehbinder, 'Rechtsschutz gegen Handlungen und Unterlassungen der Organe und Einrichtungen der Europäischen Gemeinschaft im Lichte der Aarhus-Konvention', pp. 665, 667. The Court however states that to amend the Plaumann formula the

certain margin of discretion to interpret the meaning of Art. 230, para. 4, it should explicitly grant individuals and their representing organisations 'qualified procedural guarantees' within CFP legislation as defined in the *Greenpeace* decision. The Court could also accept environmental groups' RAC-participation-rights as such qualified procedural guarantees.

Treaty would have to be amended by the Member States, Case-50/00 P, Unión de Pequeños Agiculturos v. Council [2002] ECR I-6677, paras. 44-45;

CHAPTER 7

Legal Review of Measures Adopted under the CFP

CFP measures may not be compatible with existing rules and principles provided by international or Community law, or fundamental rights (as protected by law).[1] The following chapter will assess some of the current questions on the legality of CFP law. First, incompatibilities of CFP secondary and tertiary legislation with international fisheries law will be assessed. Second, the compatibility of TAC legislation with fundamental rights protection requirements under Community law will be addressed. Finally, the legality of the sanction measure allowing the Commission to deduct quota will be reviewed.

1 Compatibility of the CFP regime with international law management requirements

The Community is party to several binding international conventions relating to fisheries, *i.e.*, particularly the 1982 United Nation Convention on the Law of the Sea (1982 UNCLOS)[2], the 1995 United Nations Agreement on Straddling Fish Stocks and Highly Migratory Fish Stocks (FSA)[3], and the Agreement to Promote Compliance with International Conservation and Management Measures by Fishing Vessels on the High Seas ('Compliance Agreement').[4] This section will examine whether the current CFP regime, and

1 D. Booß, 'Art. 230 EGV', in Grabitz and Hilf (ed.), *Das Recht der Europäischen Union – Kommentar*, (München: 2000), para. 119; Craig and De Bùrca, *EU Law*, pp. 206-213. See particularly Case C-308/06, Intertanko and others, [not yet in the ECR], paras. 53-66. Arts. 230, para. 2 and 234 of the Treaty. Under Community law basic regulations rank higher than implementing regulations, see Case 38/70, Deutsche Tradax GmbH v. Einfuhr- und Voratsstelle für Getreide und Futtermittel, [1971] ECR 145, para. 3 (Summary). Thus, lower-ranking implementing rules cannot derogate from the basic regulation, *Ibid.*, para. 10; see also Case C-103/96, Directeur général des douanes and droits indirects v. Eridania Beghin-Say [1997] ECR I-1453, para. 20. When all Member States have ratified the Treaty of Lisbon in 2009, the Charta of Fundamental Rights of the European Union (OJ 2007 No. C303/1) will be recognised by the European Union; see Art. 6 of the Treaty of Lisbon amending the Treaty on the European Union and the Treaty Establishing the European Community, OJ 2007 No. C306/01. Currently, see Art. 6(1) and (2) of the Treaty of the European Union (TEU); see also Case 29/69, Stauder v. Stadt Ulm [1969] ECR 419, para. 7; Case 11/70, ECR 1125, Case 11/70, Internationale Handelsgesellschaft mbH v. Einfuhr und Vorratsstelle für Getreide und Futtermittel [1970] ECR 1125, para. 2.; Case 4/73, Nold KG v. Commission [1974] ECR 491, Summary, para. 2.

2 See Art. 305 in combination with Annex IX (Art. 4) of 1982 UNCLOS; Council Decision (EC) 98/392 concerning the conclusion by the European Community of the United Nations Convention of 10 December 1982 on the Law of the Sea and the Agreement of 28 July 1994 relating to the implementation of Part XI thereof, OJ 1998 No. L 179/1.

3 See Council Decision (EC) 98/414 of 8 June 1998 on the ratification by the European Community of the Agreement for the implementing of the provisions of the United Nations Convention on the Law of the Sea of 10 December 1982 relating to the conservation and management of straddling stocks and highly migratory fish stocks, OJ 1998 No. L189/14.

4 Resolution 15/93 of the 27th Session of the FAO Conference.

the implementation practices of the Community and its Member States conform to the obligations established in these conventions. It will become clear that, in many cases, international law on fisheries is vague or does not provide for actions to be undertaken if its requirements are violated.[5] However, it does, in many cases, indicate management thresholds, objectives, guidelines, and best practices against which Community laws and their implementation can be assessed, *i.e.*, from a *de lege ferenda* point of view.

The following section focuses on rules and principles at international law which concern the relation between fishing and exploitation limits, ecosystem protection, the collection and sharing of scientific data, control and enforcement, cooperation, as well as subsidisation.

1.1 Harvesting limits

Both UNCLOS and FSA restrict fishing by establishing quantitative catch restrictions. The prescribed quantities, however, are matter of dispute. It will be tried here to explain how UNCLOS and FSA provisions are structured and how they should be interpreted.

1.1.1 1982 UNCLOS

1982 UNCLOS divides the seas in different sections in which different rules and laws apply. Here, conservation measures governing stocks located in or straddling through the territorial sea, the Exclusive Economic Zone and the High Seas will be analysed.[6]

1.1.1.1 Territorial sea

UNCLOS does not explicitly require coastal states to conserve or manage fisheries resources in their territorial sea;[7] Arts. 21(d) to (f) merely *entitle* coastal states to adopt laws and regulations on the conservation of the living resources of the sea, the prevention of infringements of fisheries laws and regulations, the preservation of the environment and the prevention, reduction and control of pollution. Accordingly, coastal states are, in principle, free to

---

5 Note, for example, national fisheries laws applying to the EEZ have been excluded from UNCLOS judicial review, Art. 297(3)(a) of 1982 UNCLOS. Besides, the Fish Stocks Agreement does not provide any legal implication where catch limitations are exceeded, see Jay Ellis, 'The Straddling Stocks Agreement and the Precautionary Principle as Interpretive Device and Rule of Law' (2001) 32 *Ocean Development and International Law*, p. 300.

6 Regarding sedentary species located on the *continental shelf*, see Art. 77(1) of 1982 UNCLOS. See also R.R. Churchill and A.V. Lowe, *The law of the sea*, (Manchester: 1999), p. 320; R. Wolfrum, V. Röben and Fred L. Morrison, 'Preservation of the Marine Environment', in Morrision and Wolfrum (ed.), *International, Regional and National Environmental Law*, (The Hague: 2000), p. 238.

7 Arts. 2 to 7 of 1982 UNCLOS.

exploit marine capture resources in these areas.[8] They are, however, required to conserve anadromous and catadromous stocks and ensure their ingress and egress through the different zones.[9]

1.1.1.2 Exclusive Economic Zone

Regarding marine capture resources located in an Exclusive Economic Zone (EEZ),[10] coastal states have sovereign rights for the purpose of exploiting and managing the natural resources, whether living or non-living.[11] To this end, they are required by Art. 61(1) UNCLOS to determine a TAC for living resources,[12] and allocate it to nationals and to third country fishers who are entitled to fish in their EEZ.[13] The right to exploit marine capture resources in the EEZ is subject to the following limitations.[14] Besides setting a TAC, coastal states are required by Art. 61(2) to ensure, through conservation and management measures provided in a non-exhaustive list in Art. 62(4), that the 'maintenance of the living resources in the EEZ is not endangered by overexploitation'. This requirement is qualified by Art. 61(3), which prescribes that *measures* 'maintain or restore populations of harvested species at levels which can produce the maximum sustainable yield' (MSY). MSY is, again, 'qualified by relevant environmental and economic factors, including the economic needs of coastal fishing communities and the special requirements of developing states'.

What is actually required of the parties bound by the conservation requirements in Arts. 61 and 62 UNCLOS? The main obligation is to maintain living resources[15] in the EEZ, ensuring they are used optimally and not endangered

---

8 To protect these rights, Art. 19(2)(i) of 1982 UNCLOS provides, that coastal states can forbid foreign ships from engaging in fishing activities while passing through, thus, defining more closely their right of innocent passage, granted under Art. 17. In addition, Art. 42(1)(c) allows states that border an international water strait to forbid fishing for passing ships and oblige them to stow their fishing gear.

9 Arts. 66(2) Art. 67 (1) of 1982 UNCLOS.

10 Arts. 55 and 57 of 1982 UNCLOS.

11 Art. 56(1)(a) of 1982 UNCLOS.

12 According to *Burke* and *Kaye*, this provision is not meant to limit coastal states to refer to TACs as their sole management instruments; see William T. Burke, *The New International Law of Fisheries*, (Oxford: 1994), p. 46; Stuart M. Kaye, *International Fisheries Management*, (The Hague: 2001), p. 99. However, as Markowski correctly states, such an interpretation would render Art. 61(1) meaningless. Besides, Art. 61(1) is mandatory; see also Markowski, *Allocation and management of fisheries resources: an in-depth analysis of instruments in comparative perspective*, (Bremen; forthcoming 2008), pp. 11-12.

13 Art. 61(1) of 1982 UNCLOS. If states are not able to fully exploit their TACs, Art. 62(2) requires that they give other states access to their surplus catch. They shall particularly consider the interests of landlocked, geographically disadvantaged states, developing states and states that have traditionally fished in their EEZ, see Arts. 62(2) and (3) and 69 and 70.

14 As provided in Arts. 61 and 62 of 1982 UNCLOS.

15 The meaning of the term 'living resources' remains somewhat imprecise see Burke, *The New International Law of Fisheries*, p. 51.

by overexploitation. These requirements are concretised, *i.e.*, conservation and management measures shall be designed in a way that exploitation rates do not exceed the MSY.[16] *Burke* refers to MSY as the fundamental concept in UNCLOS 'that underlies and essentially determines all other consequential decisions about access to fisheries in the EEZ'.[17] MSY is not defined in UNCLOS, but others have defined it as 'the greatest harvest that can be taken from a self-regenerating stock of animals year after year while still maintaining the average size of the stock'.[18] Parties can derogate from this obligation for social, environmental and economic reasons, pursuant to Art. 61(3). Most authors agree that under Art. 61(2) and (3,) coastal states are permitted to legitimately exceed MSY levels, adopting almost any exploitation rate as long as it does not lead to overexploitation that endangers fish stocks.[19] Thus, overexploitation is not prohibited, unless it places EEZ resources in jeopardy. An argument in favour of this position is that UNCLOS was drafted to give priority to the sovereignty of the coastal states.[20] UNCLOS also precludes national conservation laws which apply to the EEZ from legal review.[21] In addition, the wording of Art. 61(2) and (3), such as 'stocks shall not be endangered by overexploitation', 'the best scientific evidence available' and 'MSY', in practice, is vague and leaves much room for interpretation.[22]

It has been noted that this widely accepted interpretation of Art. 61 has significantly contributed to the overexploitation of commercially used species and even stock collapse under EEZ fisheries regimes by the parties to UNCLOS.[23] It is thus argued here that the interpretation of UNCLOS provisions

---

16 These norms are closely linked; see Wolff, *Fisheries and the Environment*, p. 57; and José A. de Yturriaga, *The International Regime of Fisheries – From UNCLOS 1982 to the Presidential Sea*, (The Hague: 1997), p. 116.

17 William T. Burke, *The New International Law of Fisheries*, p. 44.

18 Birnie and Boyle, *International Law & the Environment*, pp. 551-554 (at 551).

19 R.R. Churchill and A.V. Lowe, *The law of the sea* (Manchester: 1999), p. 289; José A. de Yturriaga, *The International Regime of Fisheries – From UNCLOS 1982 to the Presidential Sea*, pp. 118-119; William T. Burke, *The New International Law of Fisheries*, p. 53; Proelß, *Meeresumweltschutz*, p. 108; Richard Barnes, 'The Convention on the Law of the Sea: An Effective Framework for Domestic Fisheries Conservation', in Freestone, Barnes and Mong (ed.), *The Law of the Sea – Progress and Prospects*, (Oxford: 2006), p. 242-243.

20 Coastal states have sovereign rights with regards to the exploitation of resources in the EEZ, see Art. 56(1) and Art. 61(1) of 1982 UNCLOS.

21 Art. 297(3)(a) of 1982 UNCLOS.

22 Richard Barnes, 'The Convention on the Law of the Sea: An Effective Framework for Domestic Fisheries Conservation', p. 242; William T. Burke, *The New International Law of Fisheries*, pp. 51-52; with regards to the term TAC, see Birnie and Boyle, *International Law & the Environment*, pp. 551-554.

23 Richard Barnes, 'The Convention on the Law of the Sea: An Effective Framework for Domestic Fisheries Conservation', p. 242; Rosemary Rayfuse, 'The Interrelationship Between the Global Instruments of International Fisheries Law', in Hey (ed.), *Developments in International Fisheries Law*', (The Hague: 1999), pp. 111-112.

must be responsive to the changing circumstances in (international) fisheries.[24] If the aim is to prevent overexploitation of fish stocks, bringing them within MSY levels, a more stringent interpretation of UNCLOS must be accepted. Fisheries science has shown that exploiting species continuously beyond the MSY level will ultimately result in their depletion.[25] Therefore, a continued practice of setting fishing rates below or at the maximum sustainable yield-levels constitutes the *conditio sine qua non* for achieving the requirements of Art. 61(2) and (3). *Kaye* has stated that 'no catch limit beyond the MSY would be consistent with the Law of the Sea Convention, as to do so would prevent the maintenance or restoration of populations to levels that can produce an MSY'.[26] Other authors propose a less stringent interpretation, assuming that coastal states violate Art. 61 by *repeatedly* setting catch rates above MSY-levels, and threatening stocks.[27] This view recognises legislators' right to a margin of discretion to weigh social, economic and environmental objectives when adopting management decisions, while at the same time bringing exploitation rates within the MSY catch level.[28]

According to the second interpretation of Art. 61, and in this author's opinion, parties to UNCLOS are, in principle, obliged to adopt fishing opportunities at MSY-levels. They may exceed this level *only for compelling* social and economic reasons and only for a given time span. Reasons and time spans must be clearly predefined and thus are open for judicial review in accordance with national fisheries laws. In addition, exploitation levels must not cause stocks to fall below safe biological limits, *i.e.*, to levels where they are ultimately threatened by overexploitation.

The proposed interpretation is supported by the fact that during the 1990s many UNCLOS parties adopted binding and non-binding conservation measures, such as the 1995 Fish Stocks Agreement, the FAO Code of Conduct and agreements concerned with the protection of the marine environment. These measures increasingly support the principle that economic and social

---

24 Jay Ellis, 'The Straddling Stocks Agreement and the Precautionary Principle as Interpretive Device and Rule of Law', pp. 291, 295 and 296.

25 Kura *et al.*, *Fishing for Answers – Making Sense of the Global Fish Crisis*, p. 91; Suzanne Ludicello, Michael Weber and Robert Wieland, *Fish, Markets, and Fishermen – The Economics of Overfishing*, (London: 1999), pp. 45-47.

26 Stuart M. Kaye, *International Fisheries Management*, pp. 100-101; arguing similarly, see FAO, *Reference Points for Fisheries Management – Technical Paper 347*, (Rome: 1995), p. 1; see also Markowski, *Allocation and management of fisheries resources: an in-depth analysis of instruments in comparative perspective*, pp. 11-12.

27 Rat von Sachverständigen für Umweltfragen (SRU), *Meeresschutz für Nord- und Ostsee – Sondergutachten*, (Baden-Baden: 2004), p. 120.

28 *Ibid.* p. 101; Stuart M. Kaye, *International Fisheries Management*, p. 101. Coastal states, off course, always have the option to adopt exploitation rates at maximum economic yield-levels, *i.e.*, below the MSY, see Burke, pp. 52-58, Birnie and Boyle, *International Law & the Environment*, pp. 551-554.

factors do not justify drastic overexploitation.[29] In addition, Art. 61(3) of 1982 UNCLOS states that the management measures of coastal states shall take into account non-binding (recommended, minimum) standards, whether sub-regional, regional or global.[30] Such standards currently incorporate the precautionary and ecosystem approaches, signalling the intention to decrease fishing pressure to bring fish stocks to sustainable levels.[31]

1.1.1.3 High seas

Regarding fisheries resources located in the high seas, 87(1)(e) and Art. 116(1) of 1982 UNCLOS guarantee the freedom of fishing. Art. 116(1) states that all states have the right for their nationals to engage in fishing on the high seas. However, this right is subject to certain conditions provided in the Convention,[32] particularly UNCLOS provisions regarding the conservation of resources on the high seas as laid down in Arts. 116(c), 117, 118 and 119 of 1982 UNCLOS.[33] Art. 117 sets down the general rule that 'all states have the duty to take, or to cooperate with other states in taking, such measures for their respective nationals as may be necessary for the conservation of the living resources on the high seas'. This obligation is concretised in Art. 119 for conservation. Art. 119 spells out specific criteria that states must consider when instituting conservation measures. First, Art. 119(1) does not refer to TAC as the sole management instrument to be adopted.[34] Second, Art. 119(1)(a) and (b) provide that

---

[29] See Art. 7.1.1 of the FAO code of Conduct; Rat von Sachverständigen für Umweltfragen (SRU), *Meeresschutz für Nord- und Ostsee – Sondergutachten*, (Baden-Baden: 2004), p. 120; Rayfuse also argues that where a state is party to the FSA, the Compliance Agreement and UNCLOS, all three agreements shall be interpreted jointly; see Rosemary Rayfuse, 'The Interrelationship Between the Global Instruments of International Fisheries Law', p. 120; Markowski, *Allocation and management of fisheries resources: an in-depth analysis of instruments in comparative perspective*, pp. 11-12, 16.

[30] Wolfrum, Röben and Morrison, 'Preservation of the Marine Environment', p. 233.

[31] See, for example, standards set in Arts. 7.1.1, 7.2., 7.5. of the FAO-Code of Conduct.

[32] First, Art. 87(2) provides that the freedom of fishing on the high seas 'shall be exercised by all states with due regard for the interests of other states in the exercise of the freedom of the high seas [...]'. Second, fishing on the high seas is subject to (a) nations' treaty obligations, (b) the management obligations provided under UNCLOS with regards to specific stocks that occur both within the EEZs and the high seas, with highly migratory species, marine mammals, anadromous and catadromous stocks, see Arts. 63(2) and 64 to 67 of 1982 UNCLOS; and (c) UNCLOS provisions regarding the conservation of resources on the high seas as laid down in Arts. 116 to 119.

[33] Marine mammals are regulated by Art. 120 and 65 of 1982 UNCLOS. Art. 87(2) of 1982 UNCLOS provides that the freedom of the high seas shall be exercised by all states with *due regards* for the interests of others. This general obligation remains vague concerning its legal consequences; see Wolff, *Fisheries and the Environment*, p. 60.

[34] It provides: 'In determining the allowable catch and establishing other conservation measures for the living resources in the high seas, states shall...'; F.O. Vicuna, *The Changing International Law of High Seas Fisheries* (Cambridge: 1999), p. 46; Birnie and Boyle, *International Law & the Environment*, p. 663.

any adopted management measures shall be designed to 'maintain or restore populations of harvested species at levels which can produce the maximum sustainable yield, as qualified by relevant environmental and economic factors'. The management requirements of Art. 119(1) are similar to Art. 61. Given that Art. 119, just like Art. 61, is mainly concerned with the conservation of, in many cases, the same fisheries resources, the term 'maximum sustainable yield' as well as the other criteria should be interpreted in the same way as under Art. 61(3) of 1982 UNCLOS.[35]

1.1.2 1995 Fish Stocks Agreement (FSA)

Arts. 5, 6, and 7 of the FSA set out harvesting limitations for straddling and highly migratory species located in the high seas and the EEZ.

Article 5(a) and (b) basically reiterate conservation obligations under Arts. 61 to 64 of 1982 UNCLOS, requiring states to adopt measures to ensure long-term sustainability and optimum utilisation of fish stocks.[36] However, these conditions are imbedded in a set of more ambitious principles such as, for example, the precautionary approach. Compared to UNCLOS, the Fish Stocks Agreement substantially bolsters the MSY concept as the basis for adopting conservation measures.

Art. 6(1) of the Fish Stocks Agreement provides that states 'shall apply the precautionary approach widely to conservation, management and exploitation of straddling fish stocks and highly migratory species'.[37] The remaining paragraphs of Art. 6 and Annex II expand on this obligation. Basically, the Fish Stocks Agreement lays down a three-step procedure for adopting conservation and management strategies: (a) states complete a scientific assessment that considers biological as well as socio-economic factors; (b) they shall adopt two types of reference points, as well as (c) a system of management actions which automatically apply in cases where stocks fall under such reference points.[38]

The level of exploitation is particularly based on the reference point system. After assessing stock conditions, Art. 6(2)(b) requires states to establish a system

[35] Wolfrum, Röben and Morrison, 'Preservation of the Marine Environment', p. 235; Wolff, Fisheries and the Environment, p. 60.

[36] Parties to the Fish Stocks Agreement apply the general principles on the high seas, set out in Art. 5, *mutatis mutandis* to their EEZ.

[37] Art. 6(2) of the Fish Stocks Agreement declares: 'states shall be more cautious when information is uncertain, unreliable or inadequate. The absence of scientific information shall not be used as a reason for postponing or failing to take conservation and management measures'. The provision resembles the even more general definition included in Principle 15 of the Rio Declaration. According to the Rio Declaration, the precautionary principle provides that 'where there are threats of serious or irreversible damage, lack of scientific certainty shall not be used as a reason for postponing cost-effective measures to prevent environmental degradation'.

[38] O.S. Stokke and C. Coffey, 'Precaution, ICES and the common fisheries policy: a study of regime interplay' (2004) 28 *Marine Policy*, p. 118.

of stock-specific reference points considering the detailed criteria in Annex II (step two).[39] Section 2 of Annex II states that two types of reference points should be used: 'conservation, or limit, reference points *and* management, or target, reference points'.[40] While the so called *limit reference points* set boundaries that are intended to keep harvesting within safe biological limits, the 'target reference points' are intended to meet more ambitious management objectives.[41] Section (7) of Annex II of the FSA states that *limit reference points* should use the MSY as a minimum standard. These limits must not be exceeded for socio-economic reasons.[42]

The third consideration states must take into account when implementing Art. 6(1) is that, when stocks approach reference points, states must take predetermined management measures to ensure that such levels are not exceeded, or, if they are, take measures to restore respective stocks.[43] However, the FSA does not explicitly require specific actions if limit reference points are exceeded. In particular, it does not require parties to automatically cease all fishing activities, nor does it provide any sanctions for non-compliance. The quality of the response is defined by the state, or where competent, by the relevant international fisheries organisations.[44]

1.1.3 Compatibility of CFP measures

The process leading to the adoption of TAC regulations under the CFP was described in detail above. It has been explained that the Council regularly increases the TACs, which are recommended to it by ICES and the Commission. This practice also concerns those recommendations which advocate moratoria for stocks outside safe biological limits. In fact, many important stocks are currently exploited outside safe biological limits. If the 1982 UNCLOS and FSA conservation requirements are interpreted as suggested here, Community *exploitation rates* (as laid down in Council regulations) which exceed safe

39 See also Section (3) of Annex II.

40 Italics have been added by the author.

41 Section 2 of Annex II, see also Markowski, *Allocation and management of fisheries resources: an in-depth analysis of instruments in comparative perspective*, p. 17.

42 Markowski, *Allocation and management of fisheries resources: an in-depth analysis of instruments in comparative perspective*, p. 17.

43 Art. 6 (4) and section (4) of Annex II FSA.

44 This and other shortcomings related to the conservation obligations in Art. 6 have also been identified, see, for example, W. Edeson, D. Freestone and E. Gudmundsdottir, *Legislating for Sustainable Fisheries*, pp. 32-33; Lawrence Juda, 'The 1995 United Nations Agreement on Straddling Fish Stocks and Highly Migratory Fish Stocks: A Critique', p. 153; FAO, *Technical Guidelines for Responsible Fisheries – Precautionary Approach to Capture Fisheries and Species Introductions – Technical Paper 350/1*, (Rome: 1996), pp. 6-29; paras. 24, 25, 28, 47(c), 48(e), 49 and 50; R. Wolfrum, V. Röben and Fred L. Morrison, 'Preservation of the Marine Environment', in Morrison and Wolfrum (ed.), *International, Regional and National Environmental Law*, (The Hague: 2000), p. 244.

biological limits as well as the *continued practice* of managing stocks below the MSY level must be regarded as a violation of both UNCLOS and the FSA.

With regards to the precaution-requirements of the FSA, the Community has established a reference points system and procedures that apply where such reference levels are exceeded, *i.e.*, the recovery and management plans in Regulation 2371/02. However, these plans refer to 'safe biological limits' and not 'MSY' as their basic reference point. According to Community law, where stocks are 'outside safe biological limits', the Council must draw up plans to restore or maintain stocks at safe biological limits.[45] Recovery and management plans, again, shall include so called *'limit* and *conservation reference points'*.[46] Limit reference points refer to 'values of fish populations parameters [...] which should be avoided because they are associated with unknown population dynamics, stocks collapse or impaired recruitment'.[47] According to Art. 5(3) of Regulation 2371/02, limit reference points recommended by scientific bodies shall only be 'taken into account'. Thus, if at all, Community law's minimum standard for setting TACs is safe biological limits. Despite the fact that the FSA does not provide any legal implications for exceeding harvesting levels provided in its provision on reference points, it must be stated that Community secondary and tertiary law do not currently follow management requirement advocated by the FSA, which is to limit the exploitation of straddling and highly migratory stocks *at minimum* to MSY levels.

1.2 Interdependency of stocks with other species and ecosystems

Where coastal states exploit marine living resources, 1982 UNCLOS and the FSA require them to consider non-target species and ecosystems.

1.2.1 Requirements under 1982 UNCLOS and the FSA

Where coastal states adopt TACs for stocks located in the EEZ, they are required by 1982 UNCLOS to take into account the interdependence of stocks as well as the effects of fishing on associated or dependant species.[48] The exact implication of this norm is somewhat unclear. It has been interpreted to require states to take into account biological patterns, such as 'predator-prey relations or the mortality of other fish species caught as incidental by-catch'.[49]

---

45 See Art. 5(2) and 6(1) of Regulation 2371/02.

46 Arts. 5(2), para.2 and (3) and 6(2) and (3) of Regulation 2371/02.

47 Art. 3(j) of Regulation 2371/02.

48 Art. 61(4) of 1982 UNCLOS.

49 Wolff, *Fisheries and the Environment*, p. 65. For more elaborate analyses, see Markowski, *Allocation and management of fisheries resources: an in-depth analysis of instruments in comparative perspective*, pp. 14-15.

Nevertheless, states are to guarantee that these stocks do not fall below levels at which their reproduction is seriously threatened.[50]

Where coastal states exploit straddling and highly migratory species, the FSA requires that parties *assess* and *consider* associated ecosystems, non-target species, the marine environment and biodiversity when adopting conservation measures.[51] To prevent species belonging to the ecosystem or species associated with or dependant upon the target stocks from being seriously threatened, states must adopt conservation and management measures where necessary.[52] In sum, however, FSA provisions on the protection of non-target species, ecosystems, biodiversity and the marine environment must be regarded as vague, open to discretion, and of low mandatory force.

1.2.2 Compatibility of CFP measures and practices

Community fisheries measures do meet the very basic requirements laid down in 1982 UNCLOS and the FSA concerning the protection of non-target species and ecosystems. The management and structural measures in place increasingly take into consideration the effects of fishing on ecosystems and dependant species.[53] Measures to protect vulnerable areas have been adopted[54] and some technical measures are currently being revised.[55] In addition, the Commission has often recognised the need to improve the Community's eco-system approach to fisheries management by, *inter alia*, protecting habitats, protecting other species, coastal areas, and by increasing the knowledge base on the interdependency of fisheries resources, fishing and ecosystems.[56] It also tries to develop methods that ameliorate the problem of by-catch

---

50 Art. 63, para. 3 of 1982 UNCLOS.

51 Arts. 5 (d), (e), (f) and (g) and 6(3)(d) of the FSA.

52 Arts. 5(e), 6(3)(d) of the FSA.

53 See, for example, Arts. 2(1), second paragraph, 4(1) and (2)(g)(iv), 5(2), 6(2), 7(1), 8(1) of Regulation 2371/02; structural measures also increasingly take into account environmental issues, see Art. 25(6)(c)-(d), (7)-(8), Arts. 30(2)(a-c) and 35(1)(d), Arts. 37 and 38, Arts. 40(3)(d), (f), Arts. 43(2)(c), 44 (1)(b) and (f), Arts. 30(2)(d), (4)(d) and 38 (2)(c) of Regulation 1198/06.

54 See, for example, Council Regulation (EC) 602/04 amending Regulation 850/98 as regards the protection of deepwater coral reefs from the effects of trawling in an area northwest of Scotland, OJ 2004 No. L97/30.

55 WWF, *WWF-Mid-Term Review of the EU Common Fisheries Policy*, p. 37.

56 Commission Communication Com(1999) 363 final on Fisheries Management and Nature Conservation in the Marine Environment, pp. 9-16; Commission Communication Com(2001) 162 final Biodiversity Action Plan for Fisheries, pp. 11-29. Other measures mentioned are the improvement of the selectivity of gear, developing new technical conservation measures to reduce fishing impact, vocational training, information initiatives and consultation activities, improve the level of coherence between CFP and environmental measures, Commission Communication Com(2001), 143 final, pp. 13-14.

in its multi-species fisheries.[57] However, despite such initiatives, some critical issues continue to exist. For example, Community fisheries law does not clearly define what is required for the legal protection of ecosystems. This can reduce the effectiveness of those measures which incorporate the ecosystem approach. Discarding also remains a major problem in many Community fisheries. Ultimately, the Community's practice of allowing fishing of target species at unsustainable levels puts pressure on ecosystems and non-target species.

1.3 Assessing, collecting, sharing and using scientific data

International measures require the Community to collect, share and use scientific information on fish stocks exploited by Community vessels.

1.3.1 Requirements under 1982 UNCLOS and FSA

Regarding stocks located in the EEZ and the high seas as well as regarding straddling and highly migratory species, UNCLOS and the FSA require conservation measures to be designed on the best scientific evidence available.[58] In addition, they also require that parties exchange this scientific information.[59] However, the obligation to base measures on the best scientific evidence available is somewhat vague, as it only requires the parties to ensure that their conservation measures are science-driven; therefore, they are only bound to obtain and consider available data within the limits of their financial resources.[60]

Nevertheless, the FSA does impose some specific data collection requirements with regards to straddling and highly migratory stocks. First, it requires state parties to base their management decisions on the best scientific evidence available.[61] It also stipulates that flag states collect data in 'sufficient detail to facilitate effective stock assessment'.[62] Annex I of the FSA explains that this data shall include, *inter alia*, the amount of fish caught by species, the amount of fish

57 Commission Communication (2007) 136 final – A policy to reduce unwanted by-catches and eliminate discards in European fisheries; Communication Com(2002) 656 final from the Commission to the Council and the European Parliament – On a Community Action Plan to reduce discards of fish, pp. 9-10.

58 Art. 61(2) and Art. 119(1)(a) of 1982 UNCLOS, according to Art. 6(3)(b) of the FSA reference points and actions to be taken shall be based on the best scientific information available.

59 Art. 61(5) and 119(2) of 1982 UNCLOS and Art. 14(2) and Annex I of the FSA.

60 See above in the chapters on the UNCLOS and FSA; see also Markowski, *Allocation and management of fisheries resources: an in-depth analysis of instruments in comparative perspective*, p. 15.

61 Art. 5(b) of the Fish Stocks Agreement.

62 Art. 14(1)(b) of the FSA.

discarded, the types of fishing methods used, and the location of the fishing vessels.[63]

1.3.2 Compatibility of CFP measures and practices

The Community and its Member States collect and assess biological, technical, environmental, and socio-economic data on the fishing sector.[64] Data concern stocks located in Community waters and the high seas, and include information on vessels, catch activities, discards, location of catches, fishing methods, etc. When drawing up fisheries legislation, the Community relies extensively on these scientific data. Information is acquired mainly from the ICES, international fisheries commissions, Member States, and third countries. The scientific data are then analysed and prepared by the Commission's Scientific, Technical and Economic Committee for Fisheries (STEFC) for the Council's legislative purposes. Where stocks are not assessed[65] or the information on fish stocks is uncertain, conservation measures are based on broad assumptions underpinned by current scientific understanding.[66] Finally, the Community exchanges information with other states through IFOs.

Despite the fact that many stocks are not assessed and scientific monitoring can always be improved, for example, in the Mediterranean fisheries, the Community has gathered extensive information on fish stocks and it readily exchanges such information with other interested states. Although, in many cases, it does not follow scientific recommendations, it does base its fisheries management decisions on the available data. Thus, it could be said that the Community adheres to the standards set in international fisheries law on collecting, using and sharing data concerning the fishing sector.

1.4 Control and enforcement

International law on control and enforcement can be found within 1982 UNCLOS, the FSA, and the Agreement to Promote Compliance with International Conservation and Management Measures by Fishing Vessels on the High Seas ('Compliance Agreement').

---

63 See Art. 3 of Annex I of the FSA.

64 See Reason (7) and Arts. 3 and 4 of Council Regulation (EC) 199/2008 concerning the establishment of a Community framework for the collection, management and use of data in the fisheries sector and support for scientific advice regarding the Common Fisheries Policy, OJ 2008 No. L60/1.

65 Communication Com(2007) 295 final from the Commission to the Council on Fishing Opportunities for 2008 – Policy Statement from the European Commission, p. 5.

66 A. Karagiannakos, 'Total Allowable Catch (TAC) and quota management system in the European Union', p. 244.

1.4.1 1982 UNCLOS

Under the provisions in UNCLOS, the Community and its Member States *may* adopt, *inter alia*, a licensing system, information requirements of fishing vessels, and enforcement procedures.[67] These provisions must be regarded as optional and vague.

1.4.2 FSA

The FSA imposes compliance and enforcement demands on parties. These are: (a) requiring flag states to enhance control mechanisms, (b) strengthening control and enforcement rights of port states, and (c) broadening control and enforcement measures on the highs seas for members of regional fisheries commissions. With regards to (a), according to Arts. 18(2), vessels shall only be authorised to fish where flag states are able to exercise their responsibilities for such vessels effectively, as set out in the Fish Stocks Agreement and 1982 UNCLOS.[68] Measures include, *inter alia*, an effective licensing system, requiring prior authorisation for vessels to fish on the high seas,[69] the establishment of a national record for vessels fishing on the high seas,[70] the establishment of an effective observation system,[71] the regulation of transhipment on the high seas,[72] etc. Art. 19 requires flag states to ensure that vessels flying their flag comply with the conservation measures adopted by the regional fisheries organisations.[73] Where there is sufficient evidence of a violation, states are to refer the case to the relevant national authorities to institute proceedings without delay in accordance with their laws. Ships concerned may, where appropriate, be detained.[74] Applicable sanctions 'shall be adequate in severity to be effective in securing compliance and discourage violations'.[75] With regards to strengthening control and enforcement rights of port states (b), port states now have the right and duty to take non-discriminatory measures in accordance with international law to promote the effectiveness of international conservation and management rules.[76] If states are members of international fisheries commissions, they may,

---

67 Art. 62(4)(a) and (k) of 1982 UNCLOS.

68 Art. 18(2) FSA.

69 Art. 18(3)(a), (b) (i) to (iii) FSA.

70 Art. 18(3)(c) FSA.

71 Art. 18(3) (d) to (g) FSA.

72 Art. 18(3)(h) FSA.

73 Art. 19(1) (a) to (e) FSA.

74 Art. 19(1)(d) FSA.

75 Art. 19(2) FSA.

76 Foreign ships that are voluntarily in their ports or offshore terminals may be inspected; and landings and transhipment of illegally caught fish can be prohibited. Art. 23(2), (3) FSA. This provision is not limited to foreign vessels flying the flag of the Fish Stocks Agreement; see Edeson, Freestone and Gudmundsdottir, *Legislating for Sustainable Fisheries*, pp. 66-70.

in the respective region of the high seas, board and inspect vessels flying the flag of any other state party to the Fish Stocks Agreement.[77] With regards to broadening control and enforcement measures on the highs seas for members of regional fisheries commissions (c), Art. 19(1) requires flag states to ensure that their flag vessels comply with the conservation measures adopted by the regional fisheries organisation.[78] This obligation applies to the flag vessels of states even where such states are not members to the IFO, but only party to the Fish Stocks Agreement.[79]

1.4.3 Compliance Agreement

The Compliance agreement applies to all stocks located and exploited in the high seas. Parties to this agreement must ensure that their flag vessels do not undermine the effectiveness of international conservation and management measures. As under the FSA, the Compliance Agreement requires that states grant their flag vessels prior authorisation before they are permitted to fish on the high seas[80] and that such authorisation only be granted if states are able to exercise their management of fishing vessels effectively.[81] Parties cannot authorise fishing vessels that were previously registered in another state and have undermined the effectiveness of international conservation and management measures.[82] To make vessels and previous violators internationally identifiable, state parties shall ensure that their vessels are marked in accordance with generally accepted standards so they can be readily identified.[83] States must also require their flag vessels to provide information on their operations, including information pertaining to the area of fishing operations and catches

---

77 Art. 21(1), (5). This development has been perceived as a 'unique and far reaching exception to the flag state principle'; see M. Hayashi 'Enforcement by Non-Flag States on the High Seas under the 1995 Agreement on Straddling and Highly Migratory Fish Stocks, *Georgetown International Environmental Law Review* 9 (1996), pp. 9-10. If there is clear evidence that illegal fishing took place, the flag state must be notified promptly, Art. 21(5) FSA. If the flag state does not respond to the notification, the inspecting state itself is to secure evidence and proceed with investigations. Art.21(8)-(11) FSA.

78 Art. 19(1)(a) to (e) FSA.

79 Bob Applebaum and Amos Donohue, 'The Role of Regional Fisheries Management Organisations', in Hey (ed.), *Developments in International Fisheries Law,* (The Hague: 1999), pp. 234-235; Consequently, Art. 21(1) allows national inspectors of states party to the Fish Stocks Agreement and members of specific IFOs to board and inspect vessels of other states who are merely parties to the Agreement. Boarding and inspections must be consistent with to the rules adopted in accordance with Arts. 21 (2) and 22 of the FSA.

80 Art. III(2) Compliance Agreement.

81 Art. III(3) Compliance Agreement.

82 Arts. III(5)(a) and (b) Compliance Agreement, see also exceptions in Art. III(5)(c) and (d) Compliance Agreement.

83 Art. III(6) Compliance Agreement.

and landings.[84] States must impose sanctions of sufficient gravity on violators of the agreement.[85] States are to maintain records on their flag vessels, ensuring that such vessels are entered in that record.[86]

1.4.4 Compatibility of CFP measures and practices

In light of all of these different obligations, the Community's fisheries control and enforcement system is comprehensive. Its control and enforcement system meets most international legal requirements: it requires prior authorisation for fishing; it has instituted a licensing system, a vessel record, a comprehensive catch control system, satellite based monitoring system, rules on transhipment, and all branches of the CFP are subject to extensive monitoring.

There are, however, two international legal requirements that are in danger of not being properly implemented. First, the FSA and Compliance Agreement both require states to authorise fishing only where they are able to ensure *effectively* that fishers will comply with international conservation measures. The term 'effectively' clearly underlines the importance of functioning control mechanisms.[87] However, as described above, since its inception, the Community has struggled with guaranteeing fishers' adherence to conservation regulations. Fishers regularly transgress the rules and the application of control and enforcement measures by the Member States is often poor. Second, both FSA and the Compliance Agreement require flag states to adopt sanctions which 'are adequate in severity' or of 'sufficient gravity'. As explained in the above section on 'Community control and enforcement measures', under the current Community system, the sanctions put in place by the Member States are an insufficient deterrent, and such measures vary greatly between Member States. Given that fishers feel pressured to violate management measures in order to remain competitive, sanctions have been even less effective.

1.5 International cooperation requirements

1982 UNCLOS, the FSA and the Compliance Agreement establish obligations to co-operate.

84 Art. III(7) Compliance Agreement.

85 Art. III(8) Compliance Agreement.

86 Art. IV Compliance Agreement.

87 Moritaka Hayashi, 'The 1995 Agreement on the conservation and management of straddling and highly migratory fish stocks: significance for the Law of the Sea Convention' (1995) 29 *Ocean & Coastal Management*, p. 60.

1.5.1 1982 UNCLOS

According to 1982 UNCLOS, states that exploit identical resources in a given high seas area are required to enter into negotiations to establish measures necessary for their conservation.[88] However, the obligation to co-operate does not entail more than an obligation to enter into negotiations to resolve conflicts in good faith (*pactum de negotiando*), and not to conclude an actual agreement (*pactum de contrahendo*).[89] Moreover, articles do not state what happens if negotiations fail.[90] Similar obligations also exist for the conservation of transboundary stocks (stocks occurring in two or more EEZs), anadromous stocks (stock spawning in inland waters and living at sea), catadromous stock (stocks spawning in the sea and living in inland freshwater rivers), straddling and highly migratory species, and marine mammals.[91] With regards to highly migratory species, states are explicitly required to cooperate through international organisations. Where such organisations do not exist, states must cooperate to establish them.[92] As regards the conservation of anadromous stocks and marine mammals, it is merely recommended that states cooperate through appropriate international organisations.[93]

1.5.2 FSA

The FSA reinforces cooperation requirements with regards to the management of straddling and highly migratory fish stocks. First, parties to the FSA have a *duty to cooperate* in abolishing inconsistencies between high seas and EEZ fisheries conservation regimes.[94] In that sense, the Fish Stock Agreement imposes a much stronger obligation than Arts. 63(2) and 64 of 1982

---

88 Art. 118 of 1982 UNCLOS concretises the general rule in Art. 117 of 1982 UNCLOS.

89 Moritka Hayashi, 'The Management of Transboundary Fish Stocks under the LOS Convention' (1993) 8 *The International Journal of Marine and Coastal Law*, 2, p. 251; Wolfrum, Röben and Morrison, 'Preservation of the Marine Environment', p. 235; Wolff, *Fisheries and the Environment*, p. 60; J. Ziemer, *Das gemeinsame Interesse an einer Regelung der Hochseefischerei*, (Berlin: 2000), p. 42.

90 Birnie and Boyle, *International Law & the Environment*, p. 664; Wolfrum, Röben and Morrison, 'Preservation of the Marine Environment', p. 235.

91 Arts. 63(1) of 1982 UNCLOS on transboundary stocks, see M. Nordquist, *UN Convention of the Law of the seas 1982 – A Commentary, Vol. II* (Dordrecht: 1991), p. 646; Art. 63(2) on straddling stocks, see Yturriaga, *The International Regime of Fisheries – From UNCLOS 1982 to the Presidential Sea*, p. 165; Art. 64(1) on highly migratory species; Art. 67(3) on catadromous stocks; Ellen Hey, 'The Fisheries Provisions of the LOS Convention' in Hey (ed.), *Developments in International Fisheries Law*, (The Hague: 1999), p. 24.

92 Art. 61(1) of 1982 UNCLOS.

93 See Art. 66(3)(a) and (b) on anadromous stocks; see Art. 65 on marine mammals.

94 Art. 7(2) FSA

UNCLOS.[95] Agreements are to be reached in a reasonable period of time.[96] If states fail to agree, any party can invoke a dispute settlement procedure.[97] Second, states are required to pursue cooperation either directly or through fisheries management organisations.[98] Where no such *fora* exist, states shall cooperate to establish them.[99] To this end, they must enter into consultation in good faith.

1.5.3 Compliance Agreements

The Compliance Agreement also requires its parties to cooperate in implementing the agreement 'as appropriate'.[100] This includes exchanging information, such as evidentiary material, on the activities of fishing vessels on the high seas. They must make such information available to the FAO.

1.5.4 Compatibility of CFP measures and practices

Structures and functions of IFOs to which the Community is a party largely differ.[101] While some IFOs possess permanent secretariats, scientific councils and monitoring capacities, others do not even provide administrative or scientific capacity and many lack funding.[102] Some organisations are concerned with the conservation of different stocks, while others apply to specific species only (*e.g.*, ICCAT (tuna)).[103] NEAFC and NASCO can be invited by their parties to recommend management strategies in the EEZ.[104] ICCAT and WCPFC measures on the conservation of different tuna species regularly

95 A. Proelß, *Meeresschutz im Völker- und Europarecht*, p. 154; Georg Dahm, Jost Delbrück and Rüdiger Wolfrum, Völkerrecht Band I/2, (Berlin: 2002), p. 396.

96 Art. 7(3) FSA.

97 Arts. 27-32 FSA.

98 Art. 8(1-3) FSA.

99 Art. 8(5) FSA; Arts. 9 to 12 FSA provide a set of requirements to be considered in establishing international fisheries arrangements or organisations.

100 Art. V(3) Compliance Agreement.

101 Elli Louka, *International Environmental Law*, pp. 264-286; Michael W. Lodge *et al.*, *Recommended Best Practices for Regional Fisheries Organisations*; Wolff, *Fisheries and the Environment*, pp. 78-92.

102 CECAF, for example, does not have a permanent secretariat nor a scientific body. See also, D. Booß, ‚Umweltschutz und Fischerei', in Rengeling (ed.), *Handbuch zum europäischen Umweltrecht 2 Teilband* (Köln: 2003), pp. 1532-1536.

103 ICCAT's measures have only recently concerned trade on fisheries products. ICCAT issued recommendations concerning import restrictions for tuna caught by vessels from countries like Belize, Honduras, Equatorial Guinea; see Council Regulation (EC) No. 2092/2000 prohibiting imports of Atlantic blue-fin tuna (Thunnus thynnus) originating in Belize, Honduras and Equatorial Guinea, OJ 2000 No. L249/1; Council Regulation (EC) No. 2093/2000 prohibiting imports of Atlantic swordfish (Xiphias gladius) originating in Belize and Honduras, OJ 2000 No. L249/3.

104 Arts. 6(1) and 8(1) NEAFC-Convention and Art. 4(3) of NASCO-Convention.

apply to parties' EEZs. Some IFOs lay down (optional) binding management measures and allocate catch opportunities among parties. Other IFOs merely have consultative or advisory functions (*e.g.*, CECAF). Finally, many organisations require participation in monitoring programmes, such as vessel registers, satellite monitoring systems, observer schemes, inspections at sea, etc.[105] The Community's co-operation efforts, are carried out within the following IFOs:

- the North-West Atlantic Fisheries Organisation (NAFO);[106]
- the North Atlantic Salmon Conservation Organisation (NASCO);[107]
- the North-East Atlantic Fisheries Commission (NEAFC);[108]
- the Commission for the Conservation of Antarctic Marine Living Resources (CCAMLR);[109]
- the General Fisheries Council for the Mediterranean (GFCM);[110]
- the Indian Ocean Tuna Commission (IOTC);[111]
- the International Commission for the Conservation of Atlantic Tuna (ICCAT);[112]
- the Western Central Atlantic Fishery Commission (WECAFC);[113]
- the Fishery Committee for the Eastern Central Atlantic (CECAF);[114]
- the South-East Atlantic Fisheries Organisation (SEAFO);[115]
- the Commission for the Conservation and Management of Highly Migratory Fish Stocks in the Western and Central Pacific Ocean (WCPFC).[116]

The EC is in the process of acceding to two more organisations, *i.e.*, the Inter-American Tropical Tuna Commission (IATTC)[117] and the South West Indian Ocean Fisheries Commission (SWIOFC).[118] Representation is conducted in close cooperation with the Member States, who have also become parties to IFOs that are not exclusively concerned with fisheries.[119]

---

105 ICCAT and CCAMLR have adopted catch documentation schemes; see Elli Louka, *International Environmental Law*, p. 267; Michael W. Lodge *et al.*, *Recommended Best Practices for Regional Fisheries Organisations*, pp. 44 et seq.

106 OJ 1978 No. L378/1.

107 OJ 1982 No. L378/24.

108 OJ 1981 No. L227/21.

109 OJ 1981 No. L252/26.

110 OJ 1998 No. L190/34.

111 OJ 1995 No. L236/24.

112 OJ 1986 No. L162/34.

113 Available online at www.fao.org/fi/body/rfb/wecafc/wecafc_home.htm.

114 Available online at www.fao.org/fi/body/rfb/cecaf/cecaf_home.htm.

115 OJ 2002 No. L234/39.

116 OJ 2005 No. L32/1.

117 Available online at www.iattc.org.

118 Available online at www.fao.org/fi/body/rfb/SWIOFC/swiofc_home.htm.

119 Such as the Commission for the Conservation of Antarctic Marine Living Resources (CCAMLR), OJ 1981 No. L252/26; General Fisheries Council for the Mediterranean (GFCM), OJ 1998 No. L190/34.

At first glance, the Community appears to follow the international cooperation requirements. The Community is engaged in international co-operations regarding the protection of transboundary, anadromous stocks as well as marine mammals. It also provisionally applies the conservation measures adopted under the International Dolphin Conservation Program (ICDP).

However, the fisheries management practices of IFOs to which the Community is a party has been criticised. Given the conflicting interests of coastal and distant water fishing states, many IFOs' management regimes are deficient and their implementation ineffective. For example, it has been observed that some IFOs set catch rates at unsustainable levels and the ecosystem and precautionary approaches have not been implemented effectively.[120] Some IFOs have hardly begun to adopt management measures.[121] In many cases, there is insufficient scientific information for proper decision making. Given that most IFO agreements allow parties to opt out of proposed conservation measures, IFO legislative powers are generally considered weak.[122] The FSA requirements under the SEAFO, GFCM and WCPFC, in particular, have not been properly implemented yet. For example, under the current GFCM regime, neither limit nor target reference points have been adopted. Besides, the Community and Member States implementation of and compliance with ICCAT recommendations on setting TACs and establishing closed seasons for Atlantic bluefin tuna by the Community and its Member States has been regarded as slow or even lacking.[123]

Despite the lack of effectiveness of many of the IFO's regimes to which the Community is party, the Community must be regarded as fulfilling its international cooperation requirements. As stated above, these mainly require state parties to enter into negotiations or co-operate through international organisations. The only case where states have a duty to co-operate concerns the abolishment of inconsistencies between high seas measures and the EEZ fisheries

Membership of Member States is based on the lack of (exclusive) competences of the EC for the regulated species, scientific research and development activities. On the lack of the exclusive competence, Member States also base their participation in the Fish Stocks Agreement and in the FAO; see OJ 1998 No. L189/4 and Proposal Com(91) 387 final for a Council decision on the accession of the European Community to the FAO at the 26th session of the FAO Conference, p. 8.

120 Michael W. Lodge *et al.*, *Recommended Best Practices for Regional Fisheries Organisations*, pp. 34-37.

121 *Ibid.*, p. xiii.

122 Elli Louka, *International Environmental Law*, p. 267; For example, SEAFO, ICCAT, NEAFC, NAFO, ICCAT all impose binding measures upon the Community if it does not object within a certain time limit. See Arts. 6 (3), 23 (1) Convention on the Conservation and Management of Fishery Resources in the South East Atlantic ('SEAFO-Convention'); Arts. VIII (1)-(3) International Convention for the Conservation of Atlantic Tunas ('ICCAT-Convention'); Arts. 4, 13 The Convention for the Conservation of Salmon in the North Atlantic Ocean ('NASCO-Convention'): objections under Art. 13 make proposals non-binding for all parties. Arts. 5 (1), 7, 12 Convention on Future Multilateral Co-operation in North-East Atlantic Fisheries ('NEAFC-Convention'); Arts. XI and XII Convention on Future Multilateral Cooperation in Northwest Atlantic Fisheries ('NAFO-Convention').

123 WWF, *WWF-Mid-Term Review of the EU Common Fisheries Policy*, pp. 47-48.

regimes. Only with regards to the failure of a respective agreement can state parties invoke a dispute settlement procedure. Nevertheless, it must be kept in mind that co-operation requirements all aim at realising those conservation requirements laid down in 1982 UNCLOS and the FSA. Thus, given that these conservation requirements are far from being attained, effectiveness of international co-operation must substantially be increased. Of course, this cannot be achieved by the Community alone.

1.6 International legal instruments on the promotion of fisheries

Only two international legal instruments provide rules that are directly concerned with fisheries subsidies: the SCM-Agreement and the *non-binding* International Plan of Action for the Management of Fishing Capacity.

1.6.1 Code of Conduct and the IPOA for the Management of Fishing Capacity

The FAO Code of Conduct for Responsible Fisheries aims at providing a non-binding framework of standards, principles and rules for all political actors involved at all levels of fisheries management. Norms provided by the Code do not directly target subsidisation of fisheries. Like the Fish Stocks Agreement, the Code requires states to manage fish stocks at sustainable levels and avoid or reduce excess capacity.[124]

Pursuant to Art. 2(d) of the Code of Conduct, the FAO has organised consultations to improve the implementation of the Code. Such consultations have, *inter alia*, led to the approval of four International Plans of Action (IPOAs), including the IPOA for the management of fishing capacity.[125] Voluntary in nature, the plans provide guidance on the most urgent matters in fisheries management. The IPOA for the management of fishing capacity explicitly addresses the contributory effects of subsidies in creating fishing overcapacity. Paragraphs 25 and 26 of the IPOA state:

---

[124] Arts. 7.2.2.a, 7.6.3, 7.1.8 of the Code of Conduct.

[125] The other three IPOAs are (a) IPOA for reducing incidental catches of seabirds in long line fisheries, (b) IPOA for the conservation and management of sharks, and (c) IPOA to prevent, deter, and eliminate illegal, unreported and unregulated fishing. On the progress of implementation, see FAO Committee on Fisheries, *Progress in the Implementation of the 1995 Code of Conduct for Responsible Fisheries, Related International Plans of Action and Strategy – COFI/2007/2*, (Rome: 2007), available online at www.fao.org; see also Penelope D. Dalton, 'Implementing the International Code of Conduct for Responsible Fisheries', in Nordquist and Moore (ed.), *Current Fisheries Issues and the Food and Agriculture Organisation of the United Nations*, (The Hague: 2000), pp. 331-339; D. Greboval, 'The International Plan of Action for the Management of Fishing Capacity: Retrospect and Prospect' in Nordquist and Moore (ed.), *Current Fisheries Issues and the Food and Agriculture Organisation of the United Nations*, (The Hague: 2000), pp. 561-582.

'When developing their national plans for the management of fishing capacity, States should assess the possible impact of all factors, including subsidies, contributing to overcapacity on the sustainable management of their fisheries, distinguishing between factors, including subsidies, which contribute to overcapacity and unsustainability and those which produce a positive effect or are neutral'.

'States should reduce and progressively eliminate all factors, including subsidies and economic incentives and other factors which contribute, directly or indirectly, to the build-up of excessive fishing capacity thereby undermining the sustainability of marine living resources, giving due regard to the needs of artisinal fisheries.'

Thus, the IPOA recommends a two-step procedure to (a) assess the effects of the subsidies granted to the fisheries sector, and (b) adjust their subsidisation practice in order to avoid excess capacity.

1.6.2 Obligations under the Agreement on Subsidies and Countervailing Measures

Under the WTO umbrella agreement, specific measures on subsidies have been adopted called the Subsidies and Countervailing Measures ('SCM' rules). To date, the existing SCM rules lack fisheries-*specific* provisions.[126] Nevertheless, fisheries are subject to the general SCM rules.[127]

Under the SCM, subsidies on *export performance* as well as *subsidies promoting the use of domestic over imported goods* are prohibited.[128] Besides, the SCM Agreement makes all subsidies that are not explicitly prohibited subject to Art. 5, which concerns so called 'actionable subsidies'.[129] Actionable subsidies can be challenged by contracting parties, who have reason to believe that the subsidies cause 'adverse effects' to their interests. Parties may request consultations to discuss the alleged violation. If these consultations fail, the challenging party may refer the case to the Dispute Settlement Body.[130] However, SCM Agree-

[126] A modification of the existing WTO subsidies regulations has been discussed; see R. Grynberg, 'WTO fisheries subsidies negotiations: implications for fisheries access arrangements and sustainable management' (2003) 27 *Marine Policy*, pp. 499-511; G. Porter, *Fisheries Subsidies and Overfishing: Towards a Structured Discussion*, (Geneva: 2003), pp. 3-9.

[127] Unlike Community law, WTO rules do not treat fisheries as part of the agriculture sector, to which a special sub-agreement applies. Agriculture is excluded from the SCM agreement, but fisheries are not, see Art. 3.1. SCM. See also F.D. Schorr, *Healthy Fisheries, Sustainable Trade: Crafting New Rules on Fishing Subsidies in the World Trade Organisations*, (Washington D.C.: 2004), p. 37.

[128] Art. 3.1.(a) and (b) of the SCM-Agreement.

[129] Mitsuo Matsushita, Thomas J. Schoenbaum and Petros C. Mavroidis, The World Trade Organization, (Oxford: 2006), p. 364; Schorr, *Healthy Fisheries, Sustainable Trade: Crafting New Rules on Fishing Subsidies in the World Trade Organisations*, p. 42.

[130] See Art. 7.4. of the SCM-Agreement.

ment measures are generally regarded as weak and ineffective instruments to discipline fisheries subsidies for several reasons.[131] First, challenging parties are required to prove that the subsidy in question had an adverse effect to its interests (*i.e.*, that it has caused injury to the domestic industry).[132] The complexity of the fisheries sector and existing subsidisation programs, however, makes it difficult to directly link subsidies with (distorted) market impacts.[133] Second, due to the comparatively small size of the fishing sector and the 'economically dependent position of many smaller economies'[134] in relation to the countries of importation, no actions against fisheries subsidies have been initiated.

Therefore, the SCM rules do not provide a clear ban on specific subsidies to the fishing sectors; and even though SCM parties could theoretically claim that they are experiencing adverse effects from subsidies and that they ought to be removed, the rules have not provided an effective legal framework which enables states to do so. Given this, several WTO parties have promoted the reform of the SCM regime.[135]

1.6.3 Compatibility of CFP measures and practices

The non-binding IPOA for the Management of Fishing Capacity explicitly calls on the Community, first, to assess the effects of its subsidies on their fisheries sectors, and, second, to adjust its subsidisation practice to avoid excess fishing capacities.[136]

Since 1983, the Community frequently evaluates its subsidies schemes. Assessments have focussed on the effects of subsidies on fishing capacity. This work has been carried out by the Commission, the Court of Auditors and external scientists (financed by the Community).[137] These assessments have contrib-

---

[131] Schorr, *Healthy Fisheries, Sustainable Trade*, p. 43.

[132] See the definition of 'adverse effects' as provided in Art. 5 SCM-Agreement.

[133] Grynberg, 'WTO fisheries subsidies negotiations: implications for fisheries access arrangements and sustainable management', p. 501.

[134] Schorr, *Healthy Fisheries, Sustainable Trade*, p. 43.

[135] Grynberg, 'WTO subsidies negotiations: implication for fisheries arrangements and sustainable management', pp. 499-511; Schorr, *Healthy Fisheries, Sustainable Trade: Crafting New Rules on Fishing Subsidies in the World Trade Organisations*, pp. 18 *et seq.*

[136] Paragraphs 25 and 26 of the IPOA for the Management of Fishing Capacity.

[137] See, for example, A. Hatcher and C. Robinson, Overcapacity, overcapitalisation and subsidies in European fisheries. Proceeding of the first Concerted Action workshop on economics and the Common Fisheries Policy, (Portsmouth: 1999); Court of Auditors Special Report No. 3/93 concerning the implementation of the measures for the restructuring, modernisation and adaptation of the capacities of fishing fleets in the Community together with Commission's replies ('Court of Auditors Report – 1993'), OJ 1994 No. C2/1, paragraphs 1.19 *et seq.*; Commission Report Sec(91) 2288 final on the Common fisheries Policy; Commission Green Paper Com(2001) 135 final on the Future of the Common Fisheries Policy, pp. 10-11 and 26-27.

uted to policy making.[138] The Community took the second step, as recommended by the IPOA, by adjusting its subsidisation practice to avoid excess fishing capacities. Since the early 1990s, it has increasingly changed its support practices. Throughout this period, the most harmful subsidies were continuously reduced and aid for the adjustment of the sector as well as support for social and environmental purposes was raised. This course has been maintained under the current EFF regime. The Community continues to support fleet exits and refuses to grant aid for the construction of vessels and the increase of fish holds on fishing vessels. It also provides funds for the adjustment of fleet parts that target the most exploited stocks. In addition, it increasingly integrates environmental objectives and local development strategies into its provisions.

However, one may argue that the Community's efforts to adjust its subsidies programs have not been effective and that subsidisation has clearly undermined the efficacy of conservation measures, contributing to overexploitation. Taking an overall view of the Community's promotion practice, the Community has for a long time ignored the effects of its subsidies on fishing capacities, international conservation obligations, and ultimately the state of fisheries resources. The link between overexploitation and overcapacity became clear in the late 1980s and early 1990s. In response, the Community began to finance measures aimed at *decreasing* fishing and overall production capacity. However, until 2004, support for *increasing* fishing and overall production capacity eclipsed the amount spent on decreasing it. This subsidisation practice largely maintained this overcapacity, which has substantially contributed to the ongoing overexploitation of fish stocks by Community fishers and decline of fisheries resources at levels advocated under 1982 UNCLOS and the FSA. Despite some important progress in the sector, the Commission, nevertheless, stated in 2007 that the current reduction of fishing capacity 'appears too modest when compared with the big reduction efforts required for some major fish stocks, the steady technological creep and the poor economic performance of large parts of the fleet'.[139] In 2007, the WWF also pointed out that it remains 'unclear how far capacities must be reduced in order to solve the chronic problem of overfishing'.[140] It is suggested here that the last argument applies to the entire fisheries sector,

[138] Song, 'The EC's Common Fisheries Policy in 1990s', pp. 40-42.

[139] Commission Com(2007) 828 final annual report from the Commission to the Council and the European Parliament on Member States' efforts to during 2006 to achieve a sustainable balance between fishing capacity and fishing opportunities, pp. 9-10; see also Commission Com(2006) 872 final annual report from the Commission to the Council and the European Parliament on Member States' efforts to during 2005 to achieve a sustainable balance between fishing capacity and fishing opportunities, p. 11. WWF, *WWF-Mid-Term Review of the EU Common Fisheries Policy*, pp. 54-61. However, see somewhat misleadingly: European Commission, *Facts and Figures on the CFP*, (Brussels: 2008), p. 11; Commission Staff Working Document SEC(2008) 2760 accompanying the Proposal for a Council Regulation establishing a Community control system for ensuring compliance with the rules of the Common fisheries Policy, p. 6; Court of Auditors, *Special Report 7/2007*, para. 114.

[140] WWF, *WWF-Mid-Term Review of the EU Common Fisheries Policy*, p. 58.

including the processing industry, the marketing sector and other segments. All of them contribute to the demand for fisheries products and thus exert pressure on fisheries resources. Accordingly, it remains unclear how far the entire fishing sector must be reduced to resolve the chronic problem of overfishing. An assessment in this regard is lacking. Thus, the overall production power of the entire fisheries sector must be adjusted to levels that provide for a sustainable balance between exploitation, production capacities and fisheries resources. Only then will the Community have a somewhat clear benchmark for subsidisation. In view of this, some subsidies granted under the EFF must be regarded as problematic. Aid is currently supplied, for example, for engine replacements, fishing ports, landing sites and shelters and the promotion of the development of markets or promotional campaigns. Such aid reduces private sector production costs, creating incentives for fishers and producers to remain in or enter the business as well as consumers to buy fish products.

Finally, the WTO rules (SCM rules) forbid both subsidising of exports of fisheries products, as well as promoting the use of domestic over imported goods.[141] The SCM provisions also provide that parties to the convention must not, through the use of subsidies, cause adverse effects to the interests of other SCM Members.[142] At present, however, the Community does not promote exports nor the use of domestic fisheries products over foreign ones. Even though one may argue that subsidies to the Community's fisheries sector have adverse effects on the interests of other SCM Parties, such a claim, for the abovementioned reasons, has not been brought before a WTO panel. Therefore, until such a claim is successful, these subsidies are not prohibited under the SCM Agreement.[143]

2 Fundamental rights implications for fishing limitations under Community law

The respect for fundamental rights forms an integral part of the *general principles* of Community law protected by the ECJ.[144] The ECJ and Community organs are to respect the constitutional traditions of the Member States, international human rights treaties and particularly the European Con-

[141] Art. 3.1.(a) and (b) of the SCM-Agreement.

[142] Art. 5 of the SCM-Agreement.

[143] Given SCM-rules widely acknowledged ineffectiveness, the question in how far Community subsidies comply with Art. 5 of the SCM-Agreement will be neglected here.

[144] Case 11/70, Internationale Handelsgesellschaft mbH v. Einfuhr und Vorratsstelle für Getreide und Futtermittel [1970] ECR 1125, (Summary) para. 2. When all Member States will have ratified the Treaty of Lisbon (hopefully) in 2009, the Charta of Fundamental Rights of the European Union (OJ 2007 No. C303/1) will be recognised by the European Union; see Art. 6 of the Treaty of Lisbon amending the Treaty on the European Union and the Treaty Establishing the European Community, OJ 2007 No. C306/01. Currently, see Art. 6 (1) and (2) of the Treaty of the European Union (TEU).

vention on Human Rights.[145] There are two important consequences of managing fishing activities that should be considered with respect to fundamental rights protection. First, fishing limitations encroach upon the freedom of fishers to act. Thus, this section will look into the extent to which fishing limitations impair fishers' fundamental rights and the justification for such infringements. Second, allowing fishers to fish at unsustainable levels allows fishers to earn higher incomes in the short term; however, this policy can reduce or even eliminate fishers' fishing opportunities in the long term. The ensuing section will assess whether fundamental rights create a general obligation on the Community to actively protect and sustain fish stocks; and, if there is such an obligation, whether the current legislative practice of setting quotas at unsustainable are in conformity with this duty.

2.1 Nature of fishing rights under international, Community and national law

In order to find out how fishing limitations affect the rights of fishers, the nature of fishing rights under international, Community and Member States law must be assessed more closely. Fish may be taken from the high seas, EEZs or the territorial seas of costal states. Their legal status differs in these areas. On the high seas, until they are caught, fish are neither owned by individuals nor states, *i.e.*, they are common property.[146] In their EEZs and their territorial seas, coastal states have sovereign rights to exploit natural living (and non-living) resources.[147] Thus, within its EEZ and territorial sea, a coastal state holds exclusive exploitation rights to capture marine resources. Subject to the conservation requirements established at international law, coastal states are free to determine the use of the marine resources found in these areas.

Most coastal states limit access to fisheries resources by virtue of these rights, but do not exclusively allocate specified stocks, fish or exploitable areas. Fishers do not have any rights in relation to specified free living fishes, but do enjoy exploitation or use rights in regards to unspecified and free floating marine capture resources. However, these use rights share important character-

[145] Arts. 6(2) and 46(d) TEU; see also Paul Craig, *EU Administrative Law*, (Oxford: 2006), p. 485. Due to the primacy of Community law over the national (constitutional) law of the Member States, the validity of Community measures can only be judged in the light of Community law, see Case 11/70, ECR 1125, Internationale Handelsgesellschaft [1970] ECR 1125, (Summary), para. 1. Thus, the general principles of Community law are to be interpreted autonomously by the ECJ (merely respecting and drawing inspiration from the constitutional traditions of the Member States, see Rudolf Streinz, *Europarecht* (Heidelberg: 2005), para. 761; Craig, *EU Administrative Law*, p. 485; Thomas Kingreen, ‚Art. 6 EU-Vertrag', in Calliess and Ruffert (ed.), Kommentar zu EU-Vertrag und EG-Vertrag' (Neuwied: 1999 – old edition), paras. 29-43.

[146] Birnie and Boyle, *International Law & The Environment*, p. 141.

[147] Art. 2(1) and Art. 56(1)(a) of 1982 UNCLOS. According to Art. 2(1), the coastal states' sovereignty extends to the territorial seas.

istics with rights usually associated with property such as, for example, exclusivity, permanence, security and transferability. Depending on the degree to which these general characteristics are associated with the use rights, such rights have a varying economic value.[148] The more property characteristics are included in use rights, the more a legal order may be inclined to regard such rights as property rights.

As explained above, under Community law, fishers are granted limited access to living marine capture resources. The Community and the Member States define and allocate these use rights.[149] Countries like Denmark and Germany, for example, require fishers to obtain a license and a fishing allowance.[150] National quotas are distributed among all fishermen who are entitled to fish. The right to fish is not transferable. Within the Community, the United Kingdom and the Netherlands are commonly viewed as granting the most flexible fishing rights.[151] In the United Kingdom's mackerel and herring fisheries, individual catch quotas are allocated to vessels.[152] Some quotas are allocated to industry groups or producer organisations. These groups are free to choose the quota allocation system for their members. Vessel operators are generally free to exchange and trade their quotas within the organisation.[153] In the Netherlands, all-important input and output rights (*i.e.*, individual catch quotas, landing quantities and engine capacities) are tradable.[154] They are either allocated to individuals, companies or industry groups.[155]

---

[148] A. Scott, 'Introducing Property in Fishing Management', in Shotton (ed.), *Use of property rights in fisheries management – FAO Fisheries Technical Paper 404/1*, (Rome: 2000), p. 6; R. Arnason, 'Property Rights as a Means of Economic Organisation', in Shotton (ed.), *Use of property rights in fisheries management – FAO Fisheries Technical Paper 404/1*, pp. 19-25.

[149] Art. 20(3) of Regulation 2371/02.

[150] W. Davidse, L.V. Mc Ewan and N. Vestergaard, 'Property rights in fishing: from state property towards private property?' (1999) 23 *Marine Policy*, 6, p. 541-542. See § 3 of the Seefischereigesetz, BGBl. I 1984, 876.

[151] D. Symes, 'Rights-Based Management: A European Union Perspective', in Shotton (ed.), *Use of property rights in fisheries management – FAO Fisheries Technical Paper 404/1*, (Rome: 2000), p. 279; Commission Staff Working Document SEC (2007) 247 on rights-based management tools in fisheries, p. 12. Spain has also been moving progressively towards the implementation of ITQ with regards to industrial fishing; see Fernando González Laxe, 'Transferability of fishing right: The Spanish case', (2006) 30 *Marine Policy*, p. 379 et seq.

[152] W. Davidse, L.V. Mc Ewan and N. Vestergaard, 'Property rights in fishing: from state property towards private property?', pp. 541-542.

[153] *Ibid.*, p. 141.

[154] Wim Davidse, 'Property Rights and Regulatory Systems: the Strategic Response of Dutch Fishermen', in Symes (ed.), *Property Rights and Regulatory Systems in Fisheries*, (Oxford: 1998), pp. 58-59; Davidse, L.V. Mc Ewan and N. Vestergaard, 'Property rights in fishing: from state property towards private property?', p. 540.

[155] G. Valatin, 'Development of property rights-based fisheries management in the United Kingdom and the Netherlands: a comparison', pp. 141-142.

In sum, quotas have economic value to fishermen, attributing individual access rights to some and excluding others.[156] Due to the different characteristics of access rights across the Member States, the economic value of fishing rights varies throughout the Community. Regarding the presence of property characteristics under Community and national law, fishing rights are exclusive once obtained. However, they are far from being permanent and secure (*i.e.*, not-revocable); and in most cases, they are not transferable. The Council changes quotas at least once a year; and, the Commission or Member States may suspend fishing activities on the grounds of serious and unforeseen threats to the conservation of living aquatic resources.[157] Moreover, some Member States do not grant a fixed share of quota to individuals, but distribute available quota among those who apply for fishing access. In such cases, the value of the individual fishing right depends on the changing number of fishermen that participate in the specific fishery.[158] Community law, therefore, does not guarantee a stable claim for Community fishers to exploit living marine capture resources. Transferability of fishing rights, as described above, exists only in the Netherlands and, to a limited extent, the United Kingdom. In conclusion, the allocation systems of the Community and Member States for fishing opportunities provides exclusive, but insecure (*i.e.*, revocable), non-durable and largely non-transferable use rights of a fluctuating economic value, and no access or exploitation rights to specified stocks or fishes.

[156] *Ibid.*, pp. 137-144; Commission Staff Working Document SEC (2007) 247 on rights-based management tools in fisheries, p. 4; Davidse, 'Property Rights and Regulatory Systems: the Strategic Response of Dutch Fishermen', p. 59; W. Davidse, L.V. Mc Ewan and N. Vestergaard, 'Property rights in fishing: from state property towards private property?', pp. 537-547.

[157] Arts. 7 and 8 of Regulation 2371/02.

[158] This applies, for example, to the Danish allocation system; see W. Davidse, L.V. Mc Ewan and N. Vestergaard, 'Property rights in fishing: from state property towards private property?', pp. 537-547, p. 540.

2.2 The protection of fundamental rights to fishing opportunities under Community Law

It now remains to be examined whether and to what extent fishing rights are protected by the corpus of fundamental rights established through Community law.

2.2.1 Fundamental rights protection against fishing limitations

Are fundamental rights such as the right of ownership, the right to choose and practice a trade or profession, fundamental environmental rights, or the principle of legitimate expectation violated by limits on fishing opportunities? If so, are such impairments justified?

2.2.1.1 Right of ownership

The right of ownership is protected under Community law.[159] However, the ECJ has abstained from giving a precise definition to this right of substantial scope. On this subject, the Court has acknowledged that some rights are protected, such as the right to hold tangible, intangible[160] and intellectual property,[161] as well as different use rights connected to such property.[162] With regards to restrictions on the *use of property*, the Court has decided that where restrictions affect the profitability or the very existence of undertakings, the right of ownership is impaired.[163]

As described above, the Community allocates non-durable, revocable, and largely non-transferable use rights to marine capture resources. Thus, the reduction of quotas does not directly restrict fishers' fundamental right to hold or use property as currently defined by Community law (and as interpreted by the ECJ). The right to take fish out of the sea should be perceived as a privileging right granted by the Community and its Member States. This privileging right can be compared to certain advantages granted by the Community in the course of

159 Cases as provided in Kingreen, 'Art. 6 EU-Vertrag', para. 94; Schilling, EuGRZ 1998, p. 177.

160 Case 44/79, Hauer v. Land Rheinland-Pfalz [1979] ECR 3727, para. 17 et seq.

161 Case 15/74, Centrafarm BV and others v. Sterling Drug [1974] ECR 1147, para. 9.

162 Cases 258/81, Metallurgiki Halyps v. Commission [1982] ECR 4261, para. 13; Case 313/99, Mulligan and others [2002] ECR I-5719, para. 36.; Case 44/79, Hauer v. Land Rheinland-Pfalz [1979] ECR 3727, para. 19; see Carsten Nowak, ‚Eigentumsgrundrecht', in Heselhaus und Nowak (ed.), *Handbuch der Europäischen Grundrechte*, (München: 2006), para. 42.

163 Case 44/79, Hauer v. Land Rheinland-Pfalz [1979] ECR 3727, para. 19; ECJ Case 258/81, Metallurgiki Halyps v. Commission [1982] ECR 4261, paras. 11 and 13; The *right to carry on an established business* is closely connected to the right of ownership. Currently, it remains unclear whether the ECJ distinguishes between these rights at all; see Nowak, ‚Eigentumsgrundrecht', para. 45. Since most authors agree that *the right to carry on an established business* does not grant protection beyond the right to property; the author abstains from discussing this any further.

a market organisation, such as financial grants or reference quantities (*e.g.* milk quotas). The ECJ has ruled that as long as such advantages are neither derived from assets nor from occupational activities of a person concerned, the latter cannot claim their right of ownership is affected.[164]

The Court also held that the fundamental right of ownership does not extend to the protection of mere commercial interests or opportunities, 'the uncertainties of which are part of the very essence of economic activity'.[165] Even though fishing rights differ in nature from financial grants and reference quantities, they 'neither derive from assets nor occupational activities of fishers'. Besides, the privileging rights of fishers are uncertain, since stocks and quotas change annually. It is thus unlikely that the ECJ will be inclined to rule that fishing rights are protected by the fundamental right of ownership.

Even though the right to a fixed share of national quota may not be protected under the right of ownership, it does protect the use of established property to the extent that the restrictions affect the profitability or existence of an undertaking.[166] The ECJ had confirmed this in cases in which the Community restricted the use of production facilities by limiting their outputs. Limitations concerned the taking of fish,[167] milk quotas,[168] the production of steel,[169] and the use of a vineyard.[170]

From these decisions, it can be inferred that the right of ownership protects the right of fishers to use their ships, gear and production facilities. Thus, establishing limits under Art. 20(1) of Regulation 2371/02 encroaches on the fundamental right of ownership of fishers.

Can this encroachment be justified? The ECJ has decided that the right of ownership is not absolute but must be viewed in relation to its social func-

---

164 Case 2/92, The Queen v. Ministry of Agriculture, Fisheries and Food, es parte Dennis Clifford Bostock [1994] ECR I-955, para 19; Case C-44/89, Von Deetzen v. Hauptzollamt Oldenburg [1991] ECR I-5119, para. 27.

165 Case 44/79, Hauer v. Land Rheinland-Pfalz [1979] ECR 3727, para. 14; Case 2/92, The Queeen v. Ministry of Agriculture, Fisheries and Food, ex parte Dennis Clifford Bostock [1994] ECR I-955, para. 19.

166 Case 44/79, Hauer v. Land Rheinland-Pfalz [1979] ECR 3727, para. 19; Cases 258/81, Metallurgiki Halyps v. Commission [1982] ECR 4261, para. 13; Case 313/99, Mulligan and others [2002] ECR I-5719, paras. 11 and 13.

167 Case C-44/94, The Queen v. Ministry of Agriculture, Fisheries and Food, ex parte Fishermen's organisations and others [1995] ECR I-3115, paras. 9-13, 55.

168 Case ECJ Case 313/99, Mulligan and other [2002] ECR I-5719, paras.33-37; Case C-22/94, Irish Farmers Association v. Ministry of Agriculture, Food and Forestry [1997] ECR I-1809, para. 29; Case C-63/93, Duff and others [1996] ECR I-569, para. 30.

169 Cases 258/81, Metallurgiki Halyps v. Commission [1982] ECR 4261, para. 13; Case 313/99, Mulligan and others [2002] ECR I-5719, para. 13.; Case 172/83, Hoogovens Group v. Commission [1985] ECR 2831, para 29.

170 Case 44/79, Hauer v. Land Rheinland-Pfalz [1979] ECR 3727, para. 19.

tion.[171] In principle, the Community is entitled to impose restrictions on the exercise of the fundamental right of ownership provided that those restrictions correspond to the general interests of the Community. In addition, restrictions must not constitute a disproportionate and intolerable interference with the aim pursued, so to impair the very substance of the right of ownership.[172] According to the ECJ, the very substance of the right of ownership is not impaired where the affected individuals are still able to continue their businesses, 'if only to a limited extent'.[173] Thus, as long as fishing is possible, quota restrictions do not impair the very substance of their right of ownership. This is even truer since reduced quotas may actually increase the price of fish products.[174] Given that Community fishers are still allowed to exploit marine capture resources, there is no appreciable violation of right to property.

It can also be inferred from the ECJ's decision that the Court will accept fishing limitations or reductions as long as they correspond to the general interests pursued by the Community and are *proportionate*. The general Community interest is 'to ensure exploitation of living aquatic resources that provide sustainable economic, environmental and social conditions'.[175] Given the open access dilemma, some form of limitation of fishing activities must be adopted to ensure sustainable exploitation, *i.e.*, prevent depletion of fisheries resources. Thus, in principle, limits on fishing opportunities correspond to the general interest. Community measures must also be proportionate. Proportionality may be assumed where acts are appropriate and necessary, *i.e.*, where measures are

[171] Case C-44/94, The Queen v. Ministry of Agriculture, Fisheries and Food, ex parte Fishermen's organisations and others [1995] ECR I-3115, para. 55; Case 44/79, Hauer v. Land Rheinland-Pfalz [1979] ECR 3727, paras. 19-30.

[172] Case 44/79, Hauer v. Land Rheinland-Pfalz [1979] ECR 3727, para. 23; Case C-44/94, The Queen v. Ministry of Agriculture, Fisheries and Food, ex parte Fishermen's organisations and others [1995] ECR I-3115, paras. 56; Case C-22/94, Irish Farmers Association v. Ministry of Agriculture, Food and Forestry [1997] ECR I-1809, para. 27; Case 177/90, Kühn v. Landwirtschaftskammer Weser-Ems [1995] ECR I-35, para 17; Case 44/79, Hauer v. Land Rheinland-Pfalz [1979] ECR 3727, para. 23.

[173] The ECJ applies this test in relation to the question whether the very substance of the right of ownership is affected; see Case 177/90, Kühn v. Landwirtschaftskammer Weser-Ems [1995] ECR I-35, para. 17; see also Reasons (6) and (7) of Commission Regulation (EC) 147/2007.

[174] The ECJ adopted the same argument in relation to the reduction of milk quotas; see Case C-22/94, Irish Farmers Association v. Ministry of Agriculture, Food and Forestry [1997] ECR I-1809, para. 29.

[175] Art. 2(1) Regulation 2371/02. These objectives concretise the more general objectives of Arts. 33 and 174 of the Treaty; see also Reason (5) of Regulation 147/2007. The ECJ has accepted on many occasions that the objectives of the Common Agricultural Policy are public interests that legitimate restrictions to fundamental rights; see Case 306/93, SMW Winzersekt GmbH v. Land Rheinland-Pfalz [1994] ECR I-5555, para. 21 et seq. Since the ECJ regularly justifies impairments of the right of ownership and the right to pursue a trade on the same grounds, justifications outlined above apply here as well. Regarding the limitation of fishing opportunities, see particularly Case C-44/94, The Queen v. Ministry of Agriculture, Fisheries and Food, ex parte Fishermen's organisations and others [1995] ECR I-3115, paras. 55-61; Kingreen ‚Art. 6 EUV', para. 130.

suitable to achieve the pursued aim and where restrictions are not disproportionate to the aims pursued, particularly not more onerous than other available measures.[176] According to Tridimas, 'appropriateness refers to the relationship between the means and the end' and necessity refers to the 'weighing of competing interests'.[177]

There is no clear indication that the limits to fishing activities established under Community law are incompatible with the aim of the sustainable exploitation of living aquatic resources in light of economic, environmental and social conditions. As stated above, due to the open access conflict, some type of restriction must be adopted. Even though it is possible to conceive of alternative management systems, and although improvements to the existing system are desirable, it is not clear that a revised or new system would impose less onerous measures to maintain exploitation rates at a sustainable level. This is particularly so, since the amount of fish which can be harvested on a sustainable basis remains the same under different management and limitation systems.

When setting concrete quotas, however, the Council must adopt the least onerous restriction necessary to achieve its conservation objectives, *i.e.*, to ensure exploitation of living aquatic resources that provides sustainable economic, environmental and social conditions. As a consequence, the Council has a wide margin of discretion when designing limitations. Thus, from this author's perspective, in order to ensure the sustainable exploitation of fisheries resources, the Council can impose extensive limits on fishing.

2.2.1.2 Right to choose and practice a trade or profession

Community citizens have a right to choose and practice a trade or profession under Community law.[178] As with property rights, the ECJ has chosen not to provide a precise definition of this right, choosing to define its limits on a case-by-case basis.[179] Nevertheless, key elements of this right can be identified. The right to practice a trade or profession allows Community citizens to carry out dependent or independent labour in order to acquire an income.[180] The ECJ has refused to extend this right to protect citizens from restrictions and cutbacks to advantages such as reference quantities or other preferences under the right of ownership. From this, one might assume that such advantages are neither protected under the right to practice a trade or profession. However, the

---

[176] Case 331/88, The Queen v. Ministry of Agriculture, Fisheries and food, ex parte FEDESA and others [1990] ECR I-4023, para 13.

[177] Takis Tridimas, *The general principles of EU law*, (Oxford: 2006), p. 139.

[178] Case 44/79, Hauer v. Land Rheinland-Pfalz [1979] ECR 3727, para. 32; Case 306/93, SMW Winzersekt GmbH v. Land Rheinland-Pfalz [1994] ECR I-5555, para. 25.

[179] Nowak, 'Eigentumsgrundrecht', p. 847.

[180] *Ibid.*, p. 848, Kingreen, 'Art. 6 EUV', para. 132.

ECJ has made clear that its decisions relate only to the right of ownership.[181] In the light of these decisions, restricting or cutting back preferences like fisheries quotas may still violate the right to pursue a trade or profession.[182] The impairment of this right, however, could likewise be justified on the same grounds as under the right of ownership.[183]

2.2.1.3 Fundamental environmental right

Currently, there is no fundamental right to environmental protection in the Community's legal order. To date, such a right has neither been laid down in Art. 174 of the Treaty, nor in Art. 37 of the Charta of Fundamental Rights of the European Union.[184] Moreover, even though there is some constitutional protection of the environment in some Member States, the ECJ has not yet acknowledged such a right under Community law.[185]

2.2.1.4 Principle of the protection of legitimate expectations

In the *Dürbeck* decision, the ECJ ruled that the principle of the protection of legitimate expectations is a fundamental principle of the Community.[186] Where the Community itself creates a situation which encourages Community citizens to form a legitimate expectation and where the citizens act on such grounds, the Community can be obliged to declare measures which controvert existing expectations void, or grant transition periods or even compensation.[187]

---

181 ECJ 280/93, Germany v. Council [1994] ECR I-4973 para. 81 and Case T-254/97, Fruchthandelsgesellschaft Chemnitz v. Commission [1999] ECR II 2743, para. 74 as interpreted by Nowak, 'Eigentumsgrundrecht', p. 849; Kingreen, 'Art. 6 EUV', para. 134.

182 Nowak, 'Eigentumsgrundrecht', p. 849.

183 The ECJ, in many cases, has not clearly distinguished between the right to property and the right to choose and practice a trade or profession. Thus, it often justifies impairments of the right to property and the right to choose and practice a trade or profession on the same grounds: see, for example, Case 44/94, The Queen v. Minister of Agriculture, fisheries and Food, ex parte national Federation of Fishermen's Organisations and others [1994] ECR, p. I-3115, paras. 4, 55-58; Case 248/95, SAM Schiffahrt GmbH and Heinz Stapf v. Bundesrepublik Deutschland [1997] p. I-4475, para. 72; Case 210/03 Swedish Match [2004], I-11893, paras. 72-74.

184 OJ 2007 No. C303/1; see Art. 6 of the Treaty of Lisbon amending the Treaty on the European Union and the Treaty Establishing the European Community, OJ 2007 No. C306/01.

185 Carsten Nowak, 'Grundrechtsdimension des Umweltschutzes', in Heselhaus and Nowak (ed.), Handbuch der Europäischen Grundrechte, (München: 2006), pp. 1667-1680.

186 Case 112/80, Dürbeck v. Hauptzollamt Frankfurt a.M. [1981] ECR 1095, para. 48.

187 Marc Bungenberg, 'Vertrauensschutz', in Hesselhaus and Nowak (ed.), *Handbuch der Europäischen Grundrechte*, (München: 2006), paras. 34-37; Case 177/90, Kühn v. Landwirtschaftskammer Weser-Ems [1995] ECR I-35, para. 14; Case C-22/94, Irish Farmers Association v. Ministry of Agriculture, Food and Forestry [1997] ECR I-1809, para. 19.

For 26 years, the Community has limited fishing activities. Prior to that, Member States had adopted similar restrictions. Quotas and technical measures adopted by the Community have regularly been amended to correspond to the available fishing opportunities. As a result, there is no way that Community actions can be construed as giving rise to a legitimate expectation among fishers to have access to a fixed amount of fishing opportunities. However, the quota setting practice has, for a long time, allowed fishing at unsustainable levels. Accordingly, fishers could claim to legitimately expect a continuation of this liberal approach to fisheries management. This view, of course, would seriously disregard the objectives and obligations in primary and secondary Community law, which require that the Community manage fisheries sustainably to ensure the continuation of fisheries in the future. Where this objective requires the adoption of conservation measures, the principle of legitimate expectation does not bar the Community from limiting fishing activities at sustainable levels.

2.2.2 Obligation to protect fish stocks

This section will examine whether the fundamental rights of Community fishers to property and the right to choose and practice a trade or profession create a general obligation for the Community to effectively protect fish stocks; and whether the Council's current fisheries management policy fulfils this obligation.

2.2.2.1 Obligation to manage fisheries resources properly

As stated above, the ECJ has acknowledged that it would respect the constitutional traditions of the Member States as well as the European Convention on Human Rights. The constitutions of the Member States,[188] their constitutional courts,[189] and the European Court of Human Rights all increasingly recognise that fundamental rights do not merely protect citizens against public actions, but create obligations for legislators to actively protect fundamental rights.[190] The ECJ has also ruled that Community organs have the duty to

[188] See Art. 40(3) of the Irish Constitution and Art. 9(2) of the Spanish Constitution.

[189] Austria, France, Germany, and the United Kingdom, Liv Jaeckel, *Schutzpflichten im deutschen und europäischen Recht: Eine Untersuchung der deutschen Grundrechte, der Menschenrechte und Grundfreiheiten der EMRK sowie der Grundrechte und Grundfreiheiten der Europäischen Gemeinschaften*, (Baden-Baden: 2001), pp. 134 et seq.; Peter Szczekalla, *Die sogenannten grundrechtlichen Schutzpflichten im deutschen und europäischen Recht*, (Berlin: 2000), pp. 909 et seq.

[190] Joachim Suerbaum, 'Die Schutzpflichtdimension der Gemeinschaftsgrundrechte' (2003) *Europarecht*, pp. 390-416; Peters Szczekalla, 'Funktionen der Grundrechte', in Heselhaus and Nowak (ed.), *Handbuch der Europäischen Grundrechte*, (München: 2006), pp. 195-204 and 350-357; Peter Szczekalla, *Die sogenannten grundrechtlichen Schutzpflichten im deutschen und europäischen Recht*, pp. 610 et seq.; Liv Jaeckel, *Schutzpflichten im deutschen und europäischen Recht*; Thomas, Kingreen, 'Art. 6 EU-Vertrag', in Calliss and Ruffert (ed.), *Kommentar zu EU-Vertrag und EG-Vertrag*, (Neuwied: 1999), pp. 66-68; Roland

protect the fundamental rights of Community citizens.[191] It has also decided that Member States have a duty to actively protect Community market freedoms.[192] Based on these decisions, some authors claim that the Community legislators are obliged to actively protect the fundamental right of Community citizens to property, as well as the right to choose and practice a trade or profession.[193]

It is argued here, with regards to fisheries management, that this obligation requires the Council to pass effective legislation to protect the environmental conditions for the use of fishing vessels, fishing gear and other production facilities, as well as freedom to choose and practice the fisheries profession, *i.e.*, to protect fisheries resources against depletion by overfishing. This view is supported by the following arguments. Given the open access nature of fisheries resources and the existing race to fish dilemma, it is very difficult, if not impossible, for fishers to manage sustainable exploitation themselves. They are reliant on a public authority to adopt and enforce an effective management regime. Given that the exclusive competence to protect fisheries resources exploited by Community fishers lies at the Community, the Council has a duty to establish such a regime to ultimately grant fundamental rights protection.[194]

Accepting that the Community is obliged to protect fish stocks in order to protect Community fishers' fundamental rights, one must consider the content of this obligation to determine whether the protected activities are being encroached upon.[195] In terms of the content of the obligation, authors agree that in order to guarantee the fundamental rights of its citizens, the Community must protect all activities and rights which are protected under them, *e.g.*, the right to hold tangible, intangible and intellectual property or different use rights, etc.[196] The activities protected by fishers' fundamental rights to property and to choose and practice their profession are threatened by the current overexploitation of fish resources. Under the current Community legislation, economically exploited stocks are in decline, overfished, or even outside safe biological limits.[197] A continuation of the fishing practice under the current law, can lead

---

Winkler, *Die Grundrechte der Europäischen Union – System und allgemeine Grundrechtslehren*, (Wien: 2006), p. 215.

[191] Case 13/94, P/S and Cornwall County Council, [1996] ECR, p. 2143, para. 19.

[192] Case C-265/95, Commission v France [1997] ECR, p. 6959, para. 31.

[193] Nowak, 'Eigentumsgrundrecht', para. 61; Suerbaum, 'Die Schutzpflichtdimension der Gemein-schafts-grundrechte', p. 409; The European Convention on Human Rights does only to a limited extent protect the right to choose and practice a trade or profession; see Art. 4 (2) ECHR.

[194] The fact that the Community manages transboundary fisheries resources jointly must not lead to a deprivation of Community fishers' fundamental rights to property and to choose and practice a trade or profession; see Suerbaum, 'Die Schutzpflichtdimension der Gemeinschaftsgrundrechte', p. 391.

[195] Szczekalla, *Die sogenannten grundrechtlichen Schutzpflichten im deutschen und europäischen Recht*, p. 171; Suerbaum, 'Die Schutzpflichtdimension der Gemeinschaftsgrundrechte', p. 408.

[196] Szczekalla, *Die sogenannten grundrechtlichen Schutzpflichten im deutschen und europäischen Recht*, p. 171.

[197] In 2007, the Commission found that 20 stocks received a zero-TAC advice from ICES (moratorium); that 26 stocks were outside safe biological limits; that only 12 stocks were found to be within safe

to (a) a continuing decline of the remaining resources, and (b) ultimately to the extinction of certain species.

2.2.2.2 Title to sustainable and effective fisheries management

Against this background, the Council must take more effective actions to conserve the fundamental rights of fishers.[198] However, the Council retains a wide margin of discretion with regards to the scope and means of protection of fundamental rights.[199] Community fishers do not have the right to claim the adoption of a particular conservation measure, *unless* only this specific measure will end the threat to the fundamental right.[200]

One particular act is paramount to protecting the fundamental rights of fishers: the limitation of fishing opportunities at sustainable levels.[201] The establishment of a sustainable upper exploitation limit for the overall use of the resource is a precondition for the proper functioning of all other managing instruments. In order to have due regard for the fundamental rights of Community fishers to property and to choose and practice a trade or profession, the Council must change its current practice. It is clear that fish stocks cannot recover to sustainable levels if the Community allows fishing to continue below safe biological limits. In addition, in areas where fishing is frequently allowed at levels exceeding the MSY, stocks will decline to below limits at which stocks are threatened by depletion. This does not mean that the Council should have no discretion as to whether it will adopt scientific recommendations regarding setting restrictions. As suggested above, a system for setting TACs could be established which would compel the Council to adopt exploitation rates at MSY-levels. Deviations from this level could be justified where there are compelling socioeconomic reasons, as clearly defined in the law. The Council, however, could not allow exploitation of stocks outside safe biological limits.

The Community basically claims to have adapted its fisheries policy to this end by establishing the multi-annual recovery and management plans. However, as of 2007, these plans have not been successful at bringing stocks outside safe

biological limits; and that for 59 stocks the status was unknown; see Communication Com(2007) 295 final from the Commission to the Council on Fishing Opportunities for 2008 – Policy Statement from the European Commission, p. 5.

198 *Ibid.*, p. 173 (neither with regards to fisheries nor this particular case).

199 Under a political system of separated powers, in general, the legislator is to assess the given societal or environmental problems and draw up adequate legislative solutions through a political discourse; see Ernst-Wolfgang Böckenförde, 'Grundrechte als Grundsatznormen' (1990) 29 *Der Staat*, pp. 25, 29.

200 Kingreen, 'Art. 6 EU-Vertrag', p. 67; Suerbaum, 'Die Schutzpflichtdimension der Gemeinschaftsgrundrechte', p. 414.

201 Commission Communication Com(2001) 135 final Green Paper on the Future of the Common Fisheries Policy, p. 4.

biological limits to safe levels;[202] and overall progress is slow.[203] Even where recovery plans have been adopted, the fear is that they will be ignored by the Council.[204] Furthermore, Arts. 5 and 6 do not forbid the Council to allow exploitation where stocks are outside safe biological limits.

2.3 Conclusion

It was found in this section that fishing rights provided under CFP can be characterised as exclusive, but insecure (*i.e.*, revocable), non-durable and largely non-transferable use rights of a changing economic value. In addition, the fundamental right to property and the right to choose and practice a trade or profession protect the rights of fishers to use their prior-owned fishing vessels, fishing gear and all other production facilities, as well as the right to choose and practice the fisheries profession. In general, limits on fishing impair these rights. Given that fishing limitations are an appropriate and necessary means to attain the Community interest to sustain fisheries resources, this infringement is justifiable. With regards to the setting of quotas, the Council is obliged to adopt the least onerous limitation necessary to attain its conservation objectives. However, the Council has a wide margin of discretion in doing so. The Council can extensively limit fishing.

The fundamental right to property and the right to choose and practice a trade or profession also require that the Council actively protect fisheries resources. Given the open access nature of fisheries resources and the existing race to fish dilemma, it is very difficult, if not impossible, for fishers to establish sustainable exploitation levels themselves. They are reliant on a public authority to adopt and enforce an effective management regime. Given that the exclusive competence to protect fisheries resources exploited by Community fishers lies with the Community, the Council must establish such a regime to guarantee fundamental rights. The obligation to actively protect fisheries resources means that the Council must deviate from the current practice of setting quotas at unsustainable levels. Of all deficiencies of CFP legislation, the abolition of this practice is the first step to bringing fish stocks to sustainable levels. In light of the management instruments introduced in the 2002 Reform, it is doubted here that currently existing measures are sufficient to attain this goal. A system should be established which forces the Council to adopt exploitation rates at least

[202] WWF, *WWF-Mid-Term Review of the Common Fisheries Policy*, pp. 4, 10-32.

[203] *Ibid.*, p. 30.

[204] The WWF has brought an action under Art. 230 of the Treaty in respect of TACs for 2007 as adopted by the Council under Council Regulation (EC) 41/2006 fixing for 2007 the fishing opportunities and associated conditions for certain fish stocks and groups of fish stocks, applicable in Community waters and, for Community vessels, in waters where catch limitations are required, OJ 2007 No. L15/1. The WWF currently challenges this on the ground that the TAC is not compatible with the TACs provided under an existing cod recovery plan in Council Regulation (EC) 423/04 establishing measures for the recovery of cod stocks, OJ 2004 No. L70/8.

at MSY-levels. Under this policy, deviations could only be justified for compelling socio-economic reasons, which would have to be clearly defined in the law. The Council, however, could not exceed exploitation levels at which stocks are outside safe biological limits.

3 Legality of Commission sanction powers under the CFP

Arts. 23(4), 26(3), 16(1) of Regulation 2371/02 allow the Commission to adopt different kinds of sanctions under the CFP. These sections set out the conditions under which the Commission can deduce Member States' future fishing opportunities, take preventive actions, and suspend financial assistance. However, these empowering provisions are somewhat vague regarding the severity of the sanctions. This is particularly true of the Commission's quota deduction power,[205] since Art. 23(4) does not precisely define the degree to which quotas can be deduced. Accordingly, a quota deduction measure adopted by the Commission in 2007 has been contested before the Court of First Instance.[206] The applicants have argued in vague terms that the Commission lacked power to adopt the respective quota deductions on the basis of Art. 23(4) of Regulation 2371/02. To find out whether and to what extent the Commission is entitled to adopt quota reductions, it must be assessed whether the Council's adoption and the Commission's implementation of Art. 23(4) are compatible with the requirements of the Treaty. This assessment of the Commission's sanctioning powers is also of particular interest, given that currently the CFP's control and enforcement system is being revised and the Commission's sanctioning powers may be extended throughout this process.[207]

[205] While preventive actions must be proportionate to the risk of the threat to the conservation of living resources (and cannot exceed three weeks in duration), financial suspension must be proportionate to the degree of non-compliance.

[206] Case T-172/07, Atlantic Dawn Ltd. et. al v. Commission, OJ 2007 No. C170/29. The case has been declared inadmissible due to a lack of 'direct concern' in the sense of Art. 230, para. 4 of the Treaty. Parties contested Commission Regulation (EC) 147/2007 adapting certain fish quotas from 2007 to 2012 pursuant to Article 23(4) of Council Regulation (EC) 2371/2002 on the conservation and sustainable exploitation of fisheries resources under the Common Fisheries Policy, OJ 2007, No. L46/10.

[207] Commission Staff Working Document SEC(2008) 2760 accompanying the Proposal for a Council Regulation establishing a Community control system for ensuring compliance with the rules of the Common fisheries Policy, pp. 11-12.

3.1 Brief summary of Art. 23(4) of Regulation 2371/02 and Commission Regulation 147/07

The content of Art. 23(4) and Commission Regulation 147/07 has been described above in more detail. Only a brief summary will be provided here.

Art. 23(4) allows the Commission to reduce the future fishing opportunities of Member States where national quotas have been exceeded. In this task, the Commission is assisted by a management Committee composed of the representatives of Member States. If the Committee disagrees with the Commission's reduction measure, the decision making power is conferred back on the Council, which has the final say. The Commission has only recently reduced the quotas of Ireland and the United Kingdom, deducting fishing opportunities equivalent to the amount overfished. In order to minimise the economic and social impact of such measures, reductions are spread out over a six-year period, a correction factor is applied and the deduction cannot exceed a limit of 15% of the annual quota. In broad terms, Art. 23(4) has been described as a non-punitive sanction primarily aimed at furthering the proper national implementation of the Community quota system. As a side effect, quota reductions disburden stocks and ecosystems, providing a remedy for the environment. Commission implementation measures based on Art. 23(4) take the form of a tertiary legislative act.

3.2 Legality of Art. 23(4) and Regulation 147/07

Art. 23(4) of Regulation 2371/02 and the implementing measure must adhere to Art. 37 of the Treaty as well as to the principles of proportionality and subsidiarity. In addition, the conferral on the Commission of a power to sanction must be compatible with the requirements established in Arts. 202, para. 3 and 211, para. 4 of the Treaty. With a view to assessing the compatibility with the proportionality principle, it is important to briefly describe the deficits in the implementation of law by the Community in general, and focusing on the CFP in particular.

3.2.1 The Community's general and the CFP's particular implementation deficits

Implementation deficits of Community law exist for manifold reasons. Community legislation itself is often viewed as being too complex and lacking transparency.[208] Member States have different administrative and

[208] H-W. Rengeling, ‚Durchführung des Europäischen Gemeinschaftsrechts – Überblick', in H-W. Rengeling (ed.), *Handbuch zum europäischen und deutschen Umweltrecht*, (Köln: 2003), pp. 939 et seq.; Magiera, 'Durchsetzung des Europarechts', para. 39.

legal systems, histories and cultures, resulting in various approaches to the implementation of measures.[209] Often the interests of Community and Member States are in conflict (*e.g.*, allocation of Community funds).[210] Furthermore, the Commission's ability to review and control the actions of the Member States is hampered by a lack of information and personnel and the inability to give direct commands to national administrators.[211] Control instruments are widely considered to be reactive, fragmentary and slow.[212] This is particularly true with regards to the ECJ's infringement procedure.[213] Concerning environmental conservation issues, the lack of a private lobby that can match the power of those existing in other policy sectors is a key issue.[214] As a result, in order to guarantee effective implementation of Community law, the Community and, in particular, the Commission, must be able to impose sanctions as an appropriate response to infringements of Community law.[215]

The implementation deficits that have just been mentioned are equally applicable to the CFP. Despite the continued overhaul and tightening of the control regimes,[216] implementation deficits persist.[217] Infringements and non-compliance refer to the actions/inaction of fishermen (*e.g.*, violate technical or

---

209 Heinrich Siedentopf and Christoph Hauschild, 'Europäische Integration und die öffentlichen Verwaltungen der Mitgliedstaaten' (1990) 43 Zeitschrift für öffentliches Recht und Verwaltungswissenschaften, 11, pp. 452-453.

210 Magiera, 'Durchsetzung des Europarechts', para. 40.

211 Pühs, *Der Vollzug von Gemeinschaftsrecht*, pp. 134-141; see also Commission Report Com (2004) 573/2 final on the Protection of the European Communities' financial interests and the fight against fraud.

212 Magiera, 'Durchsetzung des Europarechts', para. 52.

213 Pühs, *Der Vollzug von Gemeinschaftsrecht*, pp. 132 et seq.

214 L. Krämer and G. Winter, 'Umweltrecht' in Schulze and Zuleeg (ed.), Europarecht, (Baden-Baden: 2006), para. 31 (with a view to environmental conservation).

215 Pühs, *Der Vollzug von Gemeinschaftsrecht*, pp. 261 et seq., 293 et seq.; Case 240/90, Germany v. Commission [1992] ECR I-5383, paras. 10-39.; Case 176/03, Commission v. Council [2005] ECR I-7879, paras. 47-48; Commission Report Com (2004) 573/2 final.

216 The first control regulation was adopted in 1982, consolidated in 1987 and repealed by the current Control Regulation in 1993, see Council Regulation 2057/82, OJ L220/1; Council Regulation (EC) 2241/87, OJ L207 and Regulation 2847/93.

217 Report by the Commission to the Council on the Enforcement of the CFP, Com(86) 301 final; Communication from the Commission to the Council and the European Parliament on the Common Fisheries Policy, SEC (90) 2244 of 6 December 1990; Report 1991 from the Commission to the Council and the European Parliament on the Common Fisheries Policy, SEC (91) 2288 of 18 December 1991; Report on Monitoring Implementation of the Common Fisheries Policy, SEC (92) 394 final of 6 March 1992. For an explanation of the last four reports, see Long/Curran, pp. 66-72. See also Report from the Commission on the monitoring of implementation of the Common Fisheries Policy – Synthesis of the implementation of the control system applicable to the Common fisheries Policy by Member States, Com(2001) 526 final of 28 September 2001; Report from the Commission on the monitoring of the Member States' implementation of the Common Fisheries Policy from 2002 to 2004, COM (2004) 849 final of 4 January 2005; Report from the Commission to the Council and the European Parliament

quantitative restriction) and Member States (*e.g.*, ineffective implementation of the CFP rules).[218] Although minor progress has been achieved as a result of the 2002 Reform, the Commission's evaluation of CPF implementation parrots the same criticisms of twenty years ago. These centre on a few central issues. First, the legal concepts underlying CFP rules are not interpreted and applied in a uniform way.[219] Second, the enforcement practices and sanction measures of the Member States differ greatly. Finally, co-ordination across the Member States is regarded as weak. Overall, the Commission has stated that the will to implement consistently and fully CFP on the part of the Member States is often lacking.[220]

3.2.2 Community sanction competence under Art. 37(2), para. 3 of the Treaty

Sanction measures in Regulation 2371/02 have been adopted on the basis of Art. 37(2) of the Treaty. The question is whether Art. 37(2) really empowers the Council to adopt such sanctions.

As explained above, Art. 37(2) of the Treaty gives the Community the power to adopt a comprehensive fisheries policy. For sanctions imposed on the agricultural sector, the ECJ has also repeatedly found that different kinds of penalties and sanctions may be justified under Arts. 34(2) and 37(2), para. 3 of the Treaty.[221] Insofar, the Court has interpreted Art. 34(2) and 37(2), para. 3 broadly. The Court has declared that the Community is entitled to 'impose penalties *necessary* for the effective application of the rules in the sphere of the Common Agricultural Policy'.[222] With regards to necessity, the ECJ had found on several occasions that restrictions to the sovereignty of the Member States can be justified where there is a severe threat to the '*effet utile*' of Community law.[223] The

on the monitoring of the Member States' implementation of the Common Fisheries Policy Com, Com (2007) 167 final of 10 April 2007.

[218] Recently, France was required to make a penalty payment of €58,761,250 for each six month period from the date of the judgment, until it is in compliance with the obligation to properly implement fisheries control and sanction measures; see also Case 290/87, Commission v. Netherlands [1989] ECR 3083; Case 64/88, Commission v. France [1991] ECR I-2727; 244/89, Commission v. France [1991] I-163; Case 258/89, Commission v. Spain [1991] ECR I-3977; Commission v France; Case 52/95 Commission v. France, para. 38.

[219] Commission Report Com (2007) 167 final on the monitoring of the member states' implementation of the common fisheries policy 2003-2005.

[220] *Ibid.*

[221] See cases listed in para. 12 of Case C-240/90, Germany v. Commission [1992] ECR I-5383 (except Case 11/70, Internationale Handelsgesellschaft mbH v. Einfuhr und Vorratsstelle für Getreide und Futtermittel [1970] ECR 1125, para. 18).

[222] Case C-240/90, Germany v. Commission [1992] ECR I-5383, para. 11; emphasis added by the author. It should be noted that the German version of the judgement includes the word 'all' [...] penalties *necessary* for the effective application of the rules in the sphere of the Common Agricultural Policy'.

[223] Case 30/59, De Gezamenlijky Steenkolenmijnen in Limburg/ECSC High Authority, (1961) ECR, p. 50.

'effet utile' argument does not, however, establish a general competence of the Community to impose sanctions. Sanction measures can be adopted if Member States are reluctant to implement Community law, thereby ensuring its effectiveness.[224]

Other judgements have explicitly recognised the legality of various forms of Community sanctions. Community measures, for example, required members of the agricultural trade to refund *with interest* benefits unduly received or the loss of a security equivalent to that benefit.[225] The ECJ has also confirmed the Community's power to *exclude* members of the trade from subsidies schemes or to impose *surcharges* on amounts wrongly received.[226] One thing that all of these measures have in common is that they impose sanctions that go beyond compensation. In all these cases, the ECJ has regarded such measures as necessary and proportionate to combat irregularities under the CAP.[227]

3.2.3 Proportionality and subsidiarity

Given that the Community is not exclusively competent to adopt sanctions, Art. 23(4) must be compatible with the principle of subsidiarity, as laid down in the Treaty. According to Art. 5, para. 2 of the Treaty:

> [T]he Community shall take action, in accordance with the principle of subsidiarity, only if and insofar as the objectives of the proposed action cannot be sufficiently achieved by the Member States and therefore, by reason of scale or effects of the proposed action, be better achieved by the Community.

This article requires the so-called 'test of comparative efficiency'.[228] The first part of the principle basically states that Member States cannot achieve Community objectives by themselves. The second is that Community actions are better at achieving Community objectives (*i.e.*, Community actions add value to the actions of Member States). It has been explained that the Member States' implementation of CFP quota measures is in most cases ineffective. The above-mentioned reports clearly indicate that there is an ongoing conflict between the Community and the Member States over the implementation of CFP law. Member States are unable or unwilling to strictly enforce management rules and are accused of favouring their own fishing industries over those of other Member States. In particular, Member States are not able to ensure a proper implementation of CFP rules by other Member States. A broad Community

---

[224] Pühs, *Der Vollzug von Gemeinschaftsrecht*, pp. 273-274.

[225] C 288/85, Hauptzollamt Hamburg-Jonas v. Plange Kraftfutterwerke [1987] ECR 611 and C 199/90, Italtrade v. AIMA [1991] ECR I-5545.

[226] Case C-240/90, Germany v. Commission [1992] ECR I-5383 paras. 1 and 2 (Summary).

[227] Case C-240/90, Germany v. Commission [1992] ECR I-5383, para. 19 and 28; C 288/85, Hauptzollamt Hamburg-Jonas v. Plange Kraftfutterwerke [1987] ECR 611, para. 15.

[228] Craig and De Búrca, *EU Law*, p. 135.

sanction competence would counterbalance the Members States' lack of competence to sanction the fishers of other Member States by creating incentives for all Member States to improve the effectiveness of their enforcement practices.

The sanction measures must, of course, be proportionate. Proportionality may be assumed where legislative acts are appropriate and necessary, *i.e.*, where measures are suitable to achieve the pursued aim and where restrictions are not disproportionate to the aims pursued, particularly not more onerous than other available measures.[229] Art. 23(4) of Regulation 2371/02 is a suitable measure to contribute to the preservation of the quota system. It makes public the infringement of quotas, underlines the validity of the quota norms, and provides an immediate relief for the overfished stocks and burdened ecosystems. By comparison, a proceeding before the ECJ would take a long time and ultimately result in penalty payments by the violating Member State. In addition, the deduction of future quotas likely does not constitute a disproportionate response to the overfishing of quotas. However, it depends, of course, on the amount of opportunities deduced. Given that Art. 23(4) leaves open the exact amount of quota to be deduced, the Commission's implementation practice is important here. The Commission interprets Art. 23(4) in such a way that reductions do not go beyond the mere compensation of quotas taken in excess.[230] Ultimately fishers cannot be regarded as being punished, as they do not loose any of their legally obtained fishing opportunities. The Commission even adopted a correction factor to take into consideration their economic situation. This view that deductions are not disproportionate is supported by the ECJ's opinion that sanction measures that impose detriments and burdens on Community citizens which go beyond mere compensation or remedy (*argumentum a majore ad minus)* are proportionate.[231]

3.2.4 Compatibility with Arts. 202, para. 3 and 211, para. 4 of the Treaty

Arts. 202, para. 3 and 211, para. 4 of the Treaty state that Council shall confer on the Commission the powers necessary for the implementation of the rules laid down by the Council. According to them, only *implementing* measures can be adopted by the Commission.[232]

---

229 Case 331/88, The Queen v. Ministry of Agriculture, Fisheries and food, ex parte FEDESA and others [1990] ECR I-4023, para 13.

230 See Commission Regulation 147/07.

231 The wording of Art. 23(4), however, does not require that the Commission limit itself to reducing amounts equal to the amounts taken in excess. It remains to be seen what the Commission practice will be in the future. The Commission will, however, to a certain extent, be required to continue its practice due to the principle of equal treatment.

232 Case 240/90, Germany v. Commission [1992] I-5383, para. 2 (Summary).

The ECJ has ruled that the Council itself must adopt the essential elements of the subject matter laid out in the legislation.[233] Nevertheless, early on, the Court declared that the term 'implementation' would be interpreted broadly:

> '[The] powers conferred on the Commission by the Council for the implementation of the rules of the Treaty must be given a wide interpretation as appears from the general context of the Treaty and practical requirements.'[234]

The ECJ has taken a particularly broad view of matters connected to the CAP, as well as to cases where the Council subjects Commission implementation to committee procedures.[235] It has done so for two reasons. First, it regards the Commission as the sole organ that is able to keep up with the rapidly changing agricultural sector. Second, subjecting implementation to committee procedures allows the Council (*i.e.*, the Member States) to maintain control over the process; and they have the right to intervene when necessary.[236]

The Court has also explicitly confirmed the legality of the *conferral of sanction powers* on the Commission. In *Germany vs. Commission*,[237] the ECJ recognised the competence to impose penalties in response to the illegal subsidies, imposing exclusions from the scheme of subsidies and requiring pay back surcharges on amounts wrongly received. The Court thus confirmed the Commission's power to impose sanctions on private entities that go beyond the mere reversal of the unjust enrichment. The Court also stated in this decision that, in relation to the CAP, only the fundamental guidelines of Community policies may be classified as essential. Penalties that merely underpin the chosen policy options, even where they may affect the fundamental rights of Community citizens,[238] are not regarded as essential.[239] Thus, the requirement to keep essential policies to the Council, in fact, does not impose too strict requirements.[240]

Given that Art. 23(4) grant the Commission a wide margin of discretion with regards to imposing sanctions in such a sensitive regulatory area, the question arises whether Arts. 202, para. 3 and 211, para. 4 require the Council to

---

[233] Case 25/70, Einfuhr- und Vorratsstelle für Getreide und Futtermittel v. Köster [1970] ECR 1161, para. para. 6; Case 23/75, Rey Soda v. Cassa Conguaglio Zucchero [1975] ECR 1279, paras. 10-11.

[234] Case 23/75, Rey Soda v. Cassa Conguaglio Zucchero [1975] ECR 1279, para. 1 (Summary), 10; Case 41/69, Chemiefarma v. Commission [1970] ECR 661, para. 65.

[235] Case 23/75, Rey Soda v. Cassa Conguaglio Zucchero [1975] ECR 1279, paras. 10-14; Case 25/70, Einfuhr- und Voratsstelle für Getreide und Futtermittel v. Köster [1970] ECR 1161, para. 9.

[236] Hummer, 'Art. 155 EGV', para. 53; Michael Schweitzer, ‚Art. 202 EGV', in Grabitz and hilf (ed.), *Das Recht der Europäischen Union*, (München: 2005), para. 27.

[237] Case 240/90, Germany v. Commission [1992] I-5383.

[238] Case 240/90, Germany v. Commission [1992] I-5383, paras. 32, 38.

[239] *Ibid.*, para 2 (Summary).

[240] Jan Kalbheim and Gerd Winter, 'Delegation requirements for rule-making by the Commission', in Winter (ed.), *Sources and Categories of European Union Law*, p. 591.

draft Art. 23(4) more precisely.[241] However, interpreting Arts. 202, para. 3 and 211, para. 4, the ECJ does not explicitly require the Council to specify the exact components of the delegated powers. 'For that purpose, a provision drafted in general terms provides a sufficient basis'.[242]

As explained above, while the Council must adopt the essential elements of the subject matter laid out in the legislation, the Commission can only adopt *implementing* measures. Nevertheless, early on, the Court declared that the term 'implementation' would be interpreted broadly. The ECJ also recognised the competence to impose penalties that go beyond restitution to reverse unjust enrichment. In light of this ruling, it is likely that the ECJ will find that the Commission has the power to reduce quota. As with penalties under the CAP, quota reductions cannot to be regarded as essential elements of the subject matter of the CFP, but merely as legal measures upholding a central management instrument. In addition, the Commission's power to impose sanctions is limited by a management committee.[243]

One issue is that Art. 23(4) does not specifically state how quota reductions should be characterised: they could be punitive in nature (*i.e.*, reducing fishing opportunities beyond the amount overfished), or merely compensatory (*i.e.*, stabilise the quota system by reducing the future quota to an amount equivalent to the one taken in excess). Since imposing sanctions is such a sensitive regulatory area, it would have been better if the Council had drafted Art. 23(4) more precisely.[244] However, the ECJ does not require the Council under Arts. 202, para. 3 and 211, para. 4 of the Treaty to specify the exact components of the delegated powers. 'For that purpose, a provision drafted in general terms provides a sufficient basis'.[245]

3.2.5 Hierarchy of norms

Another feature of Art. 23(4) of Regulation 2371/02 is also deserving of consideration. By means of Art. 23(4), the Council authorises the Commission to reduce Member States' future quota, *i.e.*, quota to be allocated by future annual Council regulations.[246] Can measures instituted by the Commission amend future Council measures? Two central rules have developed under Community law regarding the hierarchy of norms of Community law. First,

---

[241] Kalbheim and Winter, 'Delegation requirements for rule-making by the Commission', pp. 591-593, referring to Cases 291/86, Central-Import Münster v. Hauptzollamt Münster, para. 13; Case 240/90, Germany v. Commission [1992] I-5383, para. 41.

[242] Case 240/90, Germany v. Commission [1992] I-5383, paras. 40-41.

[243] Art. 23(4), para. 3 of Regulation 2371/02.

[244] Kalbheim and Winter, 'Delegation requirements for rule-making by the Commission', pp. 591-593, referring to Cases 291/86, Central-Import Münster v. Hauptzollamt Münster, para. 13; Case 240/90, Germany v. Commission [1992] I-5383, para. 41.

[245] Case 240/90, Germany v. Commission [1992] I-5383, paras. 40-41.

[246] As has been done by Regulation 147/07, OJ 2007, No. L46/10.

basic regulations rank higher than implementing regulations:[247] lower-ranking implementing rules thus cannot derogate from the basic regulation.[248] Second, under Community law, no Community organ outranks another. Thus, measures of the same category adopted by different organs, *i.e.*, basic measures on the one side and implementing measures on the other, are in no hierarchical order to each other.[249] As a result, implementing rules adopted by the Council do not outrank implementing rules adopted by the Commission. Therefore, the former can be amended by the latter.[250] However, taking into consideration the different effects of regulations, directives, and decisions, measures should be of the same type, *i.e.*, both should be regulations.[251]

3.3 Conclusion

Against the ECJ's current interpretation of Art. 37(2) of the Treaty, the adoption of Art. 23(4) of Regulation 2371/02 must be considered lawful. Considering the persevering implementation deficits under the CFP, quota reduction measures are obviously necessary to improve the implementation and overall functioning of the quota system. This view is supported by ECJ decisions that confirm more stringent Commission sanctions under the CAP are in conformity with Art. 37(2) of the Treaty.

Art. 23(4) of Regulation 2371/02 and the Commission's implementing measure conform with the principles of proportionality and subsidiarity. In principle, the deduction of future fishing opportunities is a proportionate measure, which is necessary to ensure the functioning of the quota system. Given that legal consequences are somewhat vague under Art. 23(4) of Regulation 2371/02, proportionality depends on the amount of fishing opportunities deduced by the Commission. However, given that under the current Commission's interpretation fishers do not lose any of the legally obtainable fishing opportunities, quota deductions under Art. 23(4) must be regarded as a proportionate action to uphold the functioning of the quota system.

---

247 Case 38/70, Deutsche Tradax GmbH v. Einfuhr- und Voratsstelle für Getreide und Futtermittel, [1971] ECR 145, para. 3 (Summary).

248 *Ibid.*, para. 10; Case C-103/96, Directeur général des douanes and droits indirects v. Eridania Beghin-Say [1997] ECR I-1453, para. 20.

249 Case 188-190/80, France, Italy and the United Kingdom v. Commission [1982] ECR 2545, para. 6. see also, R. Bieber and Isabelle Salomé, 'Hierarchy of Norms in European Union Law' (1996) 33 *Common Market Law Review*, pp. 915, 919.

250 Martin Nettesheim, ‚Art. 249 EGV', in Grabitz and Hilf (ed.), Das Recht der Europäischen Union, (München: 2002), para. 232, 234; M. Ruffer, Art. 249 EGV', in Calliess and Ruffert (ed.), Kommentar zu EU-Vertrag und EG-Vertrag, (Neuwied: 1999), para. 11; Bleckmann, *Europarecht*, p. 201; Gudrun Schmidt, ‚Art. 249 EG', in von der Groeben and Schwarze (ed.), *Kommentar zum Vertrag über die Europäische Union und zur Gründung der Europäischen Gemeinschaft*, (Baden-Baden: 2004), p. 778.

251 Nettesheim, 'Art. 249 EGV', para. 234.

The conferral of the quota deduction power on the Commission was in line with Arts. 202, para. 3 and 211, para. 4 of the Treaty. Quota reductions cannot to be regarded as essential elements of subject matter of the CFP and the Commission's sanction powers are limited by a management committee.

Summary and Conclusions

The purpose of this book is to: (a) explain the CFP's complex management and promotional regimes; (b) identify practical and legal problems and failures in both systems; and (c) assess whether CFP promotional measures are consistent with management rules and how consistency can be increased. The following section will outline the main insights arising from this analysis.

1 The environmental, socioeconomic and political background of the CFP

Overfishing has been widely identified as one of the most important factors causing the depletion of fish stocks. The EEA described in 2007 that of those stocks that had been assessed, 14% in the Arctic were outside safe biological limits, whilst for the North-East Atlantic and Baltic Seas this was 26%. Within the North-East Atlantic, the North Sea was the most severely affected with 44% of the assessed commercial stocks outside safe biological limits, followed by the Celtic Sea with 30% outside them [...]. In the Mediterranean Sea, the percentage of assessed commercial stocks outside safe biological limits in 2005 ranged from 10-20% with Aegean and Cretan being in the worst condition. [...] Bluefin tuna stocks both in the eastern Atlantic and Mediterranean Seas have been identified as being near collapse.

The Community has an undersupply of fisheries products and a negative trade balance in fishery products. In 1998, the value of the entire production chain was estimated at €20 billion, approximately 0.28 percent of the Community's gross domestic product. Despite this relatively small share, many coastal communities rely heavily on fishing. In addition, marine waters face other pressures such as traffic, tourism, industrial and agricultural activities, pollution, etc.

The CFP is concerned with the sustainable exploitation of living aquatic resources and Member States' fisheries sectors. This requires a broad range of political and legislative actions in a range of policy areas. To guarantee that Community fishermen do not fish at unsustainable levels, the Community limits fishing opportunities by adopting total allowable catches (TACs), effort limitations, technical and control measures. It also apportions the available resources among its Member States, which then allocate their share to their own fishers. When imposing catch limitations, the Community increasingly takes environmental concerns into consideration. Rules adopted under the CFP also relate to the structure of the Community's fisheries sector. The primary aim of such structural policies is to support the sector in adapting its production capacities to correspond to available resources, thereby guaranteeing efficient and sustainable production. Another area of the CFP is the common organisation of the market for fisheries products. Given the unstable supply of fisheries resources, the Community takes measures to stabilise markets. It also aims at matching supply to demand. To this end, the Community intervenes in the market by establishing common marketing standards, producer organisations, and a system of trade with third countries. To effectively limit fishing activities, structure the sector and stabilise markets, the Community is also responsible for the external relations of its fisheries policies (*i.e.*, with non-Member States). It has concluded several international agreements, which allow Community vessels to fish in the waters of third countries. Besides, the Community is currently a member of 11 international fisheries organisations (IFOs), which are concerned with the management of fisheries resources on the high seas. Finally, the Community controls the implementation of CFP measures by Member States.

The structure of the political process ultimately determines the conditions of exploitation. The relevant political actors are the three main Community organs, *i.e.*, the Council, the Commission, the Parliament, as well as the fishing sector and environmental NGOs. The Council legislates CFP policy. Composed of the political representatives of the Member States, it is mainly driven by exploitation interests of national fishing industries. As a result, laws adopted by the Council generally advance short-term economic interests over long-term environmental and economic gains. Since the inception of the common management policy in 1983, preference has influenced all measures adopted under CFP. The Commission's strong legislative power derives from its exclusive right to initiate proposals, from the fact that the Council may only overturn its proposal by unanimous vote, and by its right to adopt provisional measures in case Member States do not manage to agree on conservation measures, such as the annual TACs. Given that in practice, Member States hardly ever reach unanimity in the Council with regards to the adoption of utilisation and conservation measures, most CFP measures can only be established with the Commission giving its consent. Thus, the Commission's political responsibility for the current state of the CFP's overmodest political results should not be underestimated. Even though the Commission has acknowledged the need to incorporate a long-term vision in fisheries management, it often anticipates the Council's preferences, mirroring them in its own proposals. The overall impact of Parliament on the CFP has remained relatively small given its limited decision making power. However, Parliament does influence the CFP substantially through its powers regarding the yearly adoption of the Community budget. It must also be noted that in the course of the adoption of the Lisbon Treaty, the Parliament's legislative powers will be expanded. Parliament's fisheries committee's one-sided economical interests as well as a lack of control of its reports through the plenary make the Parliament a weak promoter of sustainability under the CFP. Finally, interest representation mainly takes place at the national level. Nevertheless, since 2002, industry and environmental NGOs have (arguably) been given a greater voice in Europe through the Regional Advisory Councils.

2 Institutional and organisational structures

For legal and political reasons the Community chose to base its fisheries policy on the provisions on agriculture in Arts. 32 to 38 of the Treaty – particularly Art. 37(2), third paragraph. According to Art. 299(1) and (2), the Treaty applies in all Member States as well as the Azores, Canary Islands, Madeira as well as French oversees departments. Overseas territories, in particular, Greenland, listed in Annex II of the Treaty, are excluded. Art. 299(1) refers to the Member States. The use of the term *Member States* implies that the Treaty applies to all territories under their sovereign control, including their inland waters, ports and territorial seas, the latter being determined by Art. 2 of 1982 UNCLOS. Since sovereignty does not extend to the marine areas beyond the territorial sea, neither the EEZ nor the high seas would be included here. Thus, Art. 299(1) of the Treaty does not justify the application of the Treaty beyond the territorial sea. The ECJ, however, has made clear that whenever rule-making competences are conferred on the Community by the Treaty, these competences extend to maritime areas insofar as Member States have similar powers under public international law. Given the fact that Member States are empowered by public international law to exploit fisheries resources on the high seas as well as to exploit fisheries resources exclusively within their own EEZ, the Community's competence to regulate their use and their conservation applies to these areas accordingly. Thus, fisheries governance in the EEZ and the high seas is not based on Art. 299, but on Art. 37(2), third paragraph of the Treaty. Based on this decision, Churchill and others have correctly concluded that the CFP rules apply to: (a) all vessels engaged in fishing activities in the territorial sea or the EEZ of one of the territories in which the CFP applies, (b) all vessels registered in the Member States fishing on the high seas (as well as to those fishing on the continental shelf for sedentary species), and (c) all vessels registered in the Member States fishing in third country waters.

Given that Member States are not generally excluded from lawmaking (both where the Community enjoys an exclusive or shared competence), it is necessary to take a close look at the scope of the competences connected to the different areas of the CFP.

The Court declared in *Commission v. United Kingdom* that from the day of the expiry of Art. 102 of the Treaty of Accession (*i.e.*, 1 January 1979), the Community had the full and definite power to adopt, 'as part of the CFP, measures relating to the conservation of the resources of the sea. Thus, the Member States are no longer entitled to institute any of their own conservation measures in the waters under their jurisdiction. The adoption of such measures, with the restrictions which they imply as regards fishing activities, is a matter, as from that date, of Community law'. In this context, the term 'conservation' includes the allocation of resources by the Community. The Court thus explicitly excluded Member States from the adoption of conservation measures in maritime waters. Insofar, the Community's competence should here be referred to as being exclusive. However, the ECJ has also confirmed that Member States have the power to fill lacunae in Community conservation legislation. Of course, the Community retains the power to fill existing lacunae, making measures taken by Member States void where it considers it appropriate. In the areas of structural policies and market organisation, Community competences are not exclusive but far reaching.

Executive powers primarily lie with Member States. In reality, Member States' implementation and enforcement rules and practices differ. Control and enforcement has not always been taken seriously, and the strength of the responses to violations of fisheries laws vary. In response, the Community has made efforts to promote the alignment of national measures to be taken in response to violations. The CFP provides a system of 'dual vigilance'. While the Member States ensure control, inspection and enforcement and provide for the judicial or administrational response to infringements of fisheries legislation, the Commission, on behalf of the Community, controls national compliance efforts and tries to ensure that they are carried out properly.

Community trade policy with regards to fisheries products basically concerns (a) trade between Member States and (b) trade between Member States and third countries. The division of competences regarding trade in fisheries products is primarily determined by the Treaty provisions establishing and governing the internal market as well as those provisions concerning the customs union. Regarding trade between Member States, Article 32(2) provides that 'save as otherwise provided in Articles 33 to 38, the rules laid down for the establishment of the common market shall apply to agricultural products'. Thus, provisions on the free movement of goods and those prohibiting quantitative restrictions determine the intra-Community trade in fisheries products. These rules abolished custom duties, charges having an equivalent effect as well as quantitative restrictions on fisheries products from other Member States, and prohibited Member States from re-introducing similar measures. Moreover, according to Art. 90 of the Treaty, Member States shall not, directly or indirectly, impose any internal taxation of any kind on fisheries products of other Member States in excess of that imposed directly or indirectly on similar domestic products, nor is it permitted to institute any internal taxation so as to afford indirect protection to other products. Finally, Member States are not allowed to circumvent the Community competition rules in Arts. 81 to 86 as well as respective implementing measures (adopted under Art. 83). The same applies to secondary legislation under Arts. 14 and 95, adopted to further the creation of the internal market. Regarding trade between the Community and third states, the Community has the competence to fix a common custom tariff and adopt its own export policy. The common custom tariff can be fixed either (a) autonomously by the Council (Art. 26 of the Treaty), (b) in the course of tariff and trade agreements with third states (Art. 133 of the Treaty), or (c) in association agreements with third states (Art. 310 of the Treaty). The Community has established integrated tariffs for the import of fisheries products. Member States must refrain from fixing individual tariffs and amending existing Community tariffs.

To demarcate *horizontal competences*, the ECJ will consider the aims, objectives and content of measures in question. The Court gave its view of the specific content of the fisheries policy in the *Mondiet* decision where it had to decide whether a certain measure had to be based on environmental or fisheries Treaty provisions. The ECJ's *Mondiet* decision makes it clear that the Court sees the content of fisheries policy as closely linked to fisheries management and the objectives laid down in the Treaty provisions on agriculture, *i.e.*, regulating fishing activities to prevent exploitation and ensure conservation, 'ensuring rational development of production' and 'assuring availability of supplies'. The decision also indicates that as long as Community measures relate to fishing activities, the structure of the production sector, and market organisation of fisheries products, they must be based on the Treaty provision on agriculture and fisheries.

Even though the Community under primary law has an exclusive competence with regards to the conservation of marine resources and comprehensive powers regarding the governance of the fisheries sector, Member States hold certain administrative and legislative powers. Dogmatically, these powers partly derive from (a) the re-delegation of Community activities in this area to Member States, (b) the fact that matters have intentionally or unintentionally remained unregulated by the Community, and (c) from the Member States' competence to adopt measures on environmental protection.

Member States are allowed to take so-called 'emergency measures' in waters falling under their sovereignty or jurisdiction. They also have the right to take measures for the conservation and management of fisheries resources and to minimise the effect of fishing on the conservation of marine ecosystems within twelve nautical miles of their baseline. These measures may, in principle, apply to all fishing vessels. However, they are allowed to do so only where the Community has not adopted specific measures addressing conservation and management in this area. Measures taken by Member States have to be non-discriminatory, compatible with the objective in Art. 2 of the Regulation 2371/2002, and must be as stringent as existing Community legislation. Member States are also entitled to unilaterally adopt conservation measures for the conservation of stocks in waters under their sovereignty or jurisdiction (*i.e.*, within their territorial sea and their EEZ). However, measures must apply solely to fishing vessels flying the flag of the particular Member State and registered in the Community. In addition, Member States may restrict fishing to vessels that 'traditionally fish in those waters from ports on the adjacent coasts' within their territorial sea. They determine the method of distributing the assigned fishing opportunities to vessels being registered within their territory and which fly their flags. In the area of market organisation, their competences are limited to the financial and administrative support of producer organisations and the ability to ensure they fulfil their tasks. Member States are, however, primarily responsible for the implementation of Community structural measures and the determination of how aid is allocated. This procedure is has been referred to as the principle of 'shared responsibility'. The Community has wilfully refrained from regulating special issues such as non-commercial fisheries and sweet- and inland fisheries.

All in all, Member States retain some competence to alter fisheries regulations for environmental purposes. However, they should neither unilaterally alter the Community principle of equal access, nor regulate access for fishers from other Member States within waters under their sovereign control or jurisdiction, if this would reduce the effectiveness of the existing conservation, distribution and access system provided in the basic regulation. Following this, one option might be to oblige Member States to co-ordinate their legislative processes (in accordance with all relevant environmental laws) with the Community and any affected Member States. Another option would be to fully confer the power to adopt fisheries regulations for the purpose of environmental protection on the Community. In such a case, the Community would be required to take into consideration and to respect Member States' environmental objectives and obligations when regulating fisheries activities.

The Community has established two different kinds of political Committees that allow stakeholders to participate in the CFP legislative process: the Consultative Committee on Fisheries and Aquaculture ('ACFA') and the Regional Advisory Councils. These committees grant stakeholders participation rights within the CFP legislative process. However, these rights are limited to the pre-decision phase, as both committees are only *consulted* by the Commission. No provision has been made for compulsory consultation and the advice and opinions provided have no binding legal effect.

3 Management – failures and success

The following section will outline the main insights concerning the Community's management regime, *i.e.*, its conservation, control and enforcement measures as well as its regime on small-scale coastal fisheries.

3.1 Conservation measures

The annual setting of TACs is the main management instrument of the CFP. Catch limitations are imposed for the commercially most important stocks and divided among Member States according to the principle of relative stability. This allocation formula guarantees each Member State a specified percentage of the TAC permitted that year. The static allocation formula is, in part, eroded by fishers' practice of relocating to other Member States to gain access to new sources of quota. Member States apply their own method for distributing their quota among their fishing operators. They decide whether fishers can exploit, market or transfer their individual quota.

There has been much criticism of the use of TACs as the central management instrument. Setting TAC requires complex and expensive scientific determinations and thus many stocks remain underassessed. TACs are also often considered as unsuitable for application to the Community's many multi-species fisheries. Ultimately, TACs have proven difficult to enforce. Another critique of the CFP management system is that despite the sustainability criteria laid down in the basic Regulation, the Council often favours short-term economic gain over long-term conservation (and economic) interests. Thus, the Council frequently adopts TAC above exploitation rates recommended by scientists.

The 2002 Reform upvalued effort restrictions as a management instrument, as a supplement to TACs in certain areas. The adoption of an additional management instrument to improve existing measures is a step forward towards an effective fisheries management. However, like TACs, effort limitations rely on expensive biological information on stocks, which are costly to provide. Besides, it is very difficult to precisely establish the fishing effort of a fishing vessel and to convert it to a specific rate of fishing mortality. Ultimately, the only real advantage they have over TACs is that they are easier to enforce. It remains to be seen whether the adoption of effort limitations will actually improve the TAC regime.

Technical measures such as gear restrictions, fishing zones and periods, and minimum fish sizes have also been adopted. Such measures have been criticised for being too complex to apply and difficult to control. Furthermore, the limits on mesh sizes have, in some cases, been criticised as being too small to protect juvenile fish.

It was proposed here that Council's discretion in setting TACs (or adopting other management instruments) is fettered by the sustainability criteria and the precautionary approach in Art. 2(1) of Regulation 2371/02. The approach set out under Art. 2(1) and 3(i) requires the Community to take management measures *to conserve stocks* even in the absence of adequate scientific information. The meaning of conservation under the basic regulation is, as described above, determined by the sustainability criteria. The latter requires that catch restriction should, in principle, prohibit exploitation beyond MSY-levels; it is only where there are compelling socio-economic reasons that MSY-levels may be exceeded. However, catch levels must be maintained *at a minimum* within safe biological limits. These two reference points, MSY-levels in normal circumstances and safe biological limits in exceptional ones, also serve as the reference points for the application of the precautionary approach. As a result, a *safety margin* or *precautionary buffer* that takes into account the unpredictable nature of stock levels and uncertainties in assessment must be factored into the sustainable exploitation rate (which respects the MSY- and F*lim*-level). This applies in particular to underassessed stocks. Under the CFP, TACs for such species are based on 'intelligent and educated guesswork'. In these cases, safety margins must be particularly wide, which means that exploitation rates may not exceed the limit that scientists consider to be *clearly sufficient* to protect stocks from falling below the MSY- and, under exceptional circumstances, F*lim*-level.

The Council's frequent practice of setting TACs above sustainable levels has also not substantially changed under the multi-annual approach to fisheries management, introduced in 2002. Article 5 of Regulation 2371/02 does not establish a firm obligation to ensure overexploited stocks return to safe biological limits. Not surprisingly, the Council's recovery plans have already been severely criticised and have even led to an action before the ECJ. It is claimed that the initial fishing mortality limits adopted under the cod recovery plan are much too high to put an end to overexploitation, and that they are inconsistent with the precautionary approach. Furthermore, in proceedings before the ECJ, the WWF has claimed that the Council had not adhered to its own recovery plan by subsequently adopting fishing rates well beyond the limits established in the plans. The Council's ability to change exploitation rates erodes one of their main advantages, namely, to increase planning security. Finally, the goal to promote the use of effort restrictions within recovery plans to increase their effectiveness has been weakened by the broad wording of Art. 5(4), paragraph 2 of Regulation 2371/02.

This report also states that Community fisheries management in the Mediterranean is quite distinct from that of other Community waters. Most fisheries are classified as artisinal or small-scale fisheries; and only Spain and Malta have established fishing zones. With the exception of Tuna, the Community does not adopt TACs for Mediterranean fisheries, instead relying on gear and area restrictions to maintain fish populations at acceptable levels. Given that CFP measures only cover a comparatively small area in the Mediterranean, GFCM management is of major relevance to Community fisheries. The GFCM has made recommendations, for example, in regards to TAC for tuna and minimum mesh sizes. Members of the Commission maintain the right to object within 120 days from the date of notification to a recommendation. In such a case, the objecting party is not obliged to give effect to the recommendation. The newness of this regime to fisheries governance in international waters means it is too early to evaluate CFCM practices and their effectiveness in the Mediterranean. However, the Commission has stated in 2002 that the scientific advice on Mediterranean fisheries is insufficient and that the state of many fisheries stocks in the Mediterranean is critical.

3.2 Control and enforcement measures

The competence to control and enforce fisheries rules basically lies with the Member States. The Community, however, controls the application, control and enforcement of laws made pursuant to the CFP by Member States. It has also adopted a comprehensive control regime which requires Member States to fulfil specific requirements. These requirements concern all fishing and associated activities, including the control of management, structural and market measures.

Member States are required to monitor fishing activities of vessels exceeding 15 metres in total length by satellite vessel monitoring system. Member States must also ensure that catching activities are documented in detail. To this end, Community laws require masters of vessels to document catches in logbooks and submit landing declarations to the local authorities. In addition, persons authorised to market fishery products are obliged to submit upon the first sale, sales notes, etc. Other control requirements concern:

- the monitoring of effort limitations;
- Community vessels' fishing activities in third country waters and on the high seas;
- monitoring of third country vessels in Community waters;
- the control of technical measures;
- the control of structural measures;
- the control of the market organisation.

Throughout the history of the CFP, Member States have often shown themselves to be unable or unwilling to fulfil these obligations. A uniform and coherent application of the Community fisheries regime is, nevertheless, a prerequisite for effective fisheries management. Against this background, the Community has developed a particular interest in monitoring Member States' control and enforcement practices. The Commission controls the application, control and enforcement of national laws made pursuant to the CFP by means of inspections, audits, inquiries, verifications carried out with the help of 24 inspectors. Where Member States do not properly apply, control and enforce CFP rules, the Commission may initiate infringement procedures before the ECJ, as well as take preventive measures, cut financial aid or, in case of quota overfishing, reduce a violating state's future quota. The Commission is willing to make use of its competences. It has, for example, recently reduced the quotas of Ireland and the United Kingdom in the upcoming years for exceeding their given quotas. With regards to infringement procedures, the ECJ is also increasingly willing to sanction Member States for transgressing Community fisheries law. In 2005, the ECJ decided a groundbreaking case. The Court found in 1991 that France had failed to enforce Community technical measures for the conservation of fish resources. Subsequently, the Commission claimed that France had not complied with this judgment and referred this matter, again, to

ECJ. As a result, the Court ordered France to pay the Commission a lump sum of €20,000,000, as well as an additional penalty payment of €57,761,250 for each six-month period from the delivery of the 2005 judgement on, in which France has failed to comply. To increase transparency, the uniform application of the CFP and an effective and co-ordinated use of available control capacities, Member States are increasingly required to co-operate on control and enforcement. In addition, the Community has established a common Fisheries Control Agency. Overall, however, control and enforcement remains a big problem under the CFP. The yearly Commission Scoreboards explain that, as it currently stands, the chance of being caught for violating the rules is very low. Moreover, sanctions imposed by Member States by and large do not serve as an effective deterrent. The Commission has stated that among Member States there 'remains a general lack of understanding of the relationship between compliance, risks, inspection activity and the effect of follow-up, including sanctions'. Ultimately, given its limited sanction competences and personnel capacity, it is difficult for the Commission to effectively control and promote the implementation of CFP measures by Member States.

3.3 Small-scale coastal fisheries

The report also examined whether fisheries management under the CFP takes into account the needs and potentials of small-scale coastal fisheries. These fisheries face difficulties that are similar to those of small-scale coastal fisheries in other parts of the world, such as immobility and difficulties in promoting their interests at a political level. This report suggests that under the current CFP management regime the needs of small-scale fisheries are not sufficiently considered. Community rules entitling Member States to take certain measures within their 12 nm zones do not impose a clear duty on national legislators to protect and promote the interests of small-scale fisheries. The same applies to the rules on the composition of the RACs. The ICZM measure is non-binding. If the Community really wants to promote small-scale fisheries, unequivocal and binding obligations are necessary. Existing measures could be modified to achieve this. First, Art. 17(2) of Regulation 2371/02 could be amended: instead of restricting fishing to vessels that traditionally fish in the 12 nm zones from entering the ports of adjacent coasts, Member States should be required to restrict fishing in these areas to small-scale vessels. Second, the Community should initiate a debate on the use of specific management instruments like exclusive fishing areas and long-term use rights in Community's small-scale fisheries. Third, the RAC operating rules should be amended to *guarantee* that small-scale fishers are included within RAC assemblies and committees.

3.4 Finding the right combination of management instruments

The development of the CFP's management regime does not necessarily allow us to assume that there is an 'ideal type of management measure' or the *'right combination of management instruments'* which can equally apply to all fisheries. To this day, the Community has clearly not managed to find 'the perfect approach' to protect stocks from overexploitation. The performance of single or combined management measures and instruments depends on innumerable factors that vary from fishery to fishery. Community fisheries management measures regulate the utilisation of the fleet (e.g., space, time, power, gear restrictions), the outtake of fish resources (e.g., TACs), as well as the initial limitation of fishing power (e.g., structural measures). The case of the CFP makes clear that in many cases management measures are only effective where they are supplemented by other instruments. For example, where TACs are not complemented by measures protecting nursery grounds or requirements on minimum-mesh-size to protect young fish, the reproduction of stocks would seriously be endangered. Moreover, where the initial number of vessels is too high, TACs, TAEs and technical measures are difficult to enforce. It is suggested here that the use of a single management instrument cannot be evaluated in isolation from other existing management measures. Success or failure of management systems in most cases depends on the right combination and co-ordination of different management measures. The CFP refers to TACs as its main management instrument. As described above, this choice, however, also created major problems. In response to these problems, the Community has continuously tried to improve and supplement its TAC-system by introducing input restrictions (TAEs and fleet reductions), reinforcing technical measures, and fortifying its control and enforcement obligations. The Community has faced different challenges with its adoption of supplemental measures.

The Community has also encountered difficulties as a consequence of introducing TAEs. As stated earlier, it is particularly difficult to determine precisely the actual fishing power of a fleet and then translate this power into fishing mortality. In addition, where TAEs are adopted in combination with TACs, scientific advice must provide the basis for the integration of these two systems. Given these difficulties, Community law requires that where TAE-quotas are not sufficient to fish up the assigned TACs, Member States are entitled to file a request to the Commission to modify the effort levels (to enable fishers to fish up to given TACs). Second, it is difficult to assess vessel engine power and the overall fishing potential of a vessel and these measurements are furthermore susceptible to fraud. These challenges have prevented the adoption of an effort regime before 2002.

Technical measures mainly aim at limiting the impact of fishing on marine ecosystem as well as on juvenile and non-target species. Adopting TACs or TAEs without imposing additional technical requirements would ultimately result in the indiscriminate exploitation of all marine resources, without recognising the specific character of the surrounding ecosystems, the conservation requirements regarding non-target species, and the reproduction patterns of fish stocks. Due to the specific features of the different marine areas of exploitation, the Community has adopted several tailor-made technical regulations covering these different areas (e.g., technical measures governing the Baltic Sea and the Mediterranean).

Further important lessons can be learned from the CFP are that management measures need to be coherent with each other and all measures must aim at sustainable exploitation. If one management measure allows for unsustainable fishing practices, the other measures are likely to be ineffective. The CFP provides at least three important examples in this regard. First, where the overall quantity of fish that can be taken from the waters is unsustainable, supplemental measures are unlikely to improve conservation. The Council's practice of setting TACs at unsustainable levels for twenty-five years is a key reason for the current state of fisheries resources in the European Community. Second, since the late 1970s, the Community's structural policies allowed too many vessels to fish in Community and third country waters. The CFP's persisting overcapacities have maintained fishing pressure at unsustainable levels and control and enforcement measures remain largely ineffective. In addition, huge investments in the sector have created political pressure to uphold high exploitation rates and subsidies programmes, which, again, have maintained overcapacities. Third, the Community's technical measures have been criticised for allowing fishing under conditions that harm the marine environment or the reproduction fish stocks, as well as for being too complex to be implemented, and too difficult to monitor. As a consequence, the overall impact of technical measures and TACs and TAEs is substantially reduced.

4 Promotional measures

Up to the early 1990s, promotion under the CFP aimed at increasing production. However, in actual fact, aid to increase production capacity has continued to exceed support for decreasing capacity until 2004.

In the early phase of the CFP, promotional measures aimed solely at increasing production. Grants for fleet renewals, the modernisation of vessels, the improvement of fishing ports, the cod-fishing sector, aquaculture projects and the processing and marketing sectors were introduced. After 1980, the Community began to finance third country access agreements. Aid was also provided for the stabilisation of markets.

From 1983 to 1986, the Community continued on this course. New types of promotional measures were introduced, such as marketing support for fisheries products, assistance for exploratory expeditions and joint ventures, and the promotion and exploitation of under- or unexploited fish species. Special financial grants funded the expansion of production capacity at the Mediterranean coasts. New access agreements were concluded with Guinea, Equatorial Guinea, Sao Tomé and Principe, Madagascar. Nonetheless, for the first time aid was provided for the temporary laying-up of vessels.

In the late 1980s and the early 1990s, the problems of resource decline and fleet overcapacities became obvious. In response, the Community adopted rules to reduce fleet size. This was the start of the contradictory spending practices for fleet increases on the one hand, and decreases on the other. Participation rates for the renewal and modernisation of the fleet were substantially lowered in 1986 and 1990. However, the total amount of aid granted for such purposes was raised substantially. In addition, subsidies which were not directly related to the fleet, but which also increased pressure on resources were raised, and support for processing and marketing as well as investments in fishing ports was bolstered. Moreover, the Community concluded new access agreements with third countries. From an overall perspective, aid granted for purposes that tend to increase production power, exceeded those amounts earmarked for conservation purposes.

From 1994 to 1999, the Community set more ambitious goals for the reduction of fishing effort. Accordingly, aid for the fleet adjustments was raised. Grants were provided for permanent reductions through scrapping vessels, the permanent re-assignment of vessels for other non-fishing purposes, and the permanent transfer to third countries. The Community also became increasingly aware of the need to offset the social consequences of the restructuring process. Assistance was made available for early retirement schemes and compensatory payments to fishers who withdrew from fishing permanently. Aid was granted to support the fisheries sector during the restructuring process. However, despite these positive developments, aid, which tends to increase pressure on marine resources, still greatly exceeded the amounts granted for sectoral adjustments.

The CFP reform in 2002 was particularly important to structural policies, and thus to promotion. The new management regulation replaced the complicated and ineffective system of MAGPs and permanent arrangements for monitoring fleet renewals with a simpler entry-exit scheme. A ceiling on the size of Member States' fleets was introduced; mandatory reduction targets were abolished. Assistance for the construction of vessels was phased out by the end of 2004, and aid for modernisation was to be granted only where it did not lead to an increase in fishing capacity. Extra funding was provided for the scrapping of vessels. Additionally, inland fisheries and environmentally friendly fishing methods have been increasingly promoted. More aid was granted to the fisheries sector to compensate for losses resulting from fishing moratoria and declining quotas. Premiums for the permanent transfer of vessels to third countries were phased out in 2004.

In general, this course of action has been maintained under the EFF. The EFF supports fleet exits and continues to refuse aid for the construction of vessels and the increase of fish holds on fishing vessels. It also provides for the adjustment of fleet parts that target the most exploited stocks. In addition, it increasingly integrates environmental objectives into its plan. Another important development is its promotion of local development strategies in areas heavily dependent on fisheries. It is particularly noteworthy that these strategies are set up and implemented by local stakeholders themselves. Despite these advances, some problems remain. The Council decided to grant some aid for engine replacement, subject to the condition that the vessel's ability to catch fish is not enhanced by such improvements. However, although these subsidies do not increase fishing capacity, they do create an incentive for fishers to remain in the business. Furthermore, the Community continues to subsidise fishing ports, landing sites and shelters and develop markets and promotional campaigns. The processing industry also receives financial support. Finally, in 2007, the Commission substantially raised the amount of national state aid, which can be provided by Member States without its consent, from €3,000 to €30,000 of aid per beneficiary per three-year period.

Recent developments under the WTO may affect the near future of CFP promotional policies. In November 2007, the WTO group on 'Rules' issued a draft proposal for a revised version of the Subsidies and Countervailing Measures Agreement. The draft agreement largely prohibits subsidies that contribute to maintaining or building overcapacities.

5 The integration of promotion and management

The integrative approach presented here is based on the assumption that a sustainable exploitation of marine capture resources ultimately depends on the level of available fish stocks and that the open access conflict inclines fishers to invest subsidies into to the fisheries sector, *i.e.*, they fuel the race to fish. Integration thus requires four times (a) assessment and (b) implementation or adjustment. First, the level of sustainable exploitation shall be identified and implemented. Second, the impact of the production sector on fish stocks shall be assessed and production capacity adjusted to levels required to attain sustainable exploitation. Third, taking into consideration the outcome of the first two steps, the effects of subsidisation schemes shall be assessed and adjusted to attain sustainable exploitation patterns. Fourth, the first three steps shall be repeated regularly and adjustments made where necessary.

Many practical lessons could be learned by tracking the development of the integration process under the CFP.

From 1970 to 1982, integration was low. Member States did not adopt and implement sustainable TACs as recommended by scientists, nor did they consider the impact of the production sector on the state of the resources. Accordingly, promotional measures did not aim at reducing of production capacity so that it was in line with sustainable management objectives; instead, they had the effect of expanding production power. The combination of an under-developed management regime and promotion measures lead to an enormous increase in fleet overcapacities.

Integration of the promotional and management regimes remained low throughout the second phase (1983 to 1986). The Council set TACs above sustainable levels. Nevertheless, the Community slowly began to attempt to link the promotion of the fleet to the available resources. Under MAGP I, Member States were required to maintain capacity at 1983 levels. Ultimately, however, the effort to adjust promotion to reduce fleet capacity failed for various reasons. From 1983 to 1986, those subsidies that tend to increase production greatly exceeded those that decrease fishing pressure. As result, from 1983 to 1987, the Community fleet had increased enormously.

From 1987 to 1993, integration of promotional and management strategies was still insufficient. Management had not earnestly committed to limiting exploitation; and even though the Community increasingly tried to adjust fleet to correspond with available exploitation opportunities, overall promotion actually lead to an increase in production capacity. In 1992, a scientific report estimated that fishing power would have to be reduced by approximately 40% in order to adjust fishing mortality to reflect available resources.

Integration of promotional and management policy improved only to a marginal extent during the fourth phase (1994 to 1999). Under the management regime, the Council had continued to set TAC levels above those recommended by scientists and new managing instruments were not used. On the other side, to improve the development and adjustment of the fisheries sectors, Member States now had to assess their sectors thoroughly under the so-called Community Programmes. In addition, reduction goals under MAGP III and IV were more ambitious than under the two prior programmes. However, the Community continued to advance subsidies for investments to increase fishing and production power. In 2001 and 2002, the Commission stated that fishing capacity of the Community fleet far exceeds that required to harvest the available fishery resources in a sustainable manner.

From 2000 to 2006, integration of promotional activities and management efforts has been substantially enhanced. Even though the Council continued to set fishing rates above sustainable levels, it seems inclined to *gradually* accept cut-backs to exploitation opportunities. Several multi-annual plans have been adopted throughout the fifth phase and effort limits are increasingly used to manage fisheries. Besides, The Community simply stopped its ineffective attempts to make Member States reduce their fishing capacity through mandatory reductions; instead, it has established ceilings on the size of Member States fleets and attempted to manage fishing effort. The fleet entry-exit scheme provides an improved and relatively simple means to link promotion to management objectives. Finally, promotion of the most harmful to the management efforts has been stopped. Money has been made available to decrease capacity and to buffer the socio-economic effects of management measures and fleet reductions suffered by fishermen. By granting aid to fishers who leave the business due to the adoption of management or recovery plans, aid is channelled to the reduction of fleet segments that target the most exploited fish stocks. In addition, public support is increasingly channelled to environmentally more friendly fishing and inland fishing. However, still the Commission is of the opinion that the reductions of the fleet are 'insufficient' and 'too modest'.

The new EFF, which will be applicable from 2007 to 2013, even improved integration. It does not provide funding for the *most harmful* subsidies and aid is made available to support fishers where they are affected by the reduction of fishing under the management regime. In addition, participation and transparency have been improved by including local stakeholders in the strategic planning of sector developments. It also increasingly supports environmentally sustainable fishing and aquaculture methods. However, in 2007, the Commission substantially raised the amount of national state aid which can be provided by Member States without its consent from €3,000 to €30,000 of aid per beneficiary per three-year period.

The integration process provided many practical lessons, which can be of value for policy makers in the future.

Policy makers must keep in mind that while it may be rather simple to agree politically on the build-up of a fishing fleet, the creation of jobs and the distribution of financial grants, it appears far more difficult to reduce existing fleets, create unemployment and reduce existing benefit schemes.

Another lesson to be gleaned from an examination of the development of the CFP is that an effective integration of promotion and management approaches requires appropriate scientific assessment and input. Scientific input is not only necessary for establishing safe exploitation limits, but must take into focus the effects of promotional measures on production sectors and the fisheries resources.

It is also important that any new programme to reduce exploitation has to be simple, precise and binding. The Community currently manages fleet entries and exits by means of a simple and binding system that uses fleet reference levels and a clear and simple reduction formula. This approach, introduced in 2002, is regarded to be more effective than earlier approaches.

The experience of the CFP also teaches that the competitiveness of the production sector should not be increased by so-called 'contrary spending'. Contrary spending grew out of the desire to pursue two opposing objectives: increase competitiveness and efficiency of the production sector, on one hand, and reduce the fleet size, on the other. As a result, for a long time the Community was financing both production increases and decreases. Aid for fleet renewals and fleet modernisation, for example, have neutralised aid for the adjustment of the sector and contributed to maintaining overcapacities. Contrary spending has been an enormous waste of public money and increased pressure on fisheries resources. Gains in efficiency and competitiveness must be achieved through other means.

An integrated approach would include grants for social and environmental purposes. The Community, for example, has provided financial aid to buffer the social consequences of limiting fishing opportunities or unforeseen natural occurrences.

It has also been learned that promotion schemes can negatively affect third country fisheries. For a long time, the Community 'externalised' its overcapacity problem by financing the conclusion of international agreements which allowed Community fishers access to third countries waters. In different cases, this has led to the shifting of the overfishing problem to these third countries. Whether the Community's current approach to fisheries access agreements will actually be effective remains to be seen.

Other factors that contribute to an effective integration have been identified such as transparency, public participation, and the fact that it is not only the catching segment of the fisheries industry which creates pressure on the marine resources. The relationship between *all factors* that contribute to rising exploitation have to be considered. This may include the production sectors' *processing* powers as well as *consumption patterns.*

Finally, to date, the Community has failed to adopt necessary measures which try to steer consumers towards sustainable products. It has been suggested that Community's should be to provide support for the establishment of ambitious labelling schemes that give credible product information, make public aid partially conditional upon sustainable production, and decide to launch marketing campaigns to foster the idea that sustainable products are value-added.

6 Access to justice

Community institutions and Member States have direct and privileged access to the European Court. However, since Community institutions and Member States are involved in the procedure leading up to the adoption of CFP legislation, they are unlikely to challenge the legality of 'their own' measures. Therefore, it is important that individuals and interest groups are allowed to have CFP measures of general application judicially reviewed. At present, however, their access to the European courts is rather limited. Direct access is granted only under the very strict conditions, established in Art. 230, para. 4 of the Treaty (as interpreted by the Court); in practice it is rarely granted. To have standing, applicants must either be addressed or be *directly* and *individually* concerned. However, many important measures adopted under the CFP are not of *direct concern* in the sense that they are not self-executing and that the institutions which implement them, have a discretion. In addition, the ECJ has devised the *Plaumann-formula*, which must be met for a claimant to be *individually concerned* by a measure that is not addressed to him; according to this test, the claimant has to be affected in a way which differentiates him from all other persons. The court has allowed very few exceptions to the *Plaumann* formula for individuals and interest groups representing private and *particularly* environmental interest.

As it is difficult for natural and legal persons as well as interest groups to meet the test for direct access pursuant to Art. 230, para. 4 of the Treaty, the main approach used by private entities to challenge CFP legislation is through requests for preliminary rulings by the national courts under Art. 234. However, indirect access does not necessarily mean a party has full access to the Community courts. For example, claimants cannot decide themselves whether a reference will be made to the ECJ. Instead, they must convince the national court to refer the case. As a result, the system has been widely criticised for not providing effective legal protection against Community legislative injustice.

It has been observed that recent developments in international and Community law should broaden the rights of fishermen, interest groups and environmental groups to access European courts. This is particularly true with a view to Arts. 9(3) and 9(4) of the Aarhus Convention. Besides, under Community law the establishment of RACs increasingly involves environmental organisations in the legislative processes under the CFP. It has been argued that these developments should be mirrored in the Court of Justice's interpretation of the 'direct and individual concern' criteria under Art. 230, para. 4, of the Treaty. However, in a recent decision, the CFI has decided that these developments do not influence the Court's interpretation of Art. 230, para. 4 of the Treaty. Ultimately, it became clear that there might be cases where an institutional imbalance within the CFP legal framework exists. The Commission and the Council may violate existing Community or international law by adopting flawed CFP legislation and by refusing to grant the right for interest groups or interested persons to have judicial review. To uphold the rule of law and to give real effect to the Community's legal obligations under the CFP and the Aarhus Convention, Community institutions should consider the following options. First, the Member States could amend the Treaty and establish procedural rights for individuals and interest groups in fisheries conservation and management matters. However, the text of the Lisbon Treaty, which has not yet been ratified, indicates that such an amendment will not occur in the near future. Second, the Court should reconsider its restrictive interpretation of both the terms 'directly and individually concerned' under Art. 230, para. 4. Third, the Council should explicitly grant individuals and their representing organisations 'qualified procedural guarantees' within CFP legislation as defined in the *Greenpeace* decision.

7 Legal review of measures adopted under the CFP

CFP measures may not be compatible with existing rules, principles, objectives, and guidelines provided by international or Community law. The following chapter will summarise some answers to current questions on the legality of CFP law.

7.1 Compatibility of the CFP with international management requirements

The Community is party to several binding international conventions relating to fisheries. This book has made clear that international law on fisheries is often vague or does not provide for actions to be undertaken if its requirements are violated. However, it also became clear that, in many cases, international measures indicate management thresholds, objectives, guidelines, and best practices against which Community laws and their implementation can be assessed.

Both UNCLOS and FSA restrict fishing by establishing quantitative catch restrictions. The prescribed quantities, however, are matter of dispute.

UNCLOS does not explicitly require coastal states to conserve or manage fisheries resources in their territorial sea; it merely *entitles* coastal states to adopt laws and regulations on the conservation of the living resources of the sea. With a view to the conservation requirement for fisheries resources located in the EEZ, it has been argued that the interpretation of UNCLOS provisions must be responsive to the changing circumstances in (international) fisheries. Accordingly, Art. 61 of 1982 UNCLOS must be read to oblige UNCLOS parties to adopt fishing opportunities at MSY-levels. They may exceed this level *only for compelling* social and economic reasons, which must be clearly predefined and thus open for judicial review in accordance with national fisheries laws. Exploitation levels must not cause stocks to fall below safe biological limits, *i.e.*, to levels where they are ultimately threatened by overexploitation. Given that UNCLOS provisions on high seas fisheries concern, in many cases, the same resources, conservation requirements should be interpreted in the same way.

The FSA requires its parties to adopt a system of reference points. These so-called *limit reference points* should use the MSY as a minimum standard for determining exploitation rates. These limits must not be exceeded for socio-economic reasons. Though the FSA does not explicitly require specific actions if limit reference points are exceeded, it does require that states must take predetermined management measures to ensure that such levels are not exceeded, or, if they are, take measures to restore respective stocks.

In fact, many important stocks are currently exploited outside safe biological limits. If the 1982 UNCLOS and FSA conservation requirements are interpreted as suggested here, Community *exploitation rates* (as laid down in Council regulations) which exceed safe biological limits as well as the *continued practice* of managing stocks below the MSY level must be regarded as a violation of both UNCLOS and the FSA.

Where coastal states exploit marine living resources, 1982 UNCLOS and the FSA require them to consider non-target species and ecosystems. It has been found that despite discards and other problems concerning the impact of fishing on the surrounding ecosystems, Community fisheries measures do meet these very basic requirements.

International measures require the Community to collect, share and use scientific information on fish stocks exploited by Community vessels. Despite the fact that many stocks remain unassessed and scientific monitoring can always be improved, the Community has gathered extensive information on fish stocks and it readily exchanges such information with other interested states. Although, in many cases, it does not follow scientific recommendations, it does base its fisheries management decisions on the available data. Thus, it could be said that the Community adheres to the standards set at international fisheries law on collecting, using and sharing data concerning the fishing sector.

International law on control and enforcement can be found within 1982 UNCLOS, the FSA, and the Compliance Agreement. In light of these different obligations, the Community's fisheries control and enforcement system has been found to be comprehensive. Its control and enforcement system meets most international legal requirements: it requires prior authorisation for fishing; it has instituted a licensing system, a vessel record, a comprehensive catch control system, satellite based monitoring system, rules on transhipment, and all branches of the CFP are subject to extensive monitoring. However, the FSA and Compliance Agreement both require states to authorise fishing only where they are able to ensure *effectively* that fishers will comply with international conservation measures. Since its inception, the Community has struggled with guaranteeing fishers' adherence to conservation regulations. Fishers regularly transgress the rules and the application of control and enforcement measures by the Member States is often poor. Second, both FSA and the Compliance Agreement require flag states to adopt sanctions which 'are adequate in severity' or of 'sufficient gravity'. As explained in the above section on 'Community control and enforcement measures', under the current Community system, the sanctions put in place by the Member States are an insufficient deterrent.

1982 UNCLOS and the FSA require the Community to co-operate. At first glance, the Community appears to follow the international cooperation requirements. The Community is engaged in international co-operations regarding the protection of transboundary, anadromous stocks as well as marine mammals. It also provisionally applies the conservation measures adopted under the International Dolphin conservation Program (ICDP). However, the fisheries management practices of IFOs to which the Community is a party have been criticised. Many IFO regimes are regarded as deficient and their implementation ineffective. Despite this lack of effectiveness of many of the IFO's regimes to which the Community is party, the Community must be regarded as fulfilling its international cooperation requirements. As stated above, these mainly require state parties to enter into negotiations or co-operate through international organisations. The only case where states have a duty to co-operate concerns the abolishment of inconsistencies between high seas measures and the EEZ fisheries regimes. It must be kept in mind, however, that co-operation requirements all aim at realising those conservation requirements laid down in 1982 UNCLOS and the FSA. Thus, given that these conservation requirements are far from being attained, effectiveness of international co-operation must substantially be increased. Of course, this cannot be achieved by the Community alone.

Only two international legal instruments provide rules that are directly concerned with fisheries subsidies: the SCM-Agreement and the *non-binding* International Plan of Action for the Management of Fishing Capacity.

The non-binding IPOA for the Management of Fishing Capacity explicitly calls on the Community, first, to assess the effects of its subsidies on their fisheries sectors, and, second, to adjust its subsidisation practice to avoid excess fishing capacities. Since 1983, the Community frequently evaluates its subsidies schemes. Assessments have focussed on the effects of subsidies on fishing capacity. These assessments also have contributed to policy making and the Community increasingly adjusted its fleet subsidies schemes. However, it could be argued that focussing assessments mainly on the effects of subsidies on the fleets has neglected the effects of other types of subsidies. Besides, some subsidies granted under the EFF must still be regarded as problematic. Aid is currently supplied, for example, for engine replacements, fishing ports, landing sites and shelters and the promotion of the development of markets or promotional campaigns. Such aid reduces private sector production costs, creating incentives for fishers and producers to remain in or enter the business as well as consumers to buy fish.

Finally, the WTO rules (SCM rules) forbid both subsidising of exports of fisheries products, as well as promoting the use of domestic over imported goods. The SCM provisions also provide that parties to the convention must not, through the use of subsidies, cause adverse effects to the interests of other SCM Members. At present, however, the Community does neither promote exports nor the use of domestic fisheries products over foreign ones. Even though one may argue that subsidies to the Community's fisheries sector have adverse effects on the interests of other SCM Parties, such a claim, for the abovementioned reasons, has not been brought before a WTO panel. Therefore, until such a claim is successful, these subsidies are not prohibited under the SCM Agreement. In November 2007, the WTO group on 'Rules' issued a draft proposal for a revised version of the SCM rules. The draft agreement largely prohibits subsidies that contribute to maintaining or building overcapacities.

7.2 Fundamental rights implications for fishing limitations

It was found that fishing rights provided under CFP can be characterised as exclusive, but insecure (*i.e.*, revocable), non-durable and largely non-transferable use rights of a changing economic value. In general, limits on fishing impair the fundamental right to property and the rights to choose and practice a trade or profession protect the right of fishers to use their prior-owned fishing vessels, fishing gear and all other production facilities, as well as the right to choose and practice the fisheries profession. Given that fishing limitations are an appropriate and necessary means to attain the Community interest to sustain fisheries resources, this infringement is justifiable. With regards to the setting of quotas, the Council is obliged to adopt the least onerous limitation necessary to attain its conservation objectives. However, the Council has a wide margin of discretion in doing so. The Council can extensively limit fishing, *i.e.*, even to take less fish than could be taken to attain the MSY-level, particularly for ecosystem and precautionary reasons.

The fundamental right to property and the right to choose and practice a trade or profession also require that the Council actively protect fisheries resources. Given the open access nature of fisheries resources and the existing race to fish dilemma, it is very difficult, if not impossible, for fishers to establish sustainable exploitation levels themselves. They are reliant on a public authority to adopt and enforce an effective management regime. Given that the exclusive competence to protect fisheries resources exploited by Community fishers lies with the Community, the Council must establish such a regime to guarantee fundamental rights. The obligation to actively protect fisheries resources means that the Council must deviate from the current practice of setting quotas at unsustainable levels. Of all deficiencies of CFP legislation, the abolition of this practice is the first step to bringing fish stocks to sustainable levels. In light of the management instruments introduced in the 2002 Reform, it is doubted here that the currently existing measures are sufficient to attain this goal. A system should be established which forces the Council to adopt exploitation rates at least at MSY-levels. Under this policy, deviations could only be justified for compelling socio-economic reasons, which would have to be clearly defined in the law. The Council, however, could not exceed exploitation levels at which stocks are outside safe biological limits.

7.3 Commission quota deduction competence

Art. 23(4) of Regulation 2371/02 allows the Commission to reduce the future fishing opportunities of Member States where national quotas have been exceeded. However, Art. 23(4) does not precisely define the degree to which quotas can be deduced. Accordingly, a quota deduction measure adopted by the Commission in 2007 has been contested before the Court of First Instance. The applicants have argued in vague terms that the Commission lacked power to adopt the respective quota deductions on the basis of Art. 23(4) of Regulation 2371/02. To find out whether and to which extent the Commission is entitled to adopt quota reductions, the Community's competence to impose sanction measures had been assessed. First of all, it had been found that the adoption of Art. 23(4) of Regulation 2371/02 was lawfully based on Art. 37 of the Treaty. Art. 23(4) of Regulation 2371/02 and the Commission's implementing measure also conform to the principles of proportionality and subsidiarity.

Considering the persevering implementation deficits under the CFP, quota reduction measures are obviously necessary to improve the implementation and overall functioning of the quota system. In principle, the deduction of future fishing opportunities is a proportionate measure, which is necessary to ensure the functioning of the quota system. Given that legal consequences are somewhat vague under Art. 23(4) of Regulation 2371/02, proportionality depends on the amount of fishing opportunities deduced by the Commission. However, the Commission's interpretation, *i.e.*, fishers do not loose any of the legally obtainable fishing opportunities, must be regarded as a proportionate action to uphold the functioning of the quota system. With regards to the subsidiarity principle, it has been found that since Member States are not able to ensure the proper implementation of the quota system by other Member States, a Community sanction power is to be regarded as compatible with the principle. Finally, the conferral of the quota deduction power on the Commission was in line with Arts. 202, para. 3 and 211, para. 4 of the Treaty. Quota reductions cannot to be regarded as essential elements of subject matter of the CFP and the Commission's sanction powers are limited by a management committee.

Bibliography

Alverson, D.L., Freeberg, M.H. and Pope, J.G., *A global assessment of sisheries bycatch and discards – FAO Technical Paper No. 339*, (Rome: 1994).

Appelbaum, B. and Donohue, A., 'The Role of Regional Fisheries Management Organisations', in E. Hey (ed.), *Developments in International Fisheries Law*, (The Hague: 1999).

Arnason, R., 'Property Rights as Means of Economic Organisation', in R. Shotton (ed.), *Use of property rights in fisheries management*, (Rome: 2000).

Balton, D.A., 'Strengthening the Law of the Sea: The New Agreement on Straddling Fish Stocks and Highly Migratory Fish Stocks' (1996) 27 *Ocean Development and International Law*, pp. 125-151.

Banks, R., 'Susbidising EU fleets: capacity reduction or capital subsidisation', in A. Hatcher and C. Robinson (eds.), *Overcapacity, overcapitalisation and subsidies in European fisheries*, (Portsmouth: 1998).

Barents, R., 'The internal market unlimited: some observations on the legal basis of Community legislation' (1993) *Common Market Law Review*, 30, pp. 85-109.

Barnes, R., 'The Convention on the Law of the Sea: an Effective Framework for Domestic Fisheries Conservation', in D. Freestone, R. Barnes and D.M. Mong (eds.), *The Law of the Sea – Progress and Prospects*, (Oxford: 2006).

Baumeister, P., 'Effektiver Individualrechtsschutz im Gemeinschaftsrecht' (2005) 40 *Europarecht*, 1, pp. 1-35.

Becker, C., Spurzem, K.J., 'Brüsseler Rituale' (2005) 51 *Mare*, (available online at: www.mare.de).

Berg, A., *Implementing and Enforcing European Fisheries Law*, (The Hague: 1999).

Bertea, S., 'Looking for Coherence within the European Community Law' (2005) 11 *Euopean Law Journal*, 2, pp. 154-172.

Beyerlin, U., *Umweltvölkerrecht*, (München: 2000).

Bieber, R., Epiney, A. and Haag, M., *Die Europäische Union*, (Baden-Baden: 2004).

Bieber, R. and Salomé, I., 'Hierarchy of Norms in European Union Law' (1996) 33 *Common Market Law Review*, pp. 907-930.

Birnie, P. and Boyle, A., *International Law & The Environment*, (Oxford: 2002).

Bleeckmann, A., *Europarecht*, (Köln: 1997).

Böckenförde, E.-W., 'Grundrechte als Grundsatznormen' (1990) 29 *Der Staat*, pp. 159-199

Bogdandy, A. von, 'Art. 2 EGV' in E. Grabitz and M. Hilf (eds.), *Das Recht der Europäischen Union – Kommentar*, (München: 2000).

Bogdandy, A. von and Bast, J., 'Art. 5 EGV' in E. Grabitz and M. Hilf (eds.), *Das Recht der Europäischen Union – Kommentar*, (München: 2005).

Bogdandy, A. von, 'Art. 10 EGV' in E. Grabitz and M. Hilf (eds.), *Das Recht der Europäischen Union – Kommentar*, (München: 2005).

Bogdandy, A. von, 'Art. 12 EGV' in E. Grabitz and M. Hilf (eds.), *Das Recht der Europäischen Union – Kommentar*, (München: 2005).

Bogdandy, A. von and Bast, J., 'The European Union's Vertical Order of Competences: The Current Law and Proposals for its Reform' (2002) 39 *Common Market Law Review*, pp. 227-268.

Booß, D., 'Art. 230 EGV' in E. Grabitz and M. Hilf (eds.), *Das Recht der Europäische Union – Kommentar*, (München: 2000).

Booß, D., 'Umweltschutz und Fischerei', in H.W. Rengeling (ed.), *Handbuch zum europäischen und deutschen Umweltrecht – Band 2 Besonderes Umweltrecht 2. Teilband*, (Köln: 2003).

Booß, D., 'Art. 37 EGV', in E. Grabitz and M. Hilf (eds.), *Das Recht der Europäischen Union – Kommentar*, (München: 2003).

Bosecke, T., 'Wesen und Systematik des Integrierten Küstenzonenmanagements und daraus resultierende Anforderungen an Inhalt und Umsetzung' (2004) *Natur und Recht*, 12, pp. 777-781.

Boude, J.-P., Bonceur, J. and Bailly, D., 'Regulating the access to fisheries: learning from the European experiences' (2001) 25 *Marine Policy*, pp. 313-322.

Boyle, A., 'Problems of Compulsory Jurisdiction and the Settlement of Disputes Relating Straddling Stocks', in O.S. Stokke (ed.), *Governing High Seas Fisheries*, (Oxford: 2001).

Brownlie, I., *Principles of Public International Law*, (New York: 2003).

Bungenberg, M., 'Vertrauenschutz', in S.M. Heselhaus and C. Nowak (eds.), *Handbuch der Europäischen Grundrechte*, (München: 2006).

Búrca, C., and Witte, D., 'The Delimitation of Powers Between the EU and its Member States', in A. Arnull and D. Wincott (eds.), *Accountability and Legitimacy in the European Union*, (Oxford: 2002).

Burke, W.T., *The New International Law of Fisheries*, (Oxford: 1994).

Calliess, C., 'Art. 17 GrCh', in C. Calliess and M. Ruffert (eds.), *Kommentar zu EU-Vertrag und EG-Vertrag*, (Neuwied: 2007).

Calliess, C., 'Arts. 6 and 174 EGV', in *Kommentar zu EU-Vertrag und EG-Vertrag*, (Neuwied: 1999).

Calliess, C., 'Individualrechtsschutz im europäischen Verfassungsbund' (2004) 1 *Göttinger Online-Beiträge zum Europarecht*, pp. 1-17.

Chalmers, D., Hadjiemmanuil, C., Monti, G. and Tomkins, A., *European Union law: text and materials*, (Oxford: 2006).

Churchill, R.R., *EEC Fisheries Law*, (Dordrecht: 1987).

Churchill, R.R. and Lowe, A.V., *The law of the sea*, (Manchester: 1999).

Clorley, J., 'Relationships of the Common Fisheries Policy and the Habitats and Birds Directives', in J. Ritterhoff , S. Gubbay and C. Zucco (eds.), *Marine Protected Areas and Fisheries – BfN Skripten 122*, (Bonn: 2004).

COFAD, *Die Küstenfischerei in Niedersachsen – Stand und Perspektiven (Kurzfassung)*, (Tutzing: 2004).

Coffey, C., 'Introduction to the Common Fisheries Policy – An Environmental Perspective' (1996) 8 *International Environmental Affairs*, 4, pp. 287-307.

Coffey, C. 'Fisheries Subsidies – Will the EU turn its back on the 2002 Reforms?' (Washington, D.C.: 2006), available online at:

http://www.panda.org/about_wwf/what_we_do/marine/publications/index.cfm?uNewsID=69600.

Conceicao-Heldt, E. d., *The Common Fisheries Policy in the European Union*, (New York: 2004).

Coopers, P., *Study into the nature and extend of subsidies in the fisheries sector of APEC members economies*, (Singapore: 2000).

Cooter, R. and Ulen, T., *Law & Economics*, (London: 2000).

Cox, A. and Schmidt, C.-C., *Subsidies in the OECD Fisheries Sector: A Review of Recent Analyses and Future Directions – OECD Paper*, (Paris: 2002).

Craig, P., *EU Adminstrative Law*, (Oxford: 2006).

Craig, P. and Búrca, G. d., *EU law*, (Oxford: 2007).

Cremer, W., 'Art. 230 EGV' in C. Callies and M. Ruffert (eds.), *EUV/EGV* (?), (München: 2007).

Cross, E.D., 'Pre-emption of Member State law in the European Economic Community: a framework analyses' (1992) 29 *Common Market Law Review*, pp. 447-472.

Cullen, H. and Charlesworth, A., 'Diplomacy by other means: the use of legal basis litigation as a political strategy by the European Parliament and Member States' (1999) 36 *Common Market Law Review*, pp. 1243-1270.

Cunningham, S., 'Overview' in S. Cunningham and T. Bostock (eds.), *Successful Fisheries Management*, (Delft: 2005).

Cunningham, S. and Greboval, D., *Managing fishing capacity: A review of policy and technical issues – FAO Fisheries Technical Paper 409*, (Rome: 2001).

Czybulka, D., 'Die Geltung der FFH-Richtlinie in der Ausschließlichen Wirtschaftszone' (2001) 12 *Natur und Recht*, 1, pp. 19-27.

Czybulka, D., 'Forschungsbedarf im marinen Fischereirecht' in H. Bauer, D. Czybulka, W. Kahl and A. Vosskuhle (eds.), *Wirtschaft im offenen Verfassungsstaat*, (München: 2006).

Dahm, G., Delbrück, J. and Wolfrum, R., *Völkerrecht*, (Berlin: 2002).

Dalton, P.D., 'Implementing the International Code of Conduct for Responsible Fisheries' in M. Nordquist and J.N. Moore (eds.), *Current Fisheries Issues and the Food and Agruculture Organisation of the United Nations*, (The Hague: 2000).

Danwitz, T., 'Aarhus-Konvention: Umweltinformationen, Öffentlichkeitsbeteiligung, Zugang zu den Gerichten' (2004) *Neue Zeitschrift für Verwaltungsrecht*, 3, pp. 272-282.

Dashwood, A., 'The relationship between the Member States and the European Union/European Community' (2004) 41 *Common Market Law Review*, pp. 355-381.

Davidse, W., 'Property Rights and Regulatory Systems: the Strategic Response of Dutch Fishermen', in D. Symes (ed.), *Property Rights and Regulatory Systems in Fisheries*, (Oxford: 1998).

Davidse, W., McEwan, L.V. and Vestergaard, N., 'Property rights in fishing: from state property towards private property?' (1999) 23 *Marine Policy*, 6, pp. 537-547.

Davies, P. and Redgwell, C., 'The International Legal Regulation of Straddling Fish Stocks' (1996) *The British Yearbook of International Law*, vol. 67, pp. 199-27.

Daw, T. and Gray, T., 'Fisheries science and sustainability in international policy: a study of failure in the European Union's Common Fisheries Policy' (2005) *Marine Policy*, 29, pp. 189-197.

Delgado, C.L., Wada, N., Rosegrant, M.W., Meijer, S. and Ahmet, M., *Fish to 2020 – Supply and Demand in Changing Global Markets*, (Washington D.C.: 2003).

Downes, D.R. and Dyke, B.V., *Fisheries Conservation and Trade Rules – Ensuring that Trade Law Promotes Sustainable Fisheries*, (Washington D.C.: 1998).

Durner, W., *Common Goods*, (Baden-Baden: 2001).

Dyke, J.M.V., 'Modifying the 1982 Law of the Sea Convention: New Initiatives on Governance of High Seas Fisheries Resources: the Straddling Stocks Negotiations' (1995) 10 *International Journal of Marine and Coastal Law*, 2, pp. 219-227.

Dyke, V., 'Sharing Ocean Resources – In a Time of Scarcity and Selfishness' in H.N. Scheiber (ed.), *Law of the Sea – The Common Heritage and Emerging Challenges*, (The Hague: 2000).

Ebbesson, J., 'European Community' in J. Ebbesson (ed.), *Access to Justice in Environmental Matters in the EU*, (The Hague: 2002).

Edeson, W., 'Sustainable Use of Marine Living Resources' (2003) 63 *Zeitschrift für ausländisches öffentliches Recht und Völkerrecht*, pp. 355-375.

Edeson, W., Freestone, D. and Gudmundsdottir, E., *Legislating for Sustainable Fisheries*, (Washington D.C.: 2001).

Ellis, J., 'The Straddling Stocks Agreement and the Precautionary Principle as Interpretive Device and Rule of Law, (2001) 32 *Ocean Development and International Law*, pp. 289-311.

European Commission, 'A level playing field for better enforcement of CFP rules', *Fishing in Europe*, 19 (2003), pp. 11-12.

Epiney, A., *Umweltrecht in der Europäischen Union*, (Köln: 2002).

Epiney, A., ‚Art. 12', in C. Calliess and M. Ruffert (eds.), *Kommentar zu EU-Vertrag und EG-Vertrag*, (Neuwied: 1999).

Epiney, A., 'Zu den Anforderungen der Aarhus-Konvention an das europäische Gemeinschaftsrecht', in J. Falke and S. Schlacke (eds.), *Information – Beteiligung – Rechtsschutz, Neue Entwicklungen im Umwelt- und Verbraucherrecht*, (Berlin: 2004).

Epiney, A. and Scheyli, M., *Strukturprinzipien des Umweltvölkerrechts*, (Baden-Baden: 1998).

Erben, C., *Das Vorsorgeprinzip im Völkerrecht*, (Berlin: 2005).

Eser, U. and Potthast, T., *Naturschutzethik*, (Baden-Baden: 1999).

European Commission, *Regional Socio-economic Studies on Employment and the Level of Dependency on Fishing*, Lot No. 23, (Brussels: 2000).

European Commission, *Fisheries and Aquaculture in Europe – European Fisheries Fund: driving sustainable development*, (Brussels: 2006).

European Commission, *The European Fisheries Fund 2007 – 2013*, (Luxembourg: 2006).

European Commission, *Maritime Facts and Figures*, (Luxembourg: 2006).

European Commission, *Facts and Figures on the CFP 2006*, (Luxembourg: 2006).

European Environmental Agency, *Europe's environment – The fourth assessment* (Luxembourg: 2007).

European Environmental Agency, *The European environment – state and outlook 2005* (Luxembourg: 2005).

Eurostat, *Fishery Statistics – Data 1990 – 2006*, (Luxembourg: 2007).

Eurostat, *Eurostat Yearbook*, (Luxembourg: 2007).

Everson, M.C., 'Independent agencies' (1995) 1 *European Law Journal*, pp. 180-204.

Falke, J., 'Komitologie – Entwicklung, Rechtsgrundlagen und erste empirische Annäherung', in C. Joerges and J. Falke (eds.), Das Ausschusswesen der Europäischen Union, (Baden-Baden: 2000).

Falke, J., 'Die Aarhus-Konvention und der Zugang zu Gerichten in Umweltangelegenheiten', in J. Falke and S. Schlacke (eds.), *Information – Beteiligung – Rechtsschutz, Neue Entwicklungen im Umwelt- und Verbraucherrecht*, (Berlin: 2004).

FAO, *Reference Points for Fisheries Management – FAO Fisheries Technical Paper 347*, (Rome: 1995).

FAO, *Technical Guidelines for Responsible Fisheries – Precautionary Approach to Capture Fisheries and Species Introductions – FAO Fisheries Technical Paper 350/1*, (Rome: 1996).

FAO, *Managing fishing capacity – FAO Fisheries Technical Paper 386*, (Rome: 1999).

FAO, *Report of the Expert Consultation on Catalysing the Transition away from Overcapacity in Marine Capture Fisheries, Report No. 691*, (Rome: 2002).

FAO, *The State of World Fisheries and Aquaculture*, (Rome: 2004).

FAO, *Guide for identifying, assessing and reporting on subsidies in the fishing sector – FAO Fisheries Technical Paper 438*, (Rome: 2004).

FAO, *Fisheries Laws and the Regulations in the Mediterranean: A Comparative Study – General Fisheries Commission for the Mediterranean*, (Rome: 2005).

FAO, *Discards in the world's marine fisheries – FAO Fisheries Technical Paper 470*, (Rome: 2005).

Fischer, R.C., 'Die gemeinsame Fischereipolitik', in E. Grabitz and M. Hilf (eds.), *Kommentar zur Europäischen Union*, (München: 1996).

Freestone, D., 'The Road From Rio: International Environmental Law After the Earth Summit' (1994) 6 *Journal of Environmental Law*, 2, pp. 193-218.

Freestone, D., 'International Fisheries Law Since Rio: The Continued Rise of the Precautionary Principle', in D. Freestone and A. Boyle (eds.), *International Law and Sustainable Development*, (Oxford: 1999).

Freestone, D. and Makuch, Z., 'The New International Environmental Law of Fisheries: The 1995 United Nations Straddling Stocks Agreement' (1998) 7 *Yearbook of International Environmental Law*, 3, pp. 3-51.

Frenz, W., *Europäisches Umweltrecht*, (München: 1997).

Froese, R. and Pauly, D., 'Dynamik der Überfischung', in J. Lozán, E. Rachor, K. Reise, J. Sündermann and H. v. Westernhagen (eds.), *Warnsignale aus Nordsee & Wattenmeer – Eine Aktuelle Umweltbilanz*, (Hamburg: 2003).

Frost, H. and Anderson, P., 'The Common Fisheries Policy of the European Union and fisheries economics' (2006) 30 *Marine Policy*, pp. 737-746.

Furrer, A., *Die Sperrwirkung des sekundären Gemeinschaftsrechts*, (Baden-Baden: 1994).

Garcia, S.M., 'The Precautionary Principle: Its Implications in Capture Fisheries Management' (1994) 22 *Ocean & Coastal Management*, pp. 99-125.

Garret, G., 'International Co-operation and Institutional Choice: The European Community's Internal Market' (1992) 46 *International Organization*, pp. 533-560.

Garrod, D. 'The Common Fisheries Policy – Now', in M. Holden (ed.), *The Common Fisheries Policy*, (Oxford: 1996).

Gellermann, M., 'Geschützte Meeresflächen in der deutschen Ausschließlichen Wirtschaftszone (AWZ) – Rechtsfragen der Einrichtung und des Schutzes mariner Natura 2000 Gebiete', in R. Hendler, P. Marburger, M. Reinhardt and M. Schröder (eds.), *Jahrbuch des Umwelt- und Technikrechts 2005*, (Berlin: 2005).

Gezelius, S. S. 'The Problem of Implementing Policies for Sustainable Fishing' in Gezelius and Raakjær (ed.), *Making Fisheries Management Work – Implementation of Policies for Sustainable Fishing*, (Berlin: 2008).

Gezelius, S. S. 'The Arrival of Modern Fisheries management in the North Atlantic: A Historical Overview', in Gezelius and Raakjær (eds.), *Making Fisheries Management Work – Implementation of Policies for Sustainable Fishing* (Berlin: 2008), pp. 27-40

Goeffrey, G.A. and Remy, P.J., *Advanced Microeconomic Theory*, (Boston: 2001).

Gordon, H.S., 'The Economic Theory of a Common Property Resource: The Fishery' (1954) 62, *Journal of Political Economy*, pp. 124-142.

Gorez, B., *Policy Study: EU-ACP Fisheries Agreements*, (Brussels: 2005).

Gréboval, D., *Managing fishing capacity – FAO Fisheries Technical Paper 386*, (Rome: 1999).

Gréboval, D., 'The International Plan of Action for the Management of Fishing Capacity: Retrospect and Prospect', in M. Nordquist and J.N. Moore (eds.), *Current Fisheries Issues and the Food and Agriculture Organisation of the United Nations*, (The Hague: 2000).

Grynberg, R., 'WTO fisheries subsidies negotiations: implications for fisheries access arrangements and sustainable management' (2003) 27 *Marine Policy*, pp. 499-511.

Guckelberger, A., 'Die EG-Verordnung zur Umsetzung der Aarhus-Konvention auf der Gemeinschaftsebene' (2008) *Natur und Recht*, 30, pp. 78-87.

Hannesson, R., *The Privatization of the Oceans*, (Cambridge: 2004).

Hardin, G., 'Tragedy of the Commons' (1968) 162 *Science*, New Series, pp. 1243-1248.

Hatchard, J. and Gray, T., 'The 2002 Reform of the Common Fisheries Policy's System of Governance – Rethoric or Reality?' (2003) 27 *Marine Policy*, pp. 545-554.

Hatcher, A., 'Subsidies for European fishing fleets: the European Community's structural policy for fisheries 1971-1999' (2000) 24 *Marine Policy*, pp. 129-140.

Hatcher, A. and Robinson, C., *Overcapacity, overcapitalisation and subsidies in the European fisheries. Proceeding of the first Concerted Action Workshop on Economics and the Common Fisheries Policy*, (Portsmouth: 1999).

Hayashi, M., 'The Management of Transboundary Fish Stocks under the LOS Convention' (1993) 8 *The International Journal of Marine and Coastal Law*, 2, pp. 245-261.

Hayashi, M., 'The 1995 Agreement on the conservation and management of straddling and highly migratory fish stocks: significance for the Law of the Sea Convention' (1995) 29 *Ocean & Coastal Management*, pp. 51-69.

Hayashi, M., 'The Straddling and Highly Migratory Fish Stocks Agreement', in E. Hey (ed.), *Developments in International Fisheries Law*, (The Hague: 1999), pp. 55-83.

Hefendehl, R., 'Europäischer Umweltschutz: Demokratiespritze für Europa oder Brüsseler Putsch' (2006) *Zeitschrift für Internationale Strafrechtsdogmatik*, 4, pp. 161-167.

Hegland, T. J. and Raakjær, J., 'Recovery Plans and the Balancing of Fishing Capacity and Fishing Possibilities – path dependence in the Common Fisheries Policy' in S. Gezelius and Raakjær (ed.), *Making Fisheries Management Work – Implementation of Politicies for Sustainable Fishing*, (Berlin: 2008).

Heinegg, W.H. v., 'Internationales öffentliches Umweltrecht', in K. Ipsen (ed.), *Völkerrecht*, (München: 2004).

Heinz, K. and Körte, A., 'Die Ziele Umweltschutz und Binnenmarkt zwischen gemeinschaftlicher Kompetenz und nationaler Verantwortung' (1991) *Juristische Ausbildung*, pp. 41-48.

Hesse, K., *Grundzüge des Verfassungsrechts der Bundesrepublik Deutschland*, (Heidelberg: 1995).

Hey, E., 'The Fisheries Provisions of the LOS Convention', in E. Hey (ed.), *Developments in International Fisheries Law*, (The Hague: 1999).

Hickey, J.E. and Walker, V.R., 'Refining the Precautionary Principle in International Environmental Law' (1995) 14 *Virginia Environmental Law Journal*, pp. 423-436.

Hilf, M., 'Freiheit des Welthandels contra Umweltschutz' (2000) *Neue Zeitschrift für Verwaltungsrecht*, 5, pp. 481-489.

Hinds, L.O.R., 'Crisis in Canada's Atlantic sea fisheries' (1995) 19 *Marine Policy*, 4, pp. 271-283.

Holden, M., *The Common Fisheries Policy*, (Oxford: 1996).

House of Lords – European Union Committee, *The Progress of the Common Fisheries Policy*, (London: 2008).

Hubold, G., 'Wege zu einer Nachhaltigen Fischerei' (2003) 14 *Zeitschrift für Umweltrecht*, 5, pp. 338-342.

Hummer, W., 'Art. 155 EGV' in E. Grabitz and M. Hilf (eds.), *Kommentar zur Europäischen Uinon*, (München: 1999).

ICES, *Environmental Status of the European Seas*, (Copenhagen: 2003).

ICES, *Report of the ICES Advisory Committee on Fishery Management , Advisory Committee on the Marine Environment and Advisory Committee on Ecosystems – Book 6 – North Sea*, (Copenhagen: 2006).

ICES, *Report of the ICES Advisory Committee on Fishery Management, Advisory Committee on the Marine Environment and Advisory Committee on Ecosystems – Book 1*, (Copenhagen: 2007).

ICTSD, *Fisheries, International Trade and Sustainable Development*, (Geneva: 2006).

IFREMER, *Small-scale Coastal Fisheries in Europe*, (Brussels: 2007).

Ingerowski, J.B. and Salomon, M., 'Ein kritischer Blick auf die aktuellen Entwicklungen in der Gemeinsamen Fischereipolitik unter Einbeziehung der neu geschaffenen regionalen Fischereibeiräte' (2006) *Natur und Recht*, pp. 535-542.

Iudicello, S., Weber, M. and Wieland, R., *Fish, Markets, and Fishermen – the Economics of Overfishing*, (London: 1999).

Jackson, J.B.C., 'Historical overfishing and the recent collapse of coastal ecosystems' (2001) *Science*, pp. 629-638.

Jaeckel, L., *Schutzpflichten im deutschen und europäischen Recht: Eine Untersuchung der deutschen Grundrechte, der Menschenrechte und Grundfreiheiten der EMRK sowie der Grundrechte und Grundfreiheiten der Europäischen Gemeinschaften*, (Baden-Baden: 2001).

Jahns-Böhm, J. and Breier, S., 'Die umweltrechtliche Querschnittsklausel des Art. 130 r II 2 EWGV' (1992) *Europäische Zeitschrift für Wirtschaftsrecht*, pp. 49-55.

Jans, J.H. and Vedder, H.H.B., *European Environmental Law*, (Groningen: 2008).

Japanese Ministry of Forestry, Agriculture and Fisheries, *Annual Report on the Developments in the Fishery Industry 2004*, (Tokio: 2004).

Jarass, H.D., 'Die Kompetenzverteilung zwischen der Europäischen Gemeinschaft und den Mitgliedstaaten' (1996) 121 *Archiv des öffentlichen Rechts*, 2, pp. 173-199.

Jarass, H.D., *Naturschutz in der Ausschließlichen Wirtschaftszone – Völkerrechtliche, EG- rechtliche und verfassungsrechtliche Probleme der Ausweisung von Meeresschutzgebieten*, (Baden-Baden: 2001).

Jentoft, S., 'Fisheries Co-management: delegating Government Responsibility to Fishermen's Organisations' (1989) *Marine Policy*, pp. 137-154.

Jentoft, S. and McCay, B.J., 'User Participation in fisheries management: Lessons Drawn from International Experiences' (1995) 19 *Marine Policy*, 3, pp. 227-246.

Juda, L., 'The 1995 United Nations Agreement on Straddling Stocks and Highly Migratory Fish Stocks: A Critique' (1997) 28 *Ocean Development and International Law*, pp. 147-166.

Kaczinsky, V.M. and Fluharty, D.L., 'European policies in West Africa: who benefits from fisheries agreements?' (2002) 26 *Marine Policy*, pp. 75-93.

Kalbheim, J. and Winter, G., 'Delegation requirements for rule-making by the Commission', in G. Winter (ed.), *Sources and Categories of European Union Law*, (Baden-Baden: 1996).

Kämmerer, J.A., '§ 124 Subventionen', in J. Isensee and P. Kirchhof (eds.), *Hand Buch des Staats Rechts V*, (Heidelberg. 2007).

Kapteyn, P.J.G. and Themaat, P.V.v., *Introduction to the Law of the European Communities*, (The Hague: 1998).

Karrigiannakos, A., 'Total Allowable Catch (TAC) and quota management system in the European Union' (1996) 20 *Marine Policy*, pp. 235-248.

Kathrein, U., 'Entwicklung innergemeinschaftlicher Sanktionen in der Europäischen Union' (1994) 49 *Österreichische Juristenzeitung*, 23, pp. 785-794.

Kaye, S.M., *International Fisheries Management*, (The Hague: 2001).

Khan, A. S., Sumaila, U.R., ,Watson, R.; Munro, G. and Pauli, D., 'The nature and magnitude of global non-fuel fisheries subsidies', in U.R. Sumaila and D. Pauly (eds.), *Catching more bait: a bottom-up re-estimation of global fisheries subsidies*, (Vancouver: 2007).

King, M., *Fisheries Biology, Assessment and Management*, (Oxford: 2007).

Kingreen, T., 'Art. 6 EU-Vertrag', in C. Callies and M. Ruffert (eds.), *Kommentar zu EU-Vertrag und EG-Vertrag*, (Neuwied: 1999 – old edition).

Klemm, C.D., 'Fisheries Conservation and Management and the Conservation of Biological Diversity', in E. Hey (ed.), *Developments in International Fisheries Law*, (The Hague: 1999).

Klyashtorin, L.B., *Climate Change and Long-Term Fluctuations of Commercial Catches – The Possibility of Forecasting – FAO Technical Paper 410*, (Rome: 2001).

Koch, H.-J., 'Die Verbandsklage im Umweltrecht' (2007) *Neue Zeitschrift für Verwaltungsrecht*, pp. 369-379.

Koch, C., 'International Decisions – Commission of the European Community v. Jégo Quéré & Cie SA' (2004) 98 *The American Journal for International Law*, 4, pp. 814-819.

Koenig, C. and Harratsch, A., *Europarecht*, (Tübingen: 2003).

Kopp, F., 'Art. 37 EGV', in R. Streinz (ed.), *EUV/EGV*, (München: 2003).

Krämer, L., *EC Environmental Law*, (London: 2007).

Krämer, L. and Winter, G., 'Umweltrecht', in R. Schulze and M. Zuleeg (eds.), *Europarecht*, (Baden-Baden: 2006).

Kura, Y., Revenga, C., Hoshino, E. and Mock, G., *Fishing for Answers – Making Sense of the Global Fish Crisis*, (Washington D.C.: 2004).

Laxe, F.G., 'Transferability of fishing rights: The Spanish case' (2006) 30 *Marine Policy*, pp. 379-388.

Leigh, M., *European integration and the common fisheries policy*, (Kent: 1983).

Lequesne, C., *The politics of fisheries in the European Union*, (Manchester: 2004).

Lindebo, E., 'Role of Subsidies in EU Fleet Capacity Management' (2005) 20 *Marine Resource Economics*, pp. 445-466.

Lindner, B., 'Zur Klagebefugnis natürlicher und juristischer Personen für Nichtigkeitsklagen gem. Art. 230 IV EG gegen EG-Verordnungen' (2003) *Neue Zeitschrift für Verwaltungsrecht*, pp. 569-572.

Lodge, M.W., Anderson, D., Lobach, T., Munro, G., Sainsbury, K. and Willock, A., *Recommended Best Practices for Regional Fisheries Management Organisations*, (London: 2007).

Long, R. and Curran, P.A., *Enforcing the Common Fisheries Policy*, (Oxford: 2000).

Lorz, A., 'Fischerei und Naturschutz – eine rechtliche Betrachtung' (1982) *Natur und Recht*, 1, pp. 4-10.

Lorz, A., 'Naturschutz und Fischereirecht der Gegenwart' (1994) *Natur und Recht*, 2, pp. 63-68.

Louka, E., *International Environmental Law – Fairness, Effectiveness, and World Order*, (Cambridge: 2006).

Ludicello, S., Weber, M. and Wieland, R., *Fish, Markets, and Fishermen – The Economics of Overfishing*, (London: 1999).

Magiera, S., 'Durchsetzung des Europarechts', in Schulze and Zuleeg (eds.), *Europarecht*, (Baden-Baden: 2006).

Majone, G., *Dilemmas of European integration – The ambiguities and pitfalls of integration by stealth*, (Oxford: 2005).

Markowski, M., Allocation and management of fisheries resources: an in-depth analysis of instruments in comparative perspective, in IUCN (ed.), *Marine Fisheries Law. A Comparative Approach*. (Gland: 2009).

Matushita, M., Schoenbaum, T.J. and Mavroidis, P.C., *The World Trade Organization*, (Oxford: 2006).

Maurer, H., *Allgemeines Verwaltungsrecht*, (München: 2006).

McKean, M.A., 'Success on the Commons: A Comparative Examination of Institutions for Common Property Resource Management' (1992) 4 *Journal of Theoretical Politics*, pp. 247-281.

Milazzo, M., *Subsidies in World Fisheries – World Bank Technical Paper No. 406*, (Washington D.C.: 1998).

Milczewski, C.v., *Der grundrechtliche Schutz des Eigentums im Europäischen Gemeinschaftsrecht*, (Frankfurt a.M.: 1993).

Miles, E.M. and Burke, W.T., 'Pressure on the United Nations Convention on the Law of the Sea of 1982. Arising from New Fisheries Conflicts. The Problem of Straddling Fish Stocks' (1989) 20 *Ocean Development & International Law*, 4, pp. 343-357.

Möller, W., *Umweltrecht, Wald, Planung, Naturschutz, Jagd u.a. Band IV*, (Hannover: 2004).

Molenaar, E.J., 'The Concept of 'Real Interest' and Other Aspects of Co-operation through Regional Fisheries Management Mechanisms' (2000) (4) *International Journal of Marine and Coastal Law*, pp. 475-531.

Moolenar, E.J., 'Regional Fisheries Management Organisations: Issues of Participation, Allocation and Unregulated Fisheries', in A.G.O. Elferink and D.R. Rothwell (eds.), *Oceans Management in the 21st Century: Institutional Frameworks and Responses*, (Leiden: 2004), pp. 69-86.

Moltke, A.v., 'Fisheries Subsidies', in A. Najam, M. Halle and R. Meléndez-Ortiz (eds.), *Trade and Environment – A Resource Book*, (Geneva: 2007), pp. 101-114.

Mooney-Sues, M.L. and Rosenberg, A.A., *Recommended Best Practices for Regional Fisheries Management Organisations: Technical Study No. 1*, (London: 2007).

Moore, G., 'The Code of Conduct for Responsible Fisheries', in E. Hey (ed.), *Development in International Fisheries Law*, (The Hague: 1999).

Morin, M., 'The fisheries resources in the European Union. The Distribution of TACs: principle of relative stability and quota-hopping' (2000) 24 *Marine Policy*, pp. 265-273.

Müller-Michaels, O., *Grundrechtlicher Eigentumsschutz in der Europäischen Union: das Eigentumsgrundrecht in der Rechtsordnung der EU, in der EMRK und in den Verfassungen Deutschlands, Italiens und Irlands*, (Berlin: 1996).

Munro, G., Houtte, A.V. and Willmann, R., *The conservation and management of shared fish stocks: legal and economic aspects – FAO Fisheries Technical Paper T465*, (Rome: 2004).

Munro, G.R., 'The economics of overcapitalisation and fishery resource management: a review', in A. Hatcher and C. Robinson (eds.), *Overcapacity, overcapitalisation and subsidies in European fisheries. Proceedings of the first Concerted Action workshop on economics and the Common Fisheries Policy*, (Portsmouth: 1999).

Mwikya, S.M., *Fisheries Access Agrements: Trade and Development Issues, ICTSD Natural Resources, International Trade and Sustainable Development Series Issue Paper No. 2*, (Geneva: 2006).

Myers, R.A. and Worm, B., 'Rapid worldwide depletion of predatory fish communities' (2003) 423 *Nature*, pp. 280-283.

Nettesheim, M. 'Art. 249 EGV', in E. Grabitz and M. Hilf (eds.), *Das Recht der Europäischen Union*, (München: 2003).

Nielsen, J.R., 'An analytical framework for studying: compliance and legitimacy in fisheries management' (2003) 27 *Marine Policy*, pp. 425-432.

Nonnenmacher, C., 'Das EG-Fischereiregime als supranationale Ressourcenordnung' (1989) *Europarecht*, 1, pp. 24 (1989), pp. 30-53.

Nonnenmacher, C., *Die Gemeinsame Fischereipolitik der Europäischen Wirtschaftsgemeinschaft*, (Tübingen: 1990).

Nordquist, M., *UN Convention of the Law of the Seas 1982 – A Commentary*, (Dordrecht: 1991).

Nowak, C., 'Das Verhältnis zwischen zentralem und dezentralem Individualrechtsschutz im Europäischen Gemeinschaftsrecht' (2000) *Europarecht*, 5, pp. 724-742.

Nowak, C., 'Eigentumsgrundrecht', in S.M. Heselhaus and C. Nowak (eds.), *Handbuch der Europäischen Grundrechte*, (München: 2006).

Nowak, C., 'Grundrechtsdimension des Umweltschutzes', in S.M. Heselhaus and C. Nowak (eds.), *Handbuch der Europäischen Grundrechte*, (München: 2006).

OECD, *Towards Sustainable Fisheries*, (Paris: 1997).

OECD, *Transistion to Responsible Fisheries – Government Financial Transfers and Resource Sustainability: Case Studies*, (Paris: 2000).

OECD, *Fish Piracy, Combating Illegal, Unregulated and Unreported Fishing*, (Paris: 2004).

OECD, *Using Market Mechanisms to Manage Fisheries*, (Paris: 2006).

Oppermann, T., *Europarecht*, (München: 2005).

Ostrom, E., *Governing the Commons – The evolution of institutions for collective action*, (Cambridge: 1990).

P.O. Commission, *America's Living Oceans, Charting a Course for Sea Change, Supply and Demand in Changing Global Markets*, (Washington D.C.: 2003).

Pauly, D., Christensen, V., Dalsgaard, J., Froese, R. and Torres, F., 'Fishing down the marine food webs' (1998) *Science*, 279, pp. 860-863.

Pauly, D., Christensen, V., Guénette, S., Pitcher, T.J., Sumaila, U.R. and Walters, C.J., 'Towards sustainability in world fisheries' (2002) 418 *Nature*, 8, pp. 689-695.

Payne, D.C., 'Policy-Making in Nested Institutions: Explaining the Conservation Failure of the EU's Common Fisheries Policy (2000) 38 *Journal of Common Market Studies*, pp. 303-324.

Pollack, M.A., 'Delegation, agency, agenda setting in the European Community (1997) 51 *International Organizations*, pp. 99-134.

Porter, G., *Fisheries and the Environment – Fisheries Subsdies and Overfishing: Towards a Structured Discussion*, (Geneva: 2001).

Porter, G., *Fisheries Subsidies and Overfishing: Towards a Structured Discussion*, (Geneva: 2003).

Porter, G., *Analyzing the Resources Impact of Fisheries Subsidies: a Matrix Approach (UNEP)*, (Nairobi: 2004).

Porter, G., *Incorporating Impacts on Resources into Disciplines on fisheries Subsidies: Issues and Options – A Discussion Paper (UNEP)*, (Geneva: 2004).

Potts, T. and M. Haward, 'International Trade, Eco-Labelling, and Sustainable Fisheries – Recent Issues, Concepts and Practices' (2006) *Environment, Development, and Sustainability*, pp. 91-106.

Priebe, R., 'Gemeinsamer Markt für die Landwirtschaft', in E. Grabitz and M. Hilf (eds.), *Das Recht der Europäischen Union*, (München: 2004).

Primosch, E.G., 'Das Vorsorgeprinzip im internationalen Umweltrecht' (1996) 51 *Zeitschrift für öffentliches Recht*, pp. 227-241.

Proelß, A., *Meeresschutz im Völker- und Europarecht*, (Berlin: 2004).

Pühs, W., *Der Vollzug von Gemeinschaftsrecht*, (Berlin: 1997).

Rat von Sachverständigen für Umweltfragen (SRU), *Meeresschutz für Nord- und Ostsee – Sondergutachten*, (Baden-Baden: 2004).

Rat von Sachverständigen für Umweltfragen (SRU), *Rechtsschutz für die Umwelt – die altruistische Verbandsklage ist unverzichtbar, Stellungnahme, Februar 2005*, (Berlin: 2005).

Rayfuse, R., 'The Interrelationship Between the Global Instruments of International Fisheries Law', in E. Hey (ed.), *Developments in International Fisheries Law*, (The Hague: 1999).

Rehbinder, E., Rechtsschutz gegen Handlungen und Unterlassungen der Organe und Einrichtungen der Europäischen Gemeinschaft im Lichte der Aarhus-Konvention', in C. Gaitanides, S. Kadelbach, R. Iglesias and C. Gil (eds.), *Europa und seine Verfassung – Festschrift für Manfred Zuleeg*, (Baden-Baden: 2005).

Rengeling, H.-W. and Heinz, K., 'Die dänische Pfandflaschenregelung – Entscheidungsrezension' (1990) *Juristische Schulung*, No. 8, pp. 613-617.

Rengeling, H.-W., 'Durchführung des Europäischen Gemeinschaftsrechts – Überblick', in H.-W. Rengeling (ed.) *Handbuch zum europäischen und deutschen Umweltrecht*, (Köln: 2003).

Renn, O. and Köppel, S., 'Einführung: Weder Fisch noch Fleisch – Vom Scheitern der europäischen Fischereipolitik – und neue Chancen' (2005) 14 *GAIA*, 4, pp. 293-294.

Reszat, P., *Gemeinsame Naturgüter im Völkerrecht*, (München: 2004).

Rijn, T., van, 'Art. 37 EGV' in H. von der Groeben and J. Schwarze (eds.), *Kommentar zum Vertrag über die Europäische Union und zur Gründung der Europäischen Gemeinschaft*, (Baden-Baden: 2004).

Rijn, T., van, 'Fischereipolitik – Kommentar nach Art. 38 EG' in H. von der Groeben and J. Schwarze (eds.), *Kommentar zum Vertrag über die Europäische Union und zur Gründung der Europäischen Gemeinschaft*, (Baden-Baden: 2004).

Rosenne, S. and Sohn, L.B., 'Article 300', in M.H. Nordquist (ed.), *United Nations Convention on the Law of the Sea 1982 – A Commentary*, (Dordrecht: 1989).

Rossiter, T. and Stead, S., 'Days at sea: from the fisher's mouths' (2003) 27 *Marine Policy*, pp. 281-288.

Sadeleer, N.d., *Environmental Principles*, (Cambridge: 2002).

Buisman, E., Smit, J. and de Vos, B., *Employment in the Fisheries Sector*, (Brussels: 2006).

Sands, P., 'International Law in the Field of Sustainable Development: Emerging Legal Principles', in W. Lang (ed.), *Sustainable Development and International Law*, (London: 1995).

Sands, P., *Principles of International Environmental Law*, (Cambridge: 2003).

Schlacke, S., *Überindividueller Rechtsschutz* (Tübingen: 2008).

Schmidt, G., 'Art. 249', in H. von der Groeben and J. Schwarze (eds.), *Kommentar zum Vertrag über die Europäische Union und zur Gründung der Europäischen Gemeinschaft*, (Baden-Baden: 2004).

Schmidt, R. und Kahl, W., *Umweltrecht*, (München: 2006).

Schneider, H., *Die gemeinsame Fischereipolitik der Europäischen Gemeinschaften*, (Köln: 1988).

Schneider, H., 'Die gemeinsame Fischereipolitik' in A. Bleckmann (ed.), *Europarecht*, (Köln. 1997).

Schöler, Y., 'Allmendetragik in der Ostsee? Mecklenburg-Vorpommerns Kutter- und Küstenfischer in den Zwängen der EU-Fischereipolitik' in B. Glaeser (ed.), *Küste – Ökologie – Mensch*, (München: 2005).

Schorr, F.D., *Healthy Fisheries, Sustainable Trade: Crafting New Rules on Fishing Subsidies in the World Trade Organisation*, (Washington D.C.: 2004).

Schrank, W.E., 'Subsidies for Fisheries: A Review of Concepts' in FAO (ed.), *Papers presented at the Expert Consultation on Economic Incentives and Responsible Fisheries*, (Rome: 2000).

Schrank, W.E., *Introducing Fisheries Subsidies – FAO Fisheries Technical Paper 437*, (Rome: 2003).

Schreiber, S., *Verwaltungskompetenzen der Europäischen Gemeinschaft*, (Baden-Baden: 1997).

Schröder, M., 'Umweltschutz als Gemeinschaftsziel und Grundsätze des Umweltschutzes' in H.-W. Rengeling (ed.), *Handbuch zum europäischen und deutschen Umweltrecht*, (Köln: 2003).

Schröder, M., 'Art. 299 EG', in H. von der Groeben and J. Schwarze (eds.), *Kommentar zum Vertrag über die Europäische Union und zur Gründung der Europäischen Gemeinschaft*, (Baden-Baden: 2004).

Schwarz, K., 'Fishing management in the Exclusive Economic Zone (EEZ) of the Federal Republic of Germany', in J. Ritterhoff, S. Gubbay and C. Zucco (eds.), *Marine Protected Areas and Fisheries – BfN-Skript 122*, (Bonn: 2004).

Schweitzer, M., 'Art. 2002 EGV', in E. Grabitz and M. Hilf (eds.), *Das Recht der Europäischen Union*, (München: 2005).

Scott, A., 'Introducing Property in Fishing Management', in R. Shotton (ed.), *Use of property rights in fisheries management – FAO Fisheries Technical Paper 404/1*, (Rome: 2000).

Shackleton, M., 'Fishing for a Policy? The Common Fisheries Policy of the Community', in H. Wallace, W. Wallace and C. Webb (eds.), *Policy-Making in the European Community*, (Chichester: 1983).

Siedentopf, H. and Hauschild, C., 'Europäische Integration und die öffentlichen Verwaltungen der Mitgliedstaaten' (1990) 43 *Zeitschrift für öffentliches Recht und Verwaltungswissenschaften*, 11, pp. 445-455.

Sissiwine, M., 'Part 1 – Challenges, Performance and the Future', in M. Sissiwine and D. Symes (eds.), *Reflection on the Common Fisheries Policy – Report to the General Directorate for Fisheries and Maritime Affairs of the European Community* (2007). Available online at:
http://www.seas-at-risk.org/1images/Midterm%20review%20Sissenwine%20and%20Symes%202007.pdf

Smit, W., 'Common Fishery Policy and National Fisheries Management' (1997) 12 *Marine Resource Economics*, pp. 355-359.

Smyrl, M.E., 'When (and how) Do the Commission's Preferences Matter?' (1998) 36 *Journal of Common Market Studies*, pp. 79-99.

Soeter, M., 'Norway and the European Union', in L. Mile (ed.), *The European Union and the Nordic Countries*, (London: 1996).

Song, Y.-H., 'The EC's Common Fisheries Policy in the 1990s' (1995) 26 *Ocean Development and International Law*, pp. 31-55.

Song, Y.-H., 'The Common Fisheries Policy of the European Union: Restructuring of the Fishing Fleet and the Financial Instrument for Fisheries Guidance' (1998) 13 *The International Journal of Marine and Coastal Law*, 4, pp. 537-577.

Song, Y.-H., 'The EC's Common Fisheries Policy in the 1990s' (1995) 26 *Ocean Development and International Law*, pp. 31-55.

Sproul, J.T., 'Green fisheries: certification as a management toll', in Pitcher, Hart and Pauly (eds.), *Reinventing Fisheries Management*, (London: 1998), pp. 137-147.

Steenblick, R. P. and Munro, G. R., 'Current international work on subsidies in the fisheries: a survey', in A. Hatcher and C. Robinson (eds.), *Overcapacity, overcapitalisation and subsidies in European fisheries. Proceedings of the first concerted action workshop on economics and the Common Fisheries Policy*, (Portsmouth: 1998).

Stokke, O.S. and Coffey, C., 'Precaution, ICES and the common fisheries policy: a study of regime interplay' (2004) 28 *Marine Policy*, pp. 117-126.

Streinz, R., 'Art. 7 EGV', in R. Streinz (ed.), *EUV/EGV*, (München: 2003).

Streinz, R., *Europarecht*, (Heidelberg: 2005).

Suerbaum, J., 'Die Schutzpflichtdimension der Gemeinschaftsgrundrechte' (2003) *Europarecht*, pp. 390-416.

Sutinen, J., Rieser, A., and Gauvin, J.R., 'Measuring and Explaining Noncompliance in Federally Managed Fisheries' (1990) 21 *Ocean Development & International Law*, pp. 335-372.

Symes, D., 'The European Community's Fisheries Policy' (1997) 35 *Ocean & Coastal Management*, pp. 137-155.

Symes, D., 'Rights-Based Management: A European Union Perspective', in R. Shotton (ed.), *Use of property rights in fisheries management – FAO Fisheries Technical Paper 404/1*, (Rome: 2000).

Symes, D. and Pope, J.G., *An Ecosystem Based Approach to the Common Fisheries Policy. Part I: Defining the Goals*, (Petersborough: 2000).

Szczekalla, P., *Die sogenannten grundrechtlichen Schutzpflichten im deutschen und europäischen Recht*, (Berlin: 2002).

Szczekalla, P., 'Grundrechte', in H.-W. Rengeling (ed.), *Handbuch zum europäischen und deutschen Umweltrecht*, (Köln: 2003).

Szczekalla, P., 'Funktion der Grundrechte', in S.M. Heselhaus and C. Nowak (eds.), *Handbuch der Europäischen Grundrechte*, (München: 2006).

Thiele, A., *Europäisches Prozessrecht*, (München: 2007).

Thiele, G., 'Art. 37 EGV', in C. Calliess and M. Ruffert (eds.), *EUV/EGV*, (München: 2007).

Thorpe, A. and Bennett, E., 'Globalisation and the Sustainability of World Fisheries: A View from Latin America' (2001) 16 *Marine Resource Economics*, pp. 143-164.

Tridimas, T., *The general principles of EU law*, (Oxford: 2006).

Trouwborst, A., *Evolution and Status of the Precautionary Principle in International Law*, (The Hague: 2002).

UNEP, *Global Environment Outlook 2000*, (Nairobi: 1999).

Valatin, G., 'Development of property rights-based fisheries management in the United Kingdom and the Netherlands: a comparison', in R. Shotton (ed.), *Use of property rights in fisheries management – FAO Fisheries Technical Paper 404/1*, (Rome: 2000).

Vicuna, F.O., 'Coastal States' Competence over High Seas Fisheries' (1995) 55 *Zeitschrift für ausländisches öffentliches Recht und Völkerrecht*, 2, pp. 520-535.

Vicuna, F. O., *The Changing International Law of High Seas Fisheries*, (Cambridge: 1999).

Virdin, J. and Schorr, D., *Hard Facts and Hidden Problems – A Review of Current Data on Fishing Subsidies – A WWF Technical Paper*, (Washington D.C.: 2001).

Vitzthum, W. G. Graf von, ' Begriff, Geschichte und Rechtsquellen des Seerechts', in Graf Vitzthum (ed.), *Handbuch des Seerechts*, (München: 2006).

Vitzthum, W. G. Graf von, 'Raum und Umwelt im Völkerrecht', in W. G. Vitzthum (ed.), *Völkerrecht*, (Berlin: 2004).

Voß, R., 'Art. 26 EGV', in E. Grabitz and M. Hilf (eds.), *Das Recht der Europäischen Union*, (München: 1999).

Vukas, B. and Vidas, D., 'Flags of Convenience and High Seas Fishing: The Emergence of a Legal Framework', in O.S. Stokke (ed.), *Governing High seas Fisheries*, (Oxford: 2001).

Ward, A., 'Judicial Review of Environmental Misconduct in the European Community: Problems, Prospects and Strategies' (2000) *Yearbook of European Environmental Law*, pp. 137-159.

Weeramantry, C.G., 'Sustainable Development', in J.-P. Beurier, A. Kiss and S. Mahmoudi (eds.), *New Technologies and The Law of the Marine Environment*, (The Hague: 2000).

Wegener, B., 'Art. 220 EGV', in C. Calliess and M. Ruffert (eds.), EUV – EGV Kommentar (München: 2007).

Weiß, A., *'Möglichkeiten der Regelung der Fischerei, des Bergbaus und der Schifffahrt in 'Baltic Sea Protected Areas' (BSPAs) in der Ausschließlichen Wirtschaftszone (AWZ) der der Bundesrepublik Deutschland vorgelagerten Ostsee*, (Bonn: 1999).

Westlund, L., *Guide for identifying, assessing and reporting on subsidies in the fisheries sector – FAO Fisheries Technical Paper 438*, (Rome: 2004).

Weymann, J., *Umweltökonomik – Eine theorieorientierte Einführung*, (Berlin: 1995).

Winkler, R., *Die Grundrechte der Europäischen Union – System und allgemeine Grundrechtslehren*, (Wien: 2006).

Winter, G., 'Umweltrechtliche Prinzipien des Gemeinschaftsrechts' (2003) Sonderheft *Zeitschrift für Umweltrecht*, pp. 137-145.

Winter, G., 'The Legal Nature of Environmental Principles in International, EC and German Law', in R. Macrory (ed.), *Principles of European Environmental Law*, (Groningen: 2004).

Winter, G., 'A Fundament and Two Pillars', in H.C. Bugge and C. Voigt (eds.), *Sustainable Development in International and National Law*, (Groningen: 2008).

Winter, G., 'Towards a Legal Clinic for Fisheries Management', in IUCN (ed.), *Marine Fisheries Law. A Comparative Approach.* (Gland: 2009).

Wise, M., *The Common Fisheries Policy of the European Union*, (London: 1984).

Wolff, N., *Fisheries and the Environment*, (Baden-Baden: 2002).

Wolff, N., 'Erhaltung lebender Meeresressourcen im Lichte des Nachhaltigkeitsgrundsatzes: Völker- und gemeinschaftsrechtliche Voraussetzungen' (2003) 14 *Zeitschrift für Umweltrecht*, 5, pp. 356-362.

Wolfrum, R., 'Precautionary Principle', in J.-P. Beurier, A. Kiss and S. Mahmoudi (eds.), *New Technologies and Law of the Marine Environment*, (The Hague: 2000).

Wolfrum, R., Röben, V. and Morrison, F.L., 'Preservation of the Marine Environment', in F.L. Morrison and R. Wolfrum (eds.), *International, Regional and National Environmental Law*, (The Hague: 2000).

World Bank, Saving Fish and Fishers – Toward Sustainable and Equitable Governance of the Global Fisheries Sector, (Washington D.C.: 2004).

World Bank, *World Development Report, Sustainable Development in a Dynamic World*, (New York: 2002).

WWF, *Hard Facts, Hidden Problems – A Review of Current Data on Fishing Subsidies – WWF Technical Paper*, (Washington D.C.: 2001).

WWF, *WWF Mid-Term Review of the EU Common Fisheries Policy*, (Brussels: 2007).

Yturriaga, J.A.d., *The International Regime of Fisheries – from UNCLOS 1982 to the Presidential Sea*, (The Hague: 1997).

Ziemer, J., *Das gemeinsame Interesse an einer Regelung der Hochseefischerei*, (Berlin: 2000).

Table of Cases

The Court of Justice of the European Communities

Court of First Instance

T-435/93, ASPEC and others v. Commission, [1995] ECR II-1281
T-442/93, AAC and others v. Commission, [1995] ECR II-1329
T-472/93, Campo Ebro and other v. Council, [1995] ECR II-421
T-482/93, Weber v. Commission, [1996] ECR II-609
T-514/93, Cobrecaf and other v. Commission, [1995] ECR II- 621
T-585/93, Greenpeace and others v. Commission, [1995] ECR II-2205
T-115/94, Opel Austria v. Council, [1997] ECR II-39
T-254/97, Fruchthandelsgesellschaft Chemnitz v. Commission, [1999] ECR II 2743
T-38/98, Assoziazione Nazionale Bieticolltori v. Council, [1998] ECR II-44191
T-47/00, Rica Foods v. Commission, [2000] ECR II-113
T-177/01, Jégo Quéré & Cie Sa v. Commission, [2002] ECR II-02365
T-264/03, Schmoldt and others v. Commission, [2004] ECR II-1515
T-37/04, The Autonomous Region of the Azores v. Council, [2004] ECR II-2153
T-94/04, EEB and others v. Commission, [2005] ECR II-4919
T-447/05, SPM v. Commission, [2007] ECR II-0000
T-91/07, WWF-UK Ltd. v. Council, not yet in the ECR
T-172/07, Atlantic Dawn Ltd. et al. v. Commission, OJ 2007 No. C170/29

The Court of Justice

30/59, De Gezamenlijky Steenkolenmijnen in Limburg v. ECSC High Authority, [1961] ECR 95
16-17/62, Conféderation Nationale des Producteurs de fruits et Légumes v. Council, [1962] ECR I-471
25/62, Plaumann & Co v. Commission, [1963] ECR 95
6/64, Costa v. E.N.E.L, [1964] ECR 1141
7/68, Commission v. Italy, [1968] ECR 423
29/69, Stauder v. Stadt Ulm, [1969] ECR 419
41/69, Chemiefarma v. Commission, [1970] ECR 661
11/70, Internationale Handelsgesellschaft mbH v. Einfuhr und Vorratsstelle für Getreide und Futtermittel, [1970] ECR 1125
25/70, Einfuhr- und Vorratsstelle für Getreide und Futtermittel v. Köster, [1970] ECR 1161
38/70, Deutsche Tradax GmbH v. Einfuhr- und Voratsstelle für Getreide und Futtermittel, [1971] ECR 145
41/70, International Fruit Company and others v. Commission, [1971] ECR
21-24/72, International Fruit Company NV and others v. Produktschap voor Groenten en Fruit, [1972] 1219
4/73, Nold KG v. Commission, [1974] ECR 491
167/73, Commission v French Republic, [1974] ECR 359

181/73, Haegemann, [1974] ECR 449
15/74, Centrafarm BV and others v. Sterling Drug [1974] ECR 1147
23/75, Rey Soda v. Cassa Conguaglio Zucchero, [1975] ECR 1279
3, 4, 6/76, Kramer and others, [1976] ECR 1279
15, 16/76, France v. Commission, [1979] ECR 321
26/76, Metro v. Commission, [1977] ECR
61/77, Commission v. Ireland, [1978] 417
185-204/78, J. Van Dam en Zonen, [1979] ECR 2345
44/79, Hauer v. Land Rheinland-Pfalz, [1979] ECR 3727
789-790/79, Calpak and Società Emiliana Lavorazione Frutta v Commission, [1980] ECR 1949
804/79, Commission v. United Kingdom, [1981] ECR 1045
112/80, Dürbeck v. Hauptzollamt Frankfurt a.M., [1981] ECR 1095
188-190/80, France, Italy and United Kingdom v. Commission, [1982] ECR 2545
203/80, Casati, [1981] ECR 2595
104/81, Hauptzollamt Mainz v. Kupferberg & Cie., [1982] ECR 3641
258/81, Metallurgiki Halyps v. Commission, [1982] ECR 4261
11/82, Piraiki-Patraiki v. Commission, [1985] ECR 207
14/83, von Colson und Kamann v. Land Nordrhein-Westfalen, [1984] ECR 1891
172/83, Hoogovens Group v. Commission, [1985] ECR 2831
86-87/84, Bout, [1985] ECR 941
207/84, De Boer v. Produktschap voor Vis en Visprodukten, [1985] ECR 3203
67/85, Van der Kooy v. Commission, [1988] ECR 219
201-202/85, Klensch v. Secrétaire d' État [1986] ECR, 3477
288/85, Hauptzollamt Hamburg-Jonas v. Plange Kraftfutterwerke, [1987] ECR 611
12/86, Demirel, [1987] ECRI-3719
46/86, Romkes v Officier van Justitie for the District Zwolle, [1987] ECR 2671
291/86, Central-Import Münster v. Hauptzollamt Münster, [1986] ECR 3679
3/87, The Queen v Ministry of Agriculture, Fisheries and Food, ex parte Agegate Ltd, [1989] 4495
165/87, Commission v. Council, [1988] ECR 5545
216/87, The Queen v. Ministry of Agriculture, Fisheries and Food, ex parte: Jaderow Ldt. et seq., [1989] ECR I-4544
290/87, Commission v. Netherlands, [1989] ECR 3083
62/88, Greece v. Council, [1990] ECR I-1527
64/88, Commission v. French Republic, [1991] ECR I-2727
68/88, Commission v. Greece, [1989] ECR 2965
331/88, The Queen v. Ministry of Agriculture, Fisheries and food, ex parte FEDESA and others, [1990] ECR I-4023

348/88, Criminal proceedings v. Hakvoort, [1990] ECR I-1647
C-9/89, Spain v. Council, [1990] I-1383
C-44/89, Von Deetzen v. Hauptzollamt Oldenburg, [1991] ECR I-5119
C-146/89, Commission v. United Kingdom, [1991] ECR I-3533
C-244/89, Commission v. France, [1991] I-163
C-246/89, Commission v. United Kingdom, [1991] ECR I-4585
C-258/89, Commission v. Spain, [1991] ECR I-3977
C-309/89, Codorniu SA v. Council, [1994] ECR I-1853
C-385/89, Extramet Industries SA v. Council, [1991] I-2501
C-73/90, Kingdom of Spain v. Council of the European Communities, [1992] ECR I-5191
C-177/90, Kühn v. Landwirtschaftskammer Weser-Ems, [1995] ECR I-35
C-199/90, Italtrade v. AIMA, [1991] ECR I-5545
C-240/90, Germany v. Commission, [1992] ECR I-5383
C-286/90, Anklagemydigheden v. Poulsen and Diva Navigation Corp., [1992] ECR I- 6019
C-313/90, CIFRS and others v. Commission, [1991] ECR I-2257
C-103/91, Spain v. Council, [2005] ECR I-2267
C-155/91, Commission v. Council, [1993] ECR I-939
C-2/92, The Queen v. Ministry of Agriculture, Fisheries and Food, es parte Dennis Clifford Bostock, [1994] ECR I-955
C-405/92, Mondiet v. Armement Islais, [1993] ECR I-6133
C-37/93, Commission v. Belgian Kingdom, [1993] ECR 6295
C-63/93, Duff and others, [1996] ECR I-569
C-280/93, Germany v. Council, [1994] ECR I-4973
C-306/93, SMW Winzersekt GmbH v. Land Rheinland-Pfalz, [1994] ECR I-5555
C-13/94, P/S and Cornwall County Council, [1996] ECR
C-22/94, Irish Farmers Association v. Ministry of Agriculture, Food and Forestry, [1997] ECR I-1809
C-44/94, The Queen v. Ministry of Agriculture, Fisheries and Food, ex parte Fishermen's organisations and others, [1995] ECR I-3115
C-61/94, Commission v. Germany, [1996] ECR I-3989
C-209/94, Buralux and others, [1996] ECR I-615
C-290/94, Commission v. Greek Republic, [1996] ECR 3285
C-52/95, Commission v. France, [1995] ECR I-4443
C-179/95, Spain v. Council, [1999] REC I-6475
C-248/95, SAM Schiffahrt GmbH and Heinz Stapf v. Bundesrepublik Deutschland, [1997] p. I-4475
C-265/95, Commission v. France, [1997] ECR I-6959
C-341/95, Safety High-Tech v. S& T Srl, [1998] ECR I-4355
C-103/96, Directeur général des douanes and droits indirects v. Eridania Beghin-Say [1997] ECR I-1453
C-386/96, Société Louis Dreyfus & Cie v. Commission, [1998] ECR I-2309

C-73/97, France v. Comafrica and others, [1999] ECR I-185
C-164-165/97, Parliament v. Council, [1999] ECR I-1139
C-166/97, Commission v. France, [1999] ECR I-1719
C-206/97, Kingdom of Sweden v. Council of the European Union, [1999] ECR I-3885
C-226/97, Criminal proceeding against Lemmens, [1998] ECR I-3711
C-106/98, Comité d'entreprise de la Société française de production v. Commission, [2000] ECR I-3659
C-313/99, Mulligan and others, [2002] ECR I-5719
C-50/00 P, Unión de Pequeños Agiculturos v. Council, [2002] ECR I-6677
C-336/00, Huber, [2002] I-7699
C-242/02, Germany v. Commission, [2002] ECR I-5603
C-263/02 P, Commission v. Jégo Quéré SA, [2004] ECR I-3425
C-304/02, Commission v. French Republic, [2005] ECR I-6263
C-91/03, Spain v. Council, [2005] ECR I-2267
C-176/03, Commission v. Council, [2004] ECR I-11671
C-210/03, Swedish Match, [2004], I-11893
C-260/05, Sniace v. Commission, [2007] ECR I-1
C–15/06, P, Regione Siciliana v. Commission, [2007] ECR I-2591

Legal Acts and Documents

International Conventions

Aarhus Convention, Aarhus, 25 June 1998

Convention for the Conservation of Salmon in the North Atlantic Ocean (NASCO-Convention), Reykjavik, 2 March 1982

Convention for the Protection of the Marine Environment of the Baltic Sea Area (HELCOM), Helsinki, 22 March 1974

Convention for the Protection of the Marine Environment of the North-Atlantic (OSPAR), Paris, 22 September 1992

Convention for the Protection of the Mediterranean Sea against pollution, Barcelona, 16 February 1976

Convention on Biological Diversity (CBD), Rio de Janeiro, 5 June 1992

Convention on Future Multilateral Cooperation in North-East Atlantic Fisheries (NEAFC-Convention), London, 18 November 1980

Convention on Future Multilateral Cooperation in North-West Atlantic Fisheries (NAFO-Convention), Ottawa, 24 October 1978

Convention on the Conservation and Management of Fishery Resources in the South East Atlantic (SEAFO-Convention), Windhoek, 20 April 2001

European Convention on Human Rights (ECHR), Rome, 4 November 1950

FAO, Agreement to Promote Compliance with International Conservation and Management Measures by Fishing Vessels on the High Seas (`Compliance Agreement'), Rome, 24 November 1993

General Agreement on Tariffs and Trade (GATT), Geneva , 30 October 1947

General Fisheries Commission for the Mediterranean Area (CFGM) Agreement, Rome, 24 September 1949

International Convention for the Conservation of Atlantic Tunas (ICCAT-Convention), Paris, 10 July 1984

United Nations Convention on the Law of the Sea (1982 UNCLOS), Montego Bay, 10 December 1982

United Nations Agreement for the implementing of the provisions of the United Nations Convention on the Law of the Sea of 10 December 1982 relating to the conservation and management of straddling stocks and highly migratory fish stocks ('FSA'), 8 September 1995, New York

Vienna Convention on the Law of the Treaties (VCLT), Vienna, 22 May 1969

European Union/European Community

Primary law

Treaty establishing the European Community, 24 December 2002 (OJ 2006 No C 321/E39)

Treaty on European Union, 24 December 2002 (OJ 2006 No C321/E7)

Treaty of Lisbon amending the Treaty on the European Union and the Treaty Establishing the European Community (OJ 2007 No. C306/01)

Consolidated version of the Treaty on the European Union and the treaty on the Functioning of the European Union (OJ 2008 No. C115/1)

Treaty concerning the Accession of Denmark, Ireland and the United Kingdom (OJ 1972 No L 73)
Treaty concerning the Accession of Greece (OJ 1979 N L 291)
Charta of Fundamental Rights of the European Union (OJ 2007 No. C303/1)

Secondary law
Council
Council Regulation (EEC) 17/64 of 5 February 1964 on the conditions for granting aid from the European Agricultural Guidance and Guarantee Fund (OJ 1964 No. 34)
Council Regulation (EEC) No. 729/70 of 21 April 1970 on the financing of the common agricultural policy (OJ 1970 No. L94/13)
Council Regulation (EEC) 2141/70 of 20 October 1970 laying down a common structural policy for the fishing industry (OJ 1970 No. L236/1)
Council Regulation (EEC) 2142/70 of 20 October 1970 on the common organisation of the market in fishery (and aquaculture) products (OJ 1970 No. L236/5)
Council Regulation (EEC) 2722/72 of 19 December 1972 on the financing by the EAGGF, Guidance Section, of the conversion projects in the salt cod-fishing industry (OJ 1972 No. L291/30)
Council Regulation (EEC) 100/76 of 19 January 1976 on the common organization of the market in fishery products (OJ 1976 No. L020/1)
Council Regulation (EEC) 101/76 of 19 January 1976 laying down a common structural policy for the fishing industry (OJ 1976 No. L20)
Council Regulation (EEC) 355/77 of 15 February 1977 on common measures to improve the conditions under which agricultural products are processed and marketed (OJ 1977 No. L51)
Council Regulation (EEC) 1852/78 of 25 July 1978 on an interim measure for restructuring the inshore fishing industry (OJ 1978 No. L 211/30)
Council Regulation (EEC) 592/79 of 26 March 1979 amending Regulation (EEC) 1852/78 on an interim common measure for restructuring the inshore fishing industry (OJ 1979 No. L78/5)
Council Regulation (EEC) 754/80 of 26 March 1980 concerning, for certain fish stocks occurring in the Community fishing zone, the fixing for 1980 of the total allowable catches, the share available for the Community and the means of making the catches (OJ 1980 No. L84/36)
Council Regulation (EEC) 1713/80 of 27 June 1980 amending Regulation (EEC) 1852/78 on an interim common measure for restructuring the inshore fishing industry (OJ 1980 No. L167/50)
Council Regulation (EEC) 348/81 of 20 January 1981 on common rules for imports of whales or other cetaceans products (OJ 1981 No. L 39/1)
Council Regulation (EEC) 3796/81 of 29 December 1981 on the common organization of the market in fishery products (OJ 1981 No. L379/1)

Council Regulation (EEC) 2057/82 of 29 June 1982 establishing certain control measures for fishing by vessels of the Member States (OJ 1982 No. No L220/1)

Council Regulation (EEC) 31/83 of 21 December 1982 amending Regulation (EEC) 1852/78 on an interim common measure for restructuring the inshore fishing industry (OJ 1983 No. L5/1)

Council Regulation (EEC) 170/83 of 25 January 1983 establishing a Community system for the conservation and management of fishery resources (OJ 1983 No. L24/1)

Council Regulation (EEC) 171/83 of 25 January 1983 laying down certain technical measures for the conservation of fishery resources (OJ 1983 No. L24)

Council Regulation (EEC) No. 2908/83 of 4 October 1983 on a common measure for restructuring, modernising, and developing the fishing industry and for developing aquaculture (OJ 1983 No. L290/1)

Council Regulation (EEC) 2088/85 of 23 July 1985 concerning the integrated Mediterranean programmes (OJ 1985 No. L197/1)

Council Regulation (EEC) 3722/85 of 20 December 1985 on a system of structural aid for the conversion of sardine canning plants (OJ 1985 No. L361/38)

Council Regulation (EEC) 3727/85 of 20 December 1985 amending Regulation (EEC) 2909/83 in measures to encourage fishing and cooperation through joint ventures in the fishing sector (OJ 1985 No. L361/56)

Council Regulation (EEC) 3733/85 of 20 December 1985 amending Regulation (EEC) 2908/83 on a common measure for restructuring, modernizing and developing the fishing industry and developing aquaculture (OJ 1985 No. L 361/78)

Council Regulation (EEC) 3094/86 of 7 October 1986 laying down certain technical measures for the conservation of fishery resources (OJ 1986 No. L288)

Council Regulation (EEC) 4028/86 of 18 December 1986 on Community measures to improve and adapt structures in the fisheries and aquaculture sector (OJ 1986 No. L376)

Council Regulation (EEC) 2241/87 of 23 July 1988 establishing certain control measures for fishing activities (OJ 1987 No. L207/1)

Council Regulation (EEC) 2658/87 of 23 July 1987 on the tariff and statistical nomenclature and on the Common Custom Tariff (OJ 1987 No. L256/1)

Council Regulation (EEC) No. 2052/88 of 24 June 1988 on the tasks of the structural funds and their effectiveness and on the coordination of their activities between themselves and with the operations of the European Investment Bank and other existing financial instruments (OJ 1988 No. L185/9)

Council Regulation (EEC) 4253/88 of 19 December 1988 laying down provisions for implementing Regulation (EEC) 2052/88 as regards coordination of the activities of the different Structural Funds between themselves and with the operations of the European Investment Bank and the other existing financial instruments (OJ 1988 No. L374)

Council Regulation (EEC) 4042/89 of 19 December 1989 on the improvement of the conditions under which fishery and aquaculture products are processed and marketed (OJ 1989 No. L388/1)

Council Regulation (EEC) 3944/90 of 20 December 1990 amending Regulation (EEC) 4028/86 on Community measures to improve and adapt structures in the fisheries and aquaculture sector (OJ 1990 No. L380/1)

Council Regulation (EEC) 2092/91 of 24 June 1991 on organic production of agricultural products and indications referring thereto on agricultural products and foodstuff (OJ 1991 No. L198/1)

Council Regulation (EEC) 3687/91 of 28 November 1991 on the common organization of the market in fishery products (OJ 1991 No. L354/1)

Council Regulation (EEC) 3880/91 of 17 December 1991 on the submission of nominal catch statistics by Member States fishing in the Northeast Atlantic (OJ 1991 No. L365/1)

Council Regulation (EEC) 3759/92 of 17 December 1992 on the common organisation of the market in fishery and aquaculture products (OJ 1992 No. L388/1)

Council Regulation (EEC) 3760/92 of 20 December 1992 establishing a Community system for fisheries and aquaculture (OJ 1992 No. L389)

Council Regulation (EEC) No 2080/93 of 20 July 1993 laying down provisions for implementing Regulation (EEC) 2052/88 as regards the financial instruments of fisheries guidance (OJ 1993 No. L193/1)

Council Regulation (EEC) 2847/93 of 12 October 1993 establishing a control system applicable to the Common Fisheries Policy (OJ 1993 No. L261/1)

Council Regulation (EC) 3690/93 of 20 December 1993 establishing a Community system laying down rules for the minimum information to be contained in fishing licences (OJ 1993 No. L341/93)

Council Regulation 3699/93 of 21 December 1993 laying down the criteria and arrangements regarding Community structural assistance in the fisheries and aquaculture sector and the processing and marketing of its products (OJ 1993 No. L356/1)

Council Regulation (EC) 1275/94 of 30 May 1994 on adjustments to the arrangements in the fisheries chapters of the Act of Accession of Spain and Portugal (OJ 1994 No. L140/1)

Council Regulation (EC) 1626/94 of 27 July 1994 laying down certain technical measures for the conservation of fishery resources in the Mediterranean (OJ 1994 No. L171)

Council Regulation (EC) 1627/94 of 27 June 1994 laying down general provisions concerning special fishing permits (OJ 1994 No. L171/7)

Council Regulation (EC) 3317/94 of 24 December 1994 laying down general provisions concerning the authorisation of fishing in the waters of a third country under a fisheries agreement (OJ 1994 No. L350/13)

Council Regulation (EC) 685/95 of 27 March 1995 on the management of the fishing effort relating to certain Community fishing areas and resources (OJ 1995 No. L71/5)

Council Regulation (EC) 2027/95 of 15 June 1995 establishing a system for the management of fishing effort relating to certain Community fishing areas and resources (OJ 1995 No. L199/1)

Council Regulation (EC) 847/96 of 6 May 1996 introducing additional condition for year-to-year management of TACs and quotas (OJ 1996 No. L115/3)

Council Regulation (EC) 686/97 of 14 April 1997 amending Regulation 2847/93 establishing a control system applicable to the common fisheries policy (OJ 1997 No. L 102/1)

Council Regulation (EC) 779/97 of 24 April 1997 introducing arrangements for the managing of fishing effort in the Baltic Sea (OJ 1997 No. L113/1)

Council Regulation (EC) 894/97 of 29 April 1997 laying down certain technical measures for the conservation of fishery resources (OJ 1998 No. L132/1)

Council Regulation (EC) 2205/97 of 30 October 1997 amending Regulation (EEC) 2847/93 establishing a control system applicable to the common fisheries policy (OJ 1997 No. L304/1)

Council Regulation (EC) 2635/97 of 18 December 1997 amending Regulation (EC) 2847/93 establishing a control system applicable to the common fisheries policy (OJ 1997 No. L356/14)

Council Regulation (EC) 850/98 of 30 March 1998 for the conservation of fishery resources through technical measures for the protection of juveniles of marine organisms (OJ 1998, No. L125/1)

Council Regulation (EC) 2468/98 of 3 November 1998 laying down the criteria and arrangements regarding community structural assistance in the fisheries and aquaculture sector and the processing and marketing of its products (OJ 1998 No. L312/19)

Council Regulation (EC) 1258/99 of 17 May 1999 on the financing of the common agricultural policy (OJ 1999 No. L171/19)

Council Regulation (EC) 1263/99 of 21 June 1999 on the Financial Instrument of Fisheries Guidance (OJ 1999 No. L161/54)

Council Regulation (EC) 1447/99 of 24 June 1999 establishing a list of types of behaviour which seriously infringe the rules of the Common Fisheries Policy (OJ 1999 No. L167/5)

Council Regulation (EC) 2792/99 of 17 December 1999 laying down detailed rules and arrangements regarding Community structural assistance in the fisheries sector (OJ 1999 No. L337/1)

Council Regulation (EC) 104/00 of 17 December 1999 on the common organisation of the markets in fishery and aquaculture products (OJ 2000 No. L017/22)

Council Regulation (EC) No. 657/00 of 27 March 2000 on closer dialogue with the fishing sector and groups affected by the common fisheries policy (OJ 2000 No. L80/7)

Council Regulation (EC) 2092/00 of 28 September 2000 prohibiting imports of Atlantic blue-fin tuna (Thunnus thynnus) originating in Belize, Honduras and Equatorial Guinea (OJ 2000 No. L249/1)

Council Regulation (EC) 2093/00 of 28 September 2000 prohibiting imports of Atlantic swordfish (Xiphias gladius) originating in Belize and Honduras (OJ 2000 No. L249/3)

Council Regulation (EC) 1451/01 of 28 June 2001 amending Regulation (EC) 2792/99 laying down the detailed rules and arrangements regarding Community structural assistance in the fisheries sector (OJ 2001 No. L198/9)

Council Regulation (EC) 2369/02 of 20 December 2002 amending Regulation (EC) 2792/99 laying down the detailed rules and arrangements regarding Community structural assistance in the fisheries sector (OJ 2002 No. L358)

Council Regulation (EC) 2370/02 of 20 December 2002 establishing an emergency community measure for scrapping fishing vessels (OJ 2002 No. L358)

Council Regulation 2371/02 of 20 December 2002 on the conservation a sustainable exploitation of fisheries resources under the Common Fisheries Policy (OJ 2002 No. L358/59)

Council Regulation (EC) 1/03 of 16 December 2002 on the implementation of the rules on competition laid down in Articles 81 and 82 of the Treaty (OJ 2003 No. L1/1)

Council Regulation 1954/03 of 4 November 2003 on the management of the fishing effort relating to certain Community fishing areas and resources and modifying Regulation (EC) 2847/93 and repealing Regulations (EC) 685/95 and (EC) 2027/93 (OJ 2003 No. L289/1)

Commission Regulation 26/04 of 30 December 2003 regarding the fishing vessels register of the Community (OJ 2004 No. L5/25)

Council Regulation (EC) 60/04 of 14 January 2004 laying down certain technical measures applicable to fishing activities in the area covered by the Convention on the conservation of Antarctic marine living resources (OJ 2004 No. L97/1)

Council Regulation (EC) 423/04 of 26 February 2004 establishing measures for the recovery of cod stocks (OJ 2004 No. L70/8)

Council Regulation (EC) 602/04 of 22 March 2004 amending Regulation (EC) 50/98 as regards the protection of deepwater coral reefs from the effects of trawling in an area north west of Scotland (OJ 2004 No. L97/30)

Council Regulation (EC) 639/04 of 30 March 2004 on the management of fishing fleets registered in the Community's outermost regions (OJ 2004 No. L102/9)

Council Regulation (EC) 811/04 of 21 April 2004 establishing measures for the recovery of the Northern hake stock (OJ 2004 No. L150/1)

Council Regulation (EC) 1415/04 of 19 July 2004 fixing the maximum annual fishing effort for certain fishing areas and fisheries (OJ 2004 No. L258/1)

Council Regulation (EC) 1421/04 of 19 July 2004 amending Regulation (EC) 2792/99 laying down the detailed rules and arrangements regarding Community structural assistance in the fisheries sector (OJ 2004 No. L260/1)

Council Regulation (EC) 485/05 of 16 March 2005 amending Regulation (EC) 2792/99 as regards a specific action for transfers of vessels to countries hit by the tsunami in 2004 (OJ 2005 No. L81/1)

Council Regulation (EC) No 768/05 of 26 April 2005 establishing a Community Fisheries Control agency and amending Regulation (EEC) 2847/93 establishing a control system applicable to the common fisheries policy (OJ 2005 No. L128/1)

Council Regulation (EC) 860/05 of 30 May 2005 amending Regulation (EC) 27/05, as concerns fishing opportunities in Greenland, Faeroese and Icelandic waters and fishing for cod in the North sea, and amending Regulation (EC) 2270/04 as concerns fishing opportunities for deep-sea sharks and roundnose grenadier (OJ 2005 No. L144/1)

Council Regulation (EC) 2187/05 of 21 December 2005 for the conservation of fishery resources through technical measures in the Baltic Sea, the Belts and the Sound, amending Regulation (EC) 1434/98 and repealing Regulation (EC) 88/98 (OJ 2005 No. L349/1)

Council Regulation (EC) 41/06 of 12 January 2006 for cod in the zones covered by the cod recovery plan as regulated in Regulation 423/04 (OJ 2006 No. L8)

Council Regulation (EC) 51/06 of 22 December 2005 fixing for 2006 the fishing opportunities and associated conditions for certain fish stocks and groups of fish stocks applicable in Community waters and, for Community vessels, in waters where catch limitations are required (OJ 2007 No. L16/1)

Council Regulation (EC) 700/06 of 25 April 2006 repealing Regulation (EC) 3690/93 establishing a Community system laying down rules for the minimum information to be contained in fishing licences (OJ 2006 No. L122/1)

Council Regulation (EC) 1198/06 of 27 July 2006 on the European Fisheries Fund (OJ 2006 No. L223/1)

Council Regulation (EC) 1941/06 of 11 December 2006 fixing the fishing opportunities and associated conditions for certain fish stocks and groups of fish stocks applicable in the Baltic Sea for 2007 (OJ 2006 No. L367/1)

Council Regulation (EC) 1967/06 of 21 December 2006 concerning management measure for the sustainable exploitation of fishery resources in the Mediterranean Sea, amending Regulation (EEC) 2847/93 and repealing Regulation (EC) 1626/94 (OJ 2007 No. L36/6)

Council Regulation (EC) 41/07 of 21 December 2006 fixing for 2007 the fishing opportunities and associated conditions for certain fish stocks and groups of fish stocks, applicable in Community waters and, for Community vessels, in waters where catch limitations are required (OJ 2007 No. L15/1)

Council Regulation (EC) 643/07 of 11 June 2007 amending Regulation (EC) 41/07 as concerns the recovery plan for bluefin tuna recommended by the International Commission for the Conservation of Atlantic Tunas (OJ 2007 No. L151/1)

Council Regulation (EC) 1098/07 of 18 September 2007 establishing a multiannual plan for the cod in the Baltic Sea and the fisheries exploiting those stocks, amending Regulation (EEC) 2847/93 and repealing Regulation (EC) 779/97 (OJ 2007 No. L248/1)

Council Regulation (EC) 1100/07 of 18 September 2007 establishing measures for the recovery of stocks of European eel (OJ 2007 No. L248/17)

Council Regulation (EC) 40/08 of 16 January 2008 fixing for 2008 the fishing opportunities and associated conditions for certain fish stocks and groups of fish stocks, applicable in Community waters and, for Community vessels, in waters where catch limitations are required (OJ 2008 No. L19/1)

Council Regulation (EC) 199/08 of 25 January 2008 concerning the establishment of a Community framework for the collection, management and use of data in the fisheries sector and support for scientific advice regarding the Common Fisheries Policy (OJ 2008 No. L60/1)

Council Regulation (EC) 43/09 fixing for 2009 the fishing opportunities and associated conditions for certain fish stocks and groups of fish stocks, applicable in Community waters and, for Community vessels, in waters where catch limitations are required, OJ 2009 No. L22/1.

Council Directive (EEC) 409/79 of 2 April 1979 on the conservation of wild birds (OJ 1979 L103/1)

Council Directive (EEC) 515/83 of 4 October 1983 concerning certain measures to adjust capacity in the fishing sector (OJ 1983 No. L290/15)

Council Directive (EEC) 590/85 of 20 December 1985 amending Directive 515/83 concerning certain measures to adjust capacity in the fisheries sector (OJ 1985 No. 372/49)
Council Directive (EEC) 43/92 of 21 May 1992 on the conservation of natural habitats and of wild fauna and flora (OJ 1992 L206/7)
Council Directive (EC) 4/03 of 28 January 2003 on public access to environmental information and repealing Council Directive (EEC) 313/90 (OJ 2003 No. L41/26)

Council Decision (EC) 15/94 of 15 April 1994 relating to the objectives and detailed rules for restructuring the Community fisheries sector over the period 1 January 1994 to 31 December 1996 with a view to achieving a lasting balance between the resources and their exploitation (OJ 1994 No. L10/20)
Council Decision (EC) 428/96 of 25 June 1996 on acceptance by the Community of the Agreement to promote compliance with international conservation and management measures by fishing vessels on the high seas (OJ 1996 No. L177/24)
Council Decision (EC) 413/97 of 26 June 1997 concerning the objectives and detailed rules for restructuring the Community fisheries sector for the period from January 1997 to 31 December 2001 with a view to achieving a balance on a sustainable basis between resources and their exploitation (OJ 1997 No L175/27)
Council Decision (EC) 392/98 of 23 March 1998 concerning the conclusion by the European Community of the United Nations Convention of 10 December 1982 on the Law of the Sea and the Agreement of 28 July 1994 relating to the implementation of Part XI thereof (OJ 1998 No. L 179/1)
Council Decision (EC) 414/98 of 8 June 1998 on the ratification by the European Community of the Agreement for the implementing of the provisions of the United Nations Convention on the Law of the Sea of 10 December 1982 relating to the conservation and management of straddling stocks and highly migratory fish stocks (OJ 1998 No. L189/14)
Council Decision (EC) 468/99 of 28 June 1999 laying down the procedures for the exercise of implementing powers conferred on the Commission (OJ 1999 No. L184/23) as amended by Council Decision (EC) 409/07 (OJ 2007 No. L155/68)
Council Decision (EC) 70/02 of 28 January 2002 amending Decision 413/97 concerning the objectives and detailed rules for restructuring the Community fisheries sector for the period from 1 January 1997 to 31 December 2001 with a view to achieving a balance on a sustainable basis between resources and their exploitation (OJ 2002 No. L031/77)
Council Decision (EC) 585/04 of 19 July 2004 establishing Regional Advisory Councils under the Common Fisheries Policy (OJ 2004 No. L256/17)

Council Decision (EC) 370/05 of 17 February 2005 on the conclusion, on behalf of the European Community, of the Convention on access to information, public participation in decision-making and access to justice in environmental matters (OJ 2005 No. L124/1)

Council Declaration of 30 May 1980 on the common fisheries policy (OJ 1980 No. C158/1)

European Parliament and Council

European Parliament and Council Directive (EC) 42/01 of 27 June 2001 on the assessment of the effects of certain plans and programmes on the environment (OJ 2001 No. L197/30)

European Parliament and Council Directive (EC) 35/03 of 26 May 2003 providing for public participation in respect of the drawing up of certain plans and programmes relating to the environment and amending with regard to public participation and access to justice Council Directives (EEC) 337/85 and 61/96 – Statement by the Commission (OJ 2003 No. L156/17)

European Parliament and Council Regulation (EC) 761/01 of 19 March 2001 allowing voluntary participation by organisations in a Community eco-management and audit scheme (EMAS) (OJ 2001 No. L114/1)

European Parliament and Council Regulation (EC) 1059/03 of 26 May 2003 on the establishment of a common classification of territorial units for statistics (NUTS) (OJ 2003 No. L154/1)

European Parliament and Council Regulation (EC) 1882/03 of 29 September 2003 adapting to Council Decision (EC) 468/99 the provisions relating to committees which assist the Commission in the exercise of its implementing powers laid down in instruments subject to the procedure referred to in Article 251 of the EC Treaty (OJ 2003 No. L284/1)

European Parliament and Council Regulation (EC) 1367/06 of 6 September 2006 on the application of the provisions of the Aarhus Convention of Access to Information, Public Participation in Decision-making and Access to Justice in Environmental Matters to Community institutions and bodies (OJ 2006 No. L264)

European Parliament and Council Recommendation (EC) 413/02 of 30 May 2002 concerning the implementation of Integrated Coastal Zone Management in Europe (OJ 2002 No. L148/24)

Commision

Commission Regulation (EEC) 1462/73 of 16 May 1973 concerning requests for aid to the EAGGF, Guidance Section, for conversion of the cod fishing fleet (OJ 1973 No. L145/11)

Commission Regulation (EEC) 645/74 of 22 March 1974 on procedures for implementing decision by the Guidance Section of the EAGGF to grant aid to the salt cod fishing industry (OJ 1978 No.L78/19)

Commission Regulation (EEC) 2992/82 of 10 November 1982 amending Regulation (EEC) 1852/78 on an interim common measure for restructuring the inshore fishing industry (OJ 1981 No. L299/24)

Commission Regulation (EEC) 2166/83 establishing a license system for certain fisheries in an area north of Scotland (Shetland area) (OJ 1983 No. L206/71).

Commission Regulation (EEC) 2807/83 of 22 September 1983 laying down detailed rules for recording on Member States catches of fish, OJ 1983 No L276/1 as amended by Commission Regulation (EC) 1804/05 (OJ 2005 No. L290/10)

Commission Regulation (EEC) 163/89 of 24 January 1989 concerning the fishing vessel register of the Community (OJ 1989 No. L20/5)

Commission Regulation (EC) 109/94 of 19 January 1994 concerning the fishing register of the Community (OJ 1994 No. L19/5)

Commission Regulation (EC) 2945/95 of 20 December 1995 amending Regulation (EEC) 2807/83 laying down detailed rules for recording information on the Member States' catches of fish (OJ 1995 No L308/18)

Commission Regulation (EC) 493/96 of 20 March 1996 amending Regulation (EC) 109/94 concerning the fishing vessel register of the Community (OJ 1996 No. L72/12)

Commission Regulation (EC) 939/97 of 26 May 1997 laying down detailed rules concerning the implementation of Council Regulation (EC) 338/97 on the protection of species of wild fauna and flora by regulating trade therein (OJ 1997, L140/9)

Commission Regulation (EC) 1489/97 of 29 July 1997 laying down detailed rules for the application of Council Regulation (EEC) 2847/93 as regards satellite-based vessel monitoring systems (OJ 1997 No. L202/18)

Commission Regulation (EC) 2090/98 of 30 September 1998 concerning the fishing vessel register of the Community (OJ 1998 No. L266/27)

Commission Regulation (EC) 2091/98 of 30 September 1998 concerning the segmentation of the Community fishing fleet and fishing effort in relation to the multiannual guidance programmes (OJ 1998 No. L266/36)

Commission Regulation (EC) 2092/98 of 30 September 1998 concerning the declaration of fishing effort relating to certain Community fishing areas and resources (OJ 1999 No. L118/66)

Commission Regulation (EC) 2740/99 of 21 December 1999 laying down detailed rules for the application of Council Regulation (EC) 1447/99 establishing a list of types of behaviour which seriously infringe the rules of the common fisheries policy (OJ 1999 No. L328/62)

Commission Regulation (EC) 366/01 of 22 February 2001 laying down detailed rules for implementing the measures provided for in Council Regulation (EC) 2792/99 (OJ 2001 No. L55)

Commission Regulation (EC) 839/02 of 21 May 2002 amending Regulation (EC) 2090/98 (OJ 2002 No. L135/5)

Commission Regulation (EC) 1954/02 of 31 October 2002 fixing the rates of the refunds applicable to certain milk products exported in the form of goods not covered by Annex I to the Treaty (OJ 2002 No. L289/1)

Commission Regulation (EC) 1595/02 of 6 September 2002 on the application of Articles 87 and 88 of the EC Treaty to State aid to small and medium-sized enterprises active in the production, processing and marketing of fisheries products (OJ 2004 No. L291/3)

Commission Regulation (EC) 129/03 of 24 January 2003 laying down detailed rules for determining the mesh size and thickness of twine of fishing nets (OJ 2003 No. L22/5)

Commission Regulation (EC) 1438/03 of 12 August 2003 laying down implementing rules on the Community Fleet Policy as defined in Chapter III of Council Regulation 2371/02 (OJ 2003 No. L204/21)

Commission Regulation (EC) 2244/03 of 18 December 2003 laying down detailed provisions regarding satellite-based Vessel Monitoring Systems (OJ 2003 No. L230/17)

Commission Regulation (EC) 26/04 of 30 December 2003 regarding the fishing vessels register of the Community (OJ 2004 No. L5/25)

Commission Regulation (EC) 916/04 of 29 April 2004 amending Regulation (EC) 1438/03 laying down implementing rules on the Community Fleet Policy by reason of the accession of Czech Republic, Estonia, Cyprus, Latvia, Lithuania, Hungary, Malta, Poland, Slovenia and Slovakia to the European Union (OJ 2004 No. 163/81)

Commission Regulation (EC) 1595/04 of 8 September 2004 on the application of Articles 87 and 88 of the EC Treaty to State aid to small and medium-sized enterprises active in the production, processing and marketing of fisheries products (OJ 2004 No. L291/3)

Commission Regulation (EC) 2104/04 of 9 December 2004 laying down detailed implementing rules for Council Regulation (EC) 639/04 (OJ 2004 No. L365/19)

Commission Regulation (EC) 1037/05 of 1 July 2005 establishing emergency measures for the protection and recovery of the anchovy stock in ICES Sub-area VIII (OJ 2005 No. L171/24)

Commission Regulation (EC) 1281/05 of 3 August 2005 on the management of fishing licences and minimal information to be contained therein (OJ 2005 No. L203/3)

Commission Regulation (EC) 1539/05 of 22 September 2005 extending the emergency measures for the protection and recovery of the anchovy stock in ICES subs-area VIII (OJ 2005 No. L247/9)

Commission Regulation (EC) 196/06 of 3 February 2006 amending Annex I to Regulation (EC) 761/2001 of the European Parliament and of the Council to take account of the European Standard EN ISO 14001:2004, and repealing Decision 97/265/EC (OJ 2006 No. L32/4)
Commission Regulation (EC) 780/06 of 24 May 2006 amending Annex VI to Council Regulation (EEC) 2092/91 on organic production of agricultural products and indications referring thereto on agricultural products and foodstuffs (OJ 2006 No. L137/9)
Commission Regulation (EC) 1042/06 of 7 July 2006 laying down detailed rules for the implementation of Art. 28(3) and (4) of Council Regulation 2371/02 on the conservation and sustainable exploitation of fisheries resources under the Common Fisheries Policy (OJ 2006 No. L187/14)
Commission Regulation (EC) 147/07 of 15 February 2007 adapting certain fish quotas from 2007 to 2012 pursuant to Article 23(4) of Council Regulation (EC) 2371/02 on the conservation and sustainable exploitation of fisheries resources under the Common Fisheries Policy (OJ 2007 No. L46/10)

Commission Decision (EEC) 128/71 of 25 February 1971 on legislation establishing and modifying the ACFA Committee; portant création d' un comité consultative de la pêche (OJ 1971, No. L68/18)
Commission Decision (EC) 478/99 of 14 July 1999 renewing the Advisory Committee on Fisheries and Aquaculture (OJ 1999 No. L187/70-73)
Commission Decision (EC) 774/04 of 9 November 2004 declaring operational the Regional Advisory Council for the North Sea under the common fisheries policy (OJ 2004 No. L342/28)
Commission Decision (EC) 864/04 of 16 December 2004 amending Commission Decision 1999/478/EC of renewing the Advisory Committee on Fisheries and Aquaculture (OJ 2004 No. L370/91)
Commission Decision (EC) 606/05 of 5 August 2005 declaring operational the Regional Advisory Council for pelagic stocks under the common fisheries policy (OJ 2005 No. L206/21)
Commission Decision (EC) 668/05 of 22 September 2005 declaring operational the Regional Advisory Council for the north-western waters under the common fisheries policy (OJ 2005 No. L249/18)
Commission Decision (EC) 191/06 of 1 March 2006 declaring operational the Regional Advisory Council for the Baltic Sea under the common fisheries policy (OJ 2006 No. L66/50)
Commission Decision (EC) 206/07 of 29 March 2007 declaring operational the Regional Advisory Council for the High Seas/Long Distance Fleet under the common fisheries policy (OJ 2007 No. L91/52)
Commission Decision (EC) 222/07 of 4 April 2007 declaring operational the Regional Advisory Council for the south western waters under the common fisheries policy (OJ 2007 No. L95/52)

Commission Decision (EC) 266/07 of 11 April 2007 adopting the list of Community inspectors and inspection means pursuant to Art. 28(4) of Council Regulation (EC) 2371/02 on the conservation and sustainable exploitation of fisheries resources under the common fisheries policy (OJ 2007 No L76/22)

Germany

Basic Law for the Federal Republic of Germany (Grundgesetz für die Bundesrepublik Deutschland), 23 May 1949 (BGBl 1949, 1), last amended by Art. 1 of the Act of 26 July 2002 (BGBl I S. 2863)

German Fisheries Code (Seefischereigesetz), BGBl. I 1984, 876 as amended by BGBl. I 2004, 2198

Maritime Fisheries Regulation (*Seefischereiverordnung*), BGBl. I 1989, 1485 as amended by BGBl. I 2005, 3499

Lower Saxony

Fisheries Code of Lower-Saxony (*Niedersächsisches Fischereigesetz*), Niedersächsisches Gesetz- und Verordnungsblatt, 1 February 1978, as amended by Niedersächsisches Gesetz- und Verordnungsblatt 10 November 2005

Schleswig-Holstein

Act of Schleswig-Holstein on the establishment of nature reserve in Schleswig-Holstein's North-Sea territory, *i.e.*, the Wadden Sea (Gesetz zum Schutz des schleswig-holsteinischen Wattenmeeres), Gesetz- und Verordnungsblatt Schleswig-Holstein, 17 December 1999, as amended by Gesetz- und Verordnungsblatt Schleswig-Holstein, 16 September 2003

Fisheries Code of Schleswig-Holstein (*Fischereigesetz für das Land Schleswig-Holstein*) Gesetz- und Verordnungsblatt Schleswig-Holstein, 10 February 1996

Mecklenburg-Western Pomerania

Fisheries Code of Mecklenburg-Western Pomerania (*Fischereigesetz für das Land Mecklenburg-Vorpommern*), Gesetz- und Verordnungsblatt Mecklenburg-Vorpommern, 4 May 2005, Nr. 7

Documents

International

Declaration on Environment and Development, Rio de Janeiro, 3-14 June 1992

FAO

FAO, *Code of Conduct on Responsible Fisheries*, Rome, 1995

FAO Committee on Fisheries, *Progress in the Implementation of the 1995 Code of Conduct for Responsible Fisheries, Related International Plans of Action and Strategy – COFI/2007/2*, Rome, February 2007

WTO

Panel of the GATT, Report United States – Restrictions on Import of Tuna, 3 September 1991

Panel of the GATT, Report United States – Restrictions on Imports of Tuna, 16 June 1994

United Nations

United Nations General Assembly Resolution 59/25 of 17 November 2004

United Nations General Assembly, *Some High Seas Fisheries Aspects Relating to Straddling Fish Stocks And Highly Migratory Fish Stocks*, Doc. A/Conf.164/INF/4

European Union/European Community

Council

Council of the European Union, Working Paper 13595 on the Reform of the Common Fisheries Policy, Brussels 31 Oct. 2002

Council of the European Union, Press Release 15479/05 (Presse 349), regarding 2702nd Council Meeting, Brussels 20-22 December 2005

Council of the European Union Press Release, 19 June 2006, No. 10302/06 (Presse 178)

Commission (Reports)

European Commission (EEC) 196/67, Report on the Situation in the Fisheries Sector of the EEC Member States and the Basic Principles for a Common Policy (OJ 1967)

European Commission, Report by the Commission to the Council on the Enforcement of the CFP (Com(86) 301 final), 4 June 1986

European Commission, Report from the Commission to the Council and the European Parliament on the Common Fisheries Policy (SEC(91) 2288 final), 18 December 1991

European Commission, Report on Monitoring Implementation of the Common Fisheries Policy (SEC(92) 394 final), 6 March 1992

European Commission, Annual Report to the Council and to the Parliament on the Progress of the Multi-annual Guidance Programmes for the Fishing Fleet at the End of 1996 (Com(97) 352 final), 11 July 1997

European Commission, Report from the Commission on the monitoring of implementation of the Common Fisheries Policy – Synthesis of the implementation of the control system applicable to the Common fisheries Policy by Member States, (Com(01) 526 final), 28 September 2001

European Commission, Report from the Commission to the Council and the European Parliament on the intermediate results of the multi-annual guidance programmes for the fishing (Com(02) 483 final), 30 June 2002

European Commission, Report on the Protection of the European Communities' financial interests and the fight against fraud (Com(04) 573/2 final), 30 August 2004

European Commission, Report from the Commission on the monitoring of the Member States' implementation of the Common Fisheries Policy from 2002 to 2004 (Com (04) 849 final), 4 January 2005

European Commission, Annual report from the Commission to the Council and the European Parliament on Member States' efforts to during 2005 to achieve a sustainable balance between capacity and fishing opportunities (Com(06) 872 final)

European Commission, Report from the Commission to the Council and the European Parliament on the monitoring of the Member States' implementation of the Common Fisheries Policy (Com(07) 167 final), 10 April 2007

European Commission, annual report from the Commission to the Council and the European Parliament on Member States' efforts to during 2006 to achieve a sustainable balance between fishing capacity and fishing opportunities (Com(07) 828 final), 19 December 2007

European Commission Staff Working Document accompanying the Proposal for a Council Regulation establishing a Community control system for ensuring compliance with the rules of the Common fisheries Policy, SEC(2008) 2760, 14 November 2008

Commission (Proposals)

European Commission, Proposal for a Council decision on the accession of the European Community to the FAO at the 26th session of the FAO Conference (Com(91) 387 final), 18 October 1991

European Commission, Proposal for a Regulation of the European Parliament and of the Council on the application of the provisions of the Aarhus Convention on Access to Information, Public Participation in Decision-making and Access to Justice in Environmental Matters to EC institutions and bodies (Com(03) 622 final), 24 October 2003

European Commission, Proposal on the European Fisheries Fund (Com(04) 497 final), 14 July 2004

European Commission, Proposal for a Council Regulation opening and providing for the management of autonomous Community tariff quotas for certain fishery products for the period 2007 to 2009 (Com(07) 289 final), 4 June 2007

Commission (Communications)

European Commission, Communication from the Commission to the Council and the European Parliament on the Common Fisheries Policy (SEC(90) 2244), 6 December 1990

European Commission, Communication on Implementation of Technical Measures in the Common Fisheries Policy (Com(95) 669 final), 15 December 1995

European Commission, (Com(99) 363 final), 14 July 1999

European Commission, Communication from the Commission to the Council and the European Parliament – Community Participation in Regional Fisheries Organisations (RFOs) (Com(99) 613 final), 8 December 1999

European Commission, Communication on integrated coastal zone management (Com(00) 547 final), 27 September 2000

European Commission, Communication on the application of the precautionary principle and the multiannual arrangements for setting TACs (Com(00) 803 final), 1 December 2000

European Commission, Communication on Elements of a Strategy for the Integration of Environmental Protection Requirements into the Common Fisheries Policy (Com (01) 143 final), 16 March 2001

European Commission, Communication – Green Paper on the future of the Common Fisheries Policy (Com(01) 135 final), 20 March 2001

European Commission, Communication on Biodiversity Action Plan for Fisheries (Com (01) 162 final), 27 March 2001

European Commission, Communication on the Reform of the Common Fisheries Policy ('Roadmap') (Com(02) 181 final), 3 April 2001

European Commission, Communication Action Plan to integrate environmental protection requirements into the Common Fisheries Policy (Com(02) 186 final), 28 May 2002

European Commission, Communication for the Sustainable Development of European Aquaculture (Com(02) 511 final), 19 September 2002

European Commission, Communication to the Council and the European Parliament laying down a Community Action Plan for the conservation and the sustainable exploitation of fisheries resources in the Mediterranean Sea under the Common Fisheries Policy (Com(02) 535 final), 9 October 2002

European Commission, Communication on an integrated framework for fisheries partnership agreements with third countries (Com(2002) 637 final), 23 December 2002

European Commission, Communication from the Commission to the Council and the European Parliament – On a Community Action Plan to reduce discards of fish (Com(02) 656 final), 26 November 2002

European Commission, Communication towards uniform and effective implementation of the Common Fisheries Policy (Com(03) 130 final), 21 March 2003

European Commission, Communication on improving fishing capacity and effort indicators under the common fisheries policy (Com(07) 39 final), 5 February 2007

European Commission, Communication on rights-based management tools in fisheries (Com(07) 73 final), 26 February 2007

European Commission, Staff Working Document accompanying the Communication from the Commission to the Council and the European Parliament on Right-based management tools in fisheries (SEC(07) 247), 26 February 2007

European Commission, Communication – A policy to reduce unwanted by-catches and eliminate discards in European fisheries (Com(07) 136 final), 28 March 2007

European Commission, Communication to the Council on Fishing Opportunities for 2008 – Policy Statement from the European Commission (Com(07) 295 final), 6 June 2007

European Commission, Communication on the proposal for a Council Regulation establishing a Community control system for ensuring compliance with the rules of the Common Fisheries Policy, Com(2008) 718 final.

European Commission, Recommendation (EC) 361/03 concerning the definition of micro, small and medium sized enterprises (OJ 2003 No. L124/36)

European Commission, Guidelines for the Examination of State Aid to Fisheries and Aquaculture (OJ 2004 No. C229/5)

European Court of Auditors

European Court of Auditors, Special Report No. 3/93 concerning the implementation of the measures for the restructuring, modernisation and adaptation of the capacities of fishing fleets in the Community together with Commission's replies ('Court of Auditors Report – 1993') (OJ 1994 No C2/1)

European Court of Auditors, Special Report No. 7/2007 on the control, inspection and sanction systems relating to the rules on conservation of Community fisheries resources (OJ 2007 No C317/1)

European Parliament

European Parliament Report No. A5-0174/04, Recommendation on the proposal for a Council decision on the accession of the Community to the Convention on the Conservation and Management of Highly Migratory Fish Stocks in the Western and Central Pacific Ocean – Committee on Fisheries

European Parliament Report No. A4-0176/98 Recommendation on the proposal for a Council Decision on the accession of the European Community to the General Fisheries Commission for the Mediterranean – Committee on Fisheries

Miscellaneous

Interinstitutional Agreement of 6 May 1999 between the European Parliament, the Council and the Commission on budgetary discipline and improvement of the budgetary procedure (OJ 1999 No. C172/1 – 22)

Arrangement in the form of an exchange of letters between the European Economic Community and the International Council for the Exploration of the Sea (OJ 1987 No. L149/14)